Circular Economy in the Construction Sector

Circular Economy in the Construction Sector

Guest Editors

Anibal C. Maury-Ramirez
Nele De Belie

MDPI

Basel • Beijing • Wuhan • Barcelona • Belgrade • Novi Sad • Cluj • Manchester

Guest Editors
Anibal C. Maury-Ramirez
Architecture Department
University of Antwerp
Antwerp
Belgium

Nele De Belie
Department of Structural
Engineering and Building
Materials
Ghent University
Ghent
Belgium

Editorial Office
MDPI AG
Grosspeteranlage 5
4052 Basel, Switzerland

This is a reprint of the Special Issue, published open access by the journal *Sustainability* (ISSN 2071-1050), freely accessible at: https://www.mdpi.com/journal/sustainability/special_issues/8G7B06MWS1.

For citation purposes, cite each article independently as indicated on the article page online and as indicated below:

Lastname, A.A.; Lastname, B.B. Article Title. *Journal Name* **Year**, *Volume Number*, Page Range.

ISBN 978-3-7258-3962-9 (Hbk)
ISBN 978-3-7258-3961-2 (PDF)
https://doi.org/10.3390/books978-3-7258-3961-2

Contents

About the Editors

Anibal C. Maury-Ramirez

Anibal Maury-Ramirez is the Principal Research Fellow in Construction Technology at the Architecture Department of the University of Antwerp working on the development of sustainable building materials, particularly on green walls and roofs, photocatalytic materials with self-cleaning and air-purifying properties, and sustainable concrete (i.e., supplementary cementitious materials, recycled aggregates, and permeable concrete). He has also been involved in the design of circular economy models and urban metabolism studies, which demand the use of novel tools to assess the sustainability of building materials (e.g., LCA, S-LCA, and LCC). Among others, he contributes to the bachelor courses "Construction 1, 2, and 3" as well as the master elective course "Construction Materials & Sustainability Assessment". Additionally, Dr. Maury-Ramirez serves as an invited lecturer at Ghent University (Belgium) and Pontificia Universidad Javeriana Cali (Colombia).

Dr. Maury-Ramirez is a Civil Engineer (Universidad del Norte, Colombia) in the ALFA Program in Materials Engineering (Tampere University, Finland) and Doctor of Civil Engineering (Ghent University, Belgium). He completed his postdoctoral research at the Hong Kong Polytechnic University (Hong Kong, China) and pursued a Postacademic Training in Low Countries Studies at Ghent University (Belgium). Dr. Maury-Ramirez has published his research in high-impact, peer-reviewed journals, with 26 articles indexed in Scopus and over 1,000 citations from fellow researchers. In addition to serving as a reviewer for multiple academic journals, he has also edited several Special Issues and books. Since 2021, he has been appointed as an executive member of the Ghent University Alumni Chapter in Colombia.

Nele De Belie

Nele De Belie obtained her Master of Science (1994) and PhD (1997) from Ghent University (Belgium), after which she carried out postdoctoral research (1997–2000) at KULeuven (Belgium), the Mt Albert Research Centre, New Zealand, and KVL University, Copenhagen, Denmark. She is a professor in Durability of Cement Bound Materials at Ghent University since 2000. Since 2018, she has been the director of the Magnel-Vandepitte Laboratory for Structural Engineering and Building Materials. She is the head of the "Concrete and Environment" research group, and her research focuses on sustainable concrete with supplementary cementitious materials, concrete durability, biodeterioration, bioconsolidation, smart concrete with self-healing or self-cleaning properties, circular economy, and life cycle assessment. She has supervised more than 60 (inter)national projects in these areas. She is the president of RILEM (International Union of Laboratories and Experts in Construction Materials, Systems and Structures) and the outgoing chair of the Technical Activities Committee (coordinating ~40 Technical Committees), and she has chaired TC 281-CCC on the carbonation of concrete with SCM (>100 members). She is the author of >400 WoS publications and 20 book chapters, editor of 10 books, inventor in 4 patent applications, and an editorial board member of 3 SCI-indexed scientific journals. She is the laureate of several awards in recognition of her scientific work, such as the Robert L'Hermite medal (2010). Since 2022, she has been appointed as a member of the Royal Flemish Academy of Belgium for Science and the Arts.

Preface

Since the publication of the Club of Rome's report "The Limits to Growth" in 1972, the need for a non-linear economic model became clear. This notion was further emphasized with the planetary boundaries theory, underscoring the demand for innovative production and consumption systems. Recently, the circular economy (CE) concept, championed by the Ellen MacArthur Foundation, has gained global recognition. While the CE aligns with key public policies such as the United Nations' Sustainable Development Goals and the European Green Deal, significant challenges and opportunities remain in advancing the CE across value chains—especially in the construction sector, which wields substantial economic influence while confronting environmental and social issues. This reprint addresses critical topics, including the following:

- Development of eco-friendly concrete and hybrid cements from recycled materials.
- Technological innovations like 3D printing and chemical activation to optimize material reuse.
- Environmental and economic evaluations of sustainable construction practices.
- Durability and mechanical enhancements in concrete and mortars using recycled aggregates.
- Implementation of sustainable building strategies in circular cities and communities.
- Examination of CE models, regulatory frameworks, and best practices globally.

In essence, the reprint fosters collaboration among stakeholders and promotes sustainable business models to tackle pressing environmental challenges in construction with a vision to drive a global adoption of CE principles.

Anibal C. Maury-Ramirez and Nele De Belie
Guest Editors

sustainability

MDPI

Article

Environmental and Economic Assessment of Eco-Concrete for Residential Buildings: A Case Study of Santiago de Cali (Colombia)

Aníbal Maury-Ramírez [1,2,*] and Nele De Belie [1]

1 Magnel-Vandepitte Laboratory, Deparment of Structural Engineering and Building Materials, Faculty of Engineering and Architecture, Ghent University (UGent), Technologiepark-Zwijnaarde 60, 9052 Ghent, Belgium; nele.debelie@ugent.be

2 Sigma Research Group, Departamento de Ingeniería Civil e Industrial, Facultad de Ingeniería y Ciencias, Pontificia Universidad Javeriana Cali (PUJCali), Calle 18 # 118-250, Santiago de Cali 760031, Colombia

* Correspondence: anibal.mauryramirez@ugent.be; Tel.: +32-9-264-55-35; Fax: +32-9-264-58-45

Abstract: Although the circular economy principles date back to the late 1960s, only with the recent stimulus from the European Commission and the Ellen McArthur Foundation has this concept gained attention worldwide. The City Hall of Santiago de Cali (Colombia) is implementing a circular economy model through a sustainable construction handbook and its certification. Among others, these stimulate the use of eco-concrete using fly ash and blast furnace slag coming from local industries (industrial symbiosis). Although concretes with these supplementary cementitious materials have been widely investigated regarding mechanical and durability properties, the economic and environmental impacts have been scarcely and independently evaluated, making the material selection a complex process. Therefore, this article presents the environmental and economic assessment of eco-concretes using fly ash and blast furnace slag for the design of a house located in Santiago de Cali (Colombia). The environmental and economic impacts are estimated by means of the environmental life cycle assessment (LCA) and life cycle costing (LCC), which are methodologies based on the ISO and ASTM standards implemented in the online software Building for Environmental and Economic Sustainability (BEES), which was selected for this case study. The results indicate that 40% fly ash concrete or 50% blast furnace slag would be recommended for reducing acidification or global warming potential, respectively. However, considering the existing public policies, the best option for the case study is 50% slag concrete. These results are of significant importance as they allow providing data-based recommendations for designers during the selection of the different eco-concretes. Additionally, these results might help establish a national roadmap to reduce carbon dioxide emissions from the construction sector, which are projected to continue increasing until 2050.

Keywords: supplementary cementitious materials; ordinary Portland cement; eco-concrete; LCA; LCC; fly ash; blast furnace slag; planetary boundaries; industrial symbiosis; circular economy

Citation: Maury-Ramírez, A.; De Belie, N. Environmental and Economic Assessment of Eco-Concrete for Residential Buildings: A Case Study of Santiago de Cali (Colombia). *Sustainability* **2023**, *15*, 12032. https://doi.org/10.3390/su151512032

Academic Editor: Hosam Saleh

Received: 6 July 2023
Revised: 2 August 2023
Accepted: 4 August 2023
Published: 6 August 2023

1. Introduction

The implementation of sustainable development principles in the design of buildings and infrastructure brings important challenges but also huge opportunities regarding innovation in construction materials that balance environmental, social and economic impacts. A current approach to integrate these concepts is circular economy (CE), which is a development paradigm that started to be built by the Club of Rome in 1968 with their MIT report "The limits to growth" published in 1972 [1]. Later, these principles were also incorporated in the Brundtland Report "Our Common Future" published by the World Commission on Environment and Development in 1987. Moreover, the United Nations Organization implicitly included CE principles in the Millennium (2000) and Sustainable Development Goals (2015), respectively. More recently, the European Commission and

Ellen MacArthur Foundation have designed important initiatives to develop the circular economy in different sectors worldwide. This is also supported by the advances in life cycle thinking, planetary and social boundaries [2,3]. A more detailed timeline for CE can be found in Figure 1.

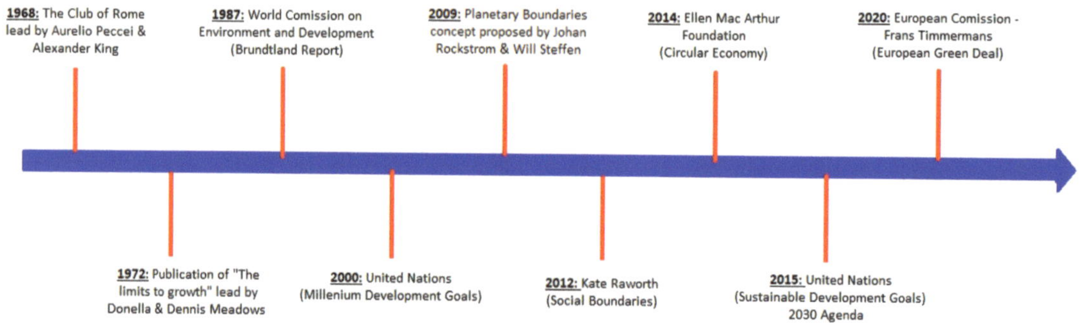

Figure 1. Circular economy timeline.

Using a public policy, consisting of a sustainability construction handbook and its certification, the City Hall of Santiago de Cali (Colombia) is implementing a CE model for the construction sector, which understands the circular economy as a consumption and production system that works for the efficiency in the consumption of materials, energy and water, considering the circular use of material flows and the resilience of ecosystems [4,5]. To do so, alliances and collaborations among different stakeholders are required for the implementation of technological innovations and the promotion of business models that respond to the fundamentals of sustainable development [5]. In addition to other strategies, the CE stimulates products such as recycled aggregates or eco-aggregates, eco-concretes and eco-mortars, eco-prefabricated products and modules, and smart construction materials [6]. These products satisfy the construction sector demand but also consider the following local waste valorization and material development experiences:

(a) Design of a system to generate air purification and self-cleaning on the surfaces of the tunnel of Colombia Avenue (Cali) using a photocatalytic mortar and UV-A lighting. This design project includes economic and environmental performances [7].
(b) Evaluation of semi-intensive green roofs with drainage layers made out of recycled and reused materials. Recycled rubber, recycled polyethylene (PET) bottles and recycled high-density polyethylene (HDPE) trays were evaluated as drainage layers regarding their hydraulic, thermal and mechanical performances in prototyped green roofs [8,9].
(c) Evaluation of the mechanical properties of concrete using recycled aggregates obtained from old paving stones coming from Almaguer (Cauca, Colombia). The results showed promising results with a mix replacing 50% of the natural fine aggregate with recycled ones [10].
(d) Evaluation of the effect in mortars of the partial replacement of Portland cement by ash from the paper industry. In particular, the fly ash was coming from the coal combustion in the manufacturing process of a local paper company [11].

Based on the previous technical results, making decisions about the best construction material for a specific application on a building or infrastructure is a complex process which is normally based on the designer's experience. Moreover, although sustainable building materials have been intensively investigated regarding their mechanical and durability properties, their environmental and economic impacts have been scarcely and independently evaluated. For example, although several articles have been recently published to evaluate the environmental impact of ordinary Portland cement (OPC) replacements with FA and BFS [12–15], few articles have applied the life cycle assessment (LCA) and life cycle costing (LCC) results to the materials selection for buildings and infrastructure.

This is because these studies used a cradle to gate approach, which does not include the construction, use and operation, and end of life of buildings or infrastructure. Moreover, few articles have integrated LCA and LCC results to the planetary boundaries and national development strategies [16].

In relation to the cases which applied LCA and LCC for materials selection with a wider system boundary, the environmental impacts and cost analysis of using copper slag as cement replacement (5%, 10%, and 15%) in concretes with compressive strengths of 20.7 MPa, 27.5 MPa, 34.5 MPa, and 41.40 MPa at 28 days for low-rise and mid-rise structures in the Philippines was performed. Based on the results, the use of copper slag was established as being beneficial to the abiotic depletion potential (fossil) and global warming potential, but it exerted damaging effects on abiotic depletion and human toxicity potentials, respectively. Moreover, the use of the copper slag as a partial cement replacement was found to reduce building costs by 1.4% and carbon emissions by 12.8% [17]. Similarly, the environmental, economic, and social impacts of using 20% fly ash as cement replacement (7FA20) in early-age high-strength concretes with compressive strengths of 55 MPa at 7 days (7OPC) and 28 days (28OPC) for two bridges in the Philippines were analyzed. The results show a higher impact of 7OPC on 17 midpoint and 3 endpoint indicators when compared to 28OPC. The most significant midpoint impacts based on normalization are fossil resource scarcity, global warming potential, ozone formation, human carcinogenicity toxicity, and terrestrial ecotoxicity. The global warming potential of 7OPC was quantified to be 636 kg CO_2 eq compared to 549 kg CO_2 eq of 28OPC concrete. Utilizing fly ash decreased the damage to resources (11%), human health (16%), and the environment (16%). Analysis of concrete constituents indicated that using chemical admixtures to obtain early high strength contributed significantly to fossil resource scarcity and resource damage [18]. In addition, the environmental impacts from building materials for the construction of a residence project in Brazil were determined. The results indicate a substantial waste of non-renewable energy, increasing global warming and harm to human health in this type of construction [19].

It is worth mentioning that FA and BFS are local wastes that have the potential of significantly reducing the current consumption of ordinary Portland cement (industrial symbiosis), which is projected to continue increasing in the global south region until 2050, as indicated by the recent Global Consensus on Sustainability in the Built Environment [20]. So, by systems to exchange waste and by-products, industrial symbiosis is a promising tool for innovative green growth of the construction sector, as this engages diverse organizations in a network to foster eco-innovation, long-term culture change and mutual benefits [21,22].

Therefore, taking into account that using life cycle assessment and life cycle costing allows consolidating, comparing, and assessing sustainability impacts [23], this article presents the selection process of eco-concretes with blast furnace slag (BSF) and fly ash (FA) to be used in a sustainable house in Santiago de Cali (Colombia).

2. Case Study

In the course of Sustainability in Construction from the Master of Civil Engineering of PUJCali, a sustainable house was designed for a family of five persons in Santiago de Cali (Colombia). To do so, a flat land of 150 m^2 and a budget of 100,000 USD were available. Considering the above indicated requirements and the local climatic conditions, students designed a one-floor house, which includes energy and water efficiency strategies, with four bedrooms, two bathrooms, a visitor´s toilet, living room, dining room and terrace. In addition, considering the high seismic risk of Santiago de Cali and high probability of an earthquake occurrence, and looking for a flexible architectural design, the following structural elements with 28 MPa compressive strength were considered: aerial beams (0.30 m × 0.35 m), columns (0.30 m × 0.30 m) and foundation beams (0.40 m × 0.3 m). Finally, for selecting the best eco-concrete, a matrix with their environmental impact categories, concrete products and associated scores that range from 1 for the best option (with lower environmental impact values), 2 for the second-best option, and so on was used. A general view of the designed house is given in Figure 2.

Figure 2. Three-dimensional (3D) view of the proposed sustainable house in Santiago de Cali (Colombia) [24].

3. Methodology

The LCA and LCC standard methods developed by the American Society for Testing and Materials (ASTM) and the International Organization for Standardization (ISO) were followed in this research, respectively. In both cases, the open access and online software Building for Environmental and Economic Sustainability (BEES) version 2.1, developed by the National Institute of Standards and Technology (NIST) from the U.S. Department of Commerce, was selected for the integrated development of LCA and LCC [25]. Using a Microsoft SQL server, BEES integrates Simapro version 8.0 and Ecoinvent 2017, which was recently ranked among the top databases for construction supplies [18]. For this case study, the following concrete mixes with a compressive strength of 28 MPa using Portland cement (reference), fly ash and blast furnace slag were assessed (Tables 1 and 2). More details about the technical challenges and limitations of using FA and BFS in concrete can be found in the recent state-of-the-art book which focus on the properties of concrete containing supplementary cementitious materials in the fresh and hardened state [26].

Table 1. Mix design for concrete with different contents of fly ash and compressive strength of 28 MPa [25].

Material	No Fly Ash or Slag (Kg/m^3)	20% Fly Ash (Kg/m^3)	30% Fly Ash (Kg/m^3)	40% Fly Ash (Kg/m^3)
Portland cement	365.2	307.3	276.5	243.2
Slag cement	0	0	0	0
Fly ash	0	77.1	118.7	162.6
Crushed coarse aggregate	678.6	678.6	678.6	678.6
Natural coarse aggregate	316.2	316.2	316.2	316.2
Crushed fine aggregate	87.4	87.4	87.4	87.4
Natural fine aggregate	656.4	656.4	656.4	656.4
Light weight aggregate	0	0	0	0
Accelerator (accel.)	0.37	0.56	0.56	0.93
Air entrainer	0.04	0.04	0.06	0.06
Water reducer and accel.	0.11	0.11	0.11	0.11
High range water red. and accel.	0	0	0	0
Water	154.8	154.8	154.8	154.8

Table 2. Mix design for concrete with different contents of blast furnace slag and compressive strength of 28 MPa [25].

Material	30% Slag (Kg/m^3)	40% Slag (Kg/m^3)	50% Slag (Kg/m^3)	20% FA, 30% Slag (Kg/m^3)
Portland cement	255.6	219.1	182.6	192.2
Slag cement	109.6	146.1	182.6	115.1
Fly ash	0	0	0	77.1
Crushed coarse aggregate	678.6	678.6	678.6	678.6
Natural coarse aggregate	316.2	316.2	316.2	316.2
Crushed fine aggregate	87.4	87.4	87.4	87.4
Natural fine aggregate	656.4	656.4	656.4	656.4
Light weight aggregate	0	0	0	0
Accelerator (accel.)	0.56	0.93	1.11	1.11
Air entrainer	0.04	0.04	0.04	0.04
Water reducer and accel.	0.11	0.11	0.11	0.11
High range water red. and accel.	0	0	0	0
Water	154.8	154.8	154.8	154.8

3.1. Life Cycle Assessment (LCA)

In particular, ISO indicates following four phases for LCAs: (a) objective and scope (ISO 14041), (b) inventory analysis (ISO 14041), (c) impact assessment (ISO 14042), and (d) results interpretation (ISO 14043) (Figure 3).

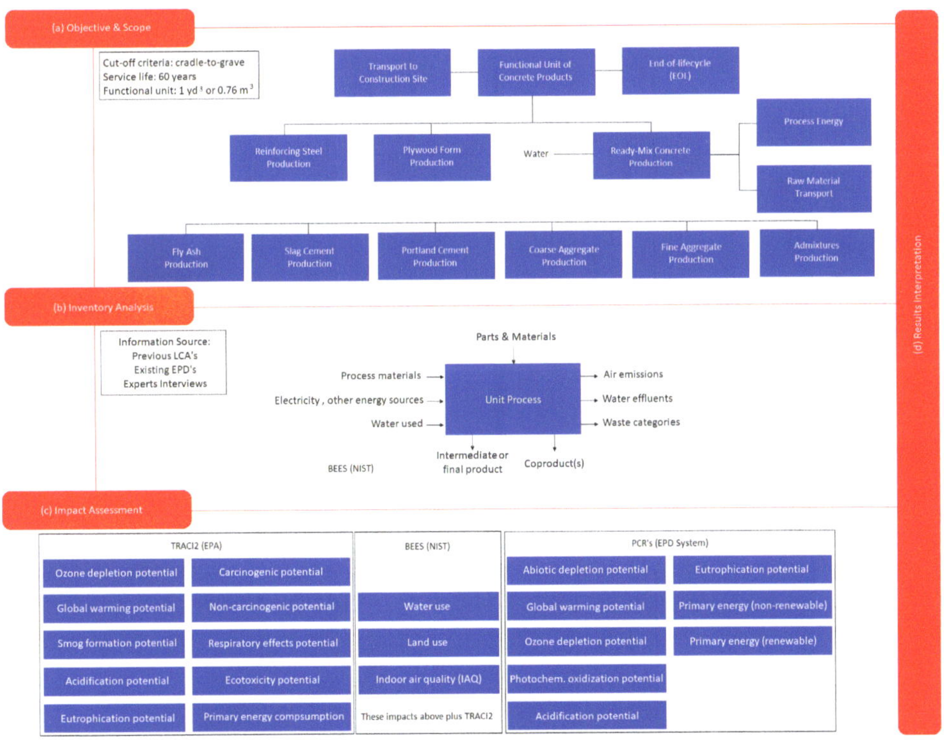

Figure 3. LCA phases (in red boxes) including system boundaries, functional unit and environmental impact categories (in blue boxes).

First, in relation to the objective and scope, the case study was performed using a cradle-to-grave approach, a service life of 60 years (without maintenance or reparation activities), and a functional unit of 1 yd^3 or 0.76 m^3 for beams and columns. Also, the system boundaries include all concrete mix components (fly ash, blast furnace slag, Portland cement, coarse and fine aggregates, admixtures), steel reinforcing and plywood forms. Second, regarding the inventory analysis, the inlet and outlets from the functional unit were estimated using several information sources such as previous LCAs, existing EPDs, and expert interviews, which were information already included and validated in the software BEES. Moreover, a distance of 120 km (74.6 mi) was set based on the concrete and the cement average delivery distance reported within Latin America and the Caribbean [27]. Third, considering the environmental impact categories developed by the U.S. Environmental Protection Agency (EPA), National Institute of Standards and Technology (NIST) and International Environmental Product Declaration System (EPD System), the BEES environmental impact categories were selected for this case study. Finally, in order to interpretate the results, the relative importance weights based on the BEES stakeholder panel were used (Table 3).

Table 3. Relative importance weights based on the BEES Stakeholder Panel [25].

Impact Category	Relative Importance Weight (%)
Climate change	29
Primary energy consumption	10
Human health criteria air	9
Human health cancer	8
Water consumption	8
Ecological toxicity	7
Eutrophication	6
Land use	6
Human health non-cancer	5
Smog formation	4
Acidification	3
Indoor air quality	3
Ozone depletion	2

It is worth mentioning that in this case study, the following conceptualization and monitoring units of the impact categories were used:

Acidification: Acidifying compounds such as sulfur and nitrogen, hydrogen chloride and ammonia can reach water, air and soil from ecosystems attacking flora and fauna. For example, a reduction in coniferous forests and an increase in fish mortality due to acidification has been reported, which is normally indicated by kilograms of SO_2 equivalents (kg SO_2 eq).

Climate change: Also identified as global warming potential (GWP), climate change refers to the increase in the average global temperature as consequence of the greenhouse gas emissions (GHGs). Major GHGs come from the combustion of fossil fuels (e.g., oil, coal and natural gas). The GWP from the GHG emissions is estimated by kilogram of carbon dioxide equivalent (kg CO_2 eq).

Ecological toxicity: This impact indicates the potential damage to land and water ecosystems, which is estimated by comparative toxic units for ecosystems (CTUe). This estimates the potentially affected fraction (PAF) of species integrated over time and volume per unit mass of a chemical released.

Eutrophication: This indicator refers to the potential toxic impacts caused by the excessive presence of mineral nutrients (i.e., nitrogen and phosphorous) on soil or water from an ecosystem. This phenomenon enhances algae growth in water, which normally block fishes' access to oxygen, causing their death. Eutrophication potential is normally estimated in kilograms of nitrogen equivalents (kg N eq).

Human health: This impact refers to potential effects on human health, which can be associated to cancer, non-cancer, or respiratory problems. The characterization factors for human toxicity are estimated by comparative toxic units for human toxicity (CTUh). These estimate the increase in morbidity (i.e., cancer or non-cancer) in the total human population per unit mass of a contaminant released. On the other hand, inhaling fine particulates from air might result in serious health problems (e.g., asthma). This impact category is reported in kilograms of particulate matter of size less than or equal to 2.5 μm equivalents ($PM_{2.5}$ eq).

Indoor air quality: This impact, also known as (IAQ), is associated with the respiratory problems caused by the excessive concentration of pollutant gases inside buildings or even infrastructure (e.g., tunnels and parking lots). Although there is no general consensus about the estimation method, this indicator is normally associated with the presence of volatile organic compounds (VOCs), which is very much controlled in building materials.

Land use: Although this is a composed factor normally measuring impacts on soil properties, BEES only considers the surface area of land occupied or transformed within the system boundaries. So, this is estimated by square meters (m^2).

Ozone depletion: This impact measures the damage potential caused by depleting gases (e.g., chlorofluorocarbons—CFCs) on the ozone layer, which protects life on earth from short wave radiation. Ozone depletion is reported in kilograms of trichlorofluormethane, also known as Freon-11, equivalents (CFC-11 eq).

Primary energy consumption: This refers to the required energy along the production system and the embodied energy in the products. Primary energy consumption can include non-renewable (e.g., coming from fossil fuels and nuclear power) and renewable energies (e.g., coming from hydropower, wind power and biomass). So, these impact categories are estimated using the cumulative energy demand method (CED) in megajoules (MJ).

Smog formation: This impact is associated with the photochemical oxidation potential caused by the formation of photochemical oxidants, including ozone (O_3), under certain environmental conditions (e.g., UV radiation) and air pollutants (e.g., NO_X, VOCs). This is estimated in kilograms of O_3 equivalents (O_3 eq) or ethylene equivalents (C_2H_4 eq).

Water consumption: Although water access and availability are different from region to region, this factor is associated with the consumption of freshwater, which controls the availability or scarcity of water in the system boundaries. The impact potential is estimated in liters (L).

3.2. Life Cycle Costing (LCC)

Determining the economic impacts of concrete products with BEES is an easier process than measuring the environmental ones. This because available data and standard methods for developing economic performance evaluations exist. The most appropriate method for determining the economic impacts of building products is the LCC method following the ASTM standard for the life cycle costing of building related investments [28].

In this case study, the total LCC of a concrete product (C_{LCC}) is the sum of the present values of first cost (C_{First}) and future costs (C_{Future}) minus the residual value (RV), as shown in the following equation:

$$C_{LCC} = C_{First} + C_{Future} - RV \tag{1}$$

Based on the LCC methodology and BEES requirements, the study period was defined as 60 years (in agreement with the LCA), installation costs of approx. 77.59 USD (343,315 COP per m^3), discount rate of 3%, and social cost of CO_2 emissions was \$12/ton (Table 4), as suggested by BEES.

Table 4. Social cost of carbon per metric ton [25].

Time (Year)	Cost (USD)
2010	12
2015	13
2020	15
2025	17
2030	19
2035	22
2040	26
2045	28
2050	32

4. Results

The online software BEES was set using the conditions previously described in the LCA methodology for the selection of the best eco-concrete using BFS and FA for a sustainable house to be constructed in Santiago de Cali (Colombia). With this aim, concrete products (i.e., beams and columns) were classified for each environmental category from 1 to 8 based on the environmental impact, where 1 indicated the smallest and 8 indicated the biggest impact, respectively. For instance, the results of the environmental impact on global warming potential from the house beams and columns can be seen in Figures 4 and 5, respectively. In this case, numbers 1 and 8 were assigned to 50% slag concrete and 100% OPC concrete (reference), respectively. These results were the same for all environmental impact factors evaluated except for acidification, in which the trend was different with number 1 assigned to 40% fly ash concrete as can be seen in Figures 6 and 7 for beams and columns, respectively.

On the other hand, the economic performance of the beams and columns for the sustainable house can be seen in Figures 8 and 9. It is evident that there are no significant differences between the cost of the evaluated concrete mixes even when including social carbon costs. So, based on this LCC, there is no evidence to conclude that using by-products in concrete might substantially increase the construction costs.

So, following the proposed methodology to select the best eco-concrete for the sustainable house, a matrix which displays the environmental and economic impact categories and different concrete products was elaborated (Table 5). Based on this, the 1st option is the 50% slag concrete, which is followed by the 20–30% fly ash–slag concrete and the 40% slag concrete (3rd option). This result does not apply for acidification, where the lowest environmental impact is obtained with the 40% fly ash (1st option), which is followed by the 30% fly ash concrete (2nd option), and the 20–30% fly ash–slag concrete (3rd option).

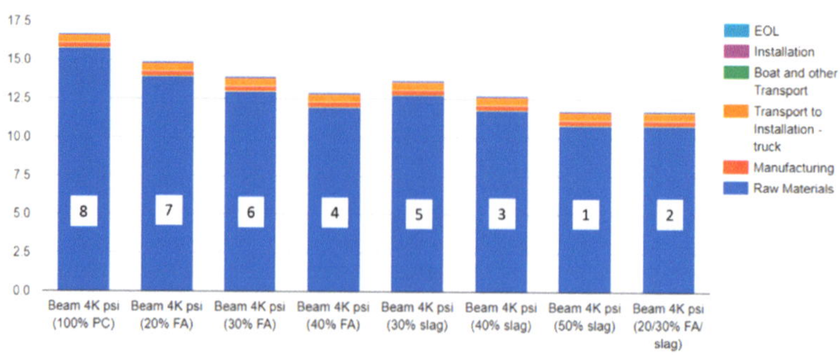

Figure 4. Comparative environmental impact based on the global warming from the different concrete products (beams). Numbers 1 to 8 were assigned from the lowest to the highest impacts, respectively.

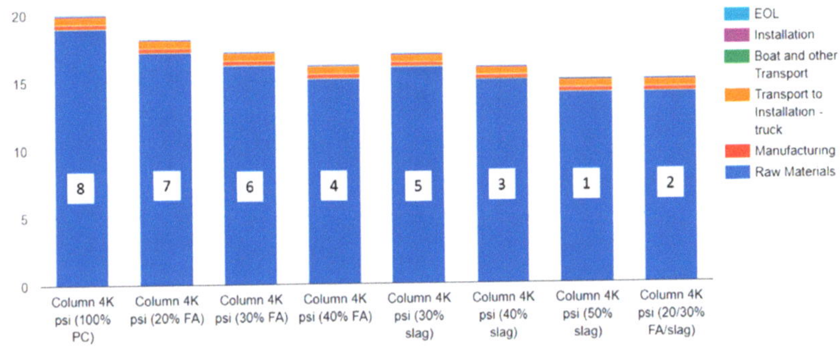

Figure 5. Comparative environmental impact based on the global warming from the different concrete products (columns). Numbers 1 to 8 were assigned from the lowest to the highest impacts, respectively.

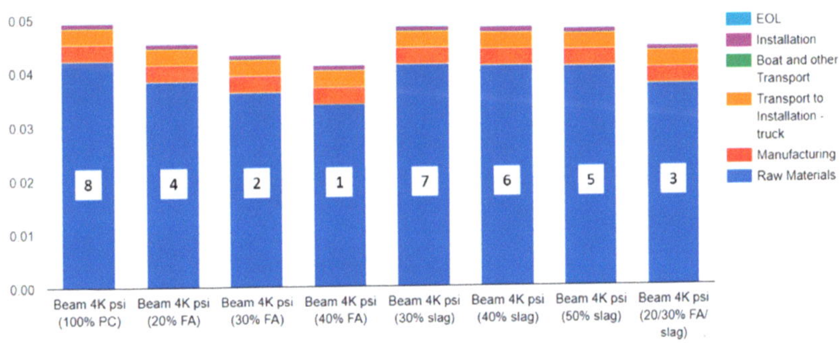

Figure 6. Comparative environmental impact based on the acidification from the different concrete products (beams). Numbers 1 to 8 were assigned from the lowest to the highest impacts, respectively.

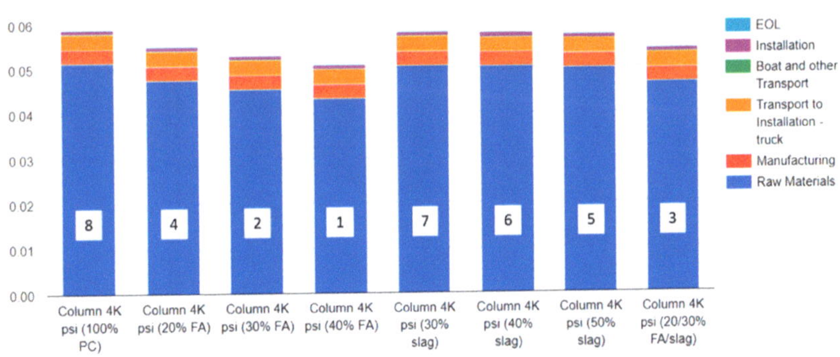

Figure 7. Comparative environmental impact based on the acidification from the different concrete products (columns). Numbers 1 to 8 were assigned from the lowest to the highest impacts, respectively.

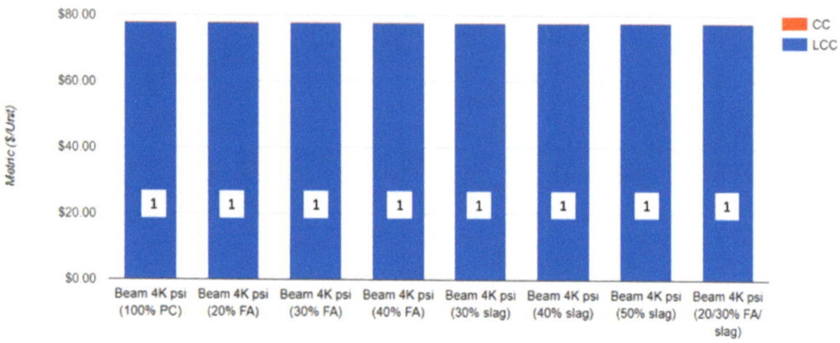

Figure 8. Comparative economic impact of beams using fly ash, blast furnace slag and OPC. As there is no difference between the costs, number 1 was assigned to all concrete mixes.

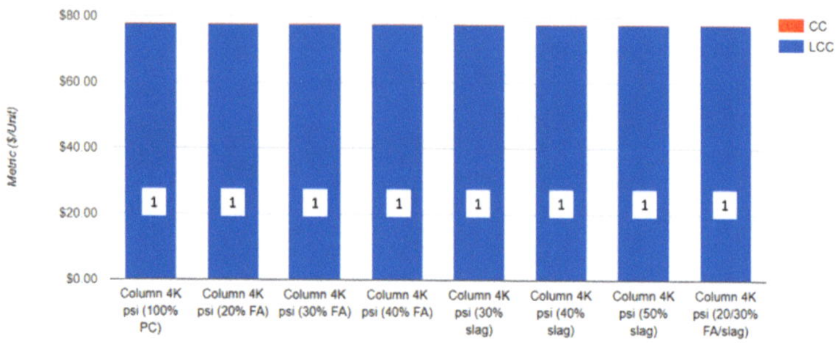

Figure 9. Comparative economic impact of columns using fly ash, blast furnace slag and OPC. As there is no difference between the costs, number 1 was assigned to all concrete mixes.

Table 5. Concrete products scores for the different environmental and economic impact categories.

Impact Category	100% PC	20% FA	30% FA	40% FA	30% Slag	40% Slag	50% Slag	20% FA, 30% Slag
Global warming	8	7	6	4	5	3	1	2
Primary energy consumption (nr)[1]	8	7	6	4	5	3	1	2
Primary energy consumption (r)[2]	8	7	6	4	5	3	1	2
Human health criteria air	8	7	6	4	5	3	1	2
Human health cancer	8	7	6	4	5	3	1	2
Water consumption	8	7	6	4	5	3	1	2
Ecological toxicity	8	7	6	4	5	3	1	2
Eutrophication	8	7	6	4	5	3	1	2
Land use	8	7	6	4	5	3	1	2
Human health non-cancer	8	7	6	4	5	3	1	2
Smog formation	8	7	6	4	5	3	1	2
Acidification	8	4	2	1	7	6	5	3
Ozone depletion	8	7	6	4	5	3	1	2
Indoor air quality	8	7	6	4	5	3	1	2
EP[3]: beams and columns	1	1	1	1	1	1	1	1

Abbreviations: (nr)[1]: non-renewable, (r)[2]: renewable, EP[3]: economic performance.

In order to make a final decision about the best eco-concrete for the selected case study, global and local scales were considered. At the global scale, the concept of planetary boundaries stated by Rockstrom et al. (2009) was considered [2]. The nine planetary boundaries are: climate change, ocean acidification, stratospheric ozone depletion, biogeochemical flows, global freshwater use, change in land use, biodiversity loss, atmospheric aerosol loading and chemical pollution. From these boundaries, climate change, biogeochemical flows (nitrogen cycle) and biodiversity loss have been already exceeded (Figure 10). As the transgression of one or more planetary boundaries may be deleterious or even catastrophic due to the risk of crossing thresholds that will trigger non-linear and abrupt environmental change within continental to planetary scale systems, the most important environmental problem for the selected case study would be climate change compared to acidification. Similarly, at the local scale, it was considered that Santiago de Cali is implementing a net zero carbon building program sponsored by the World Green Building Council [29]. Based on the previous considerations, the best option for the beams and column's project is the 50% slag concrete. However, attention has to be paid to the release of hydrogen chloride, ammonia, sulfur and nitrogen, which may, in gaseous state or dissolved in water or fixed on solid particles reach ecosystems through dissolution in rain or wet deposition, affecting trees, soil, buildings, animals and humans.

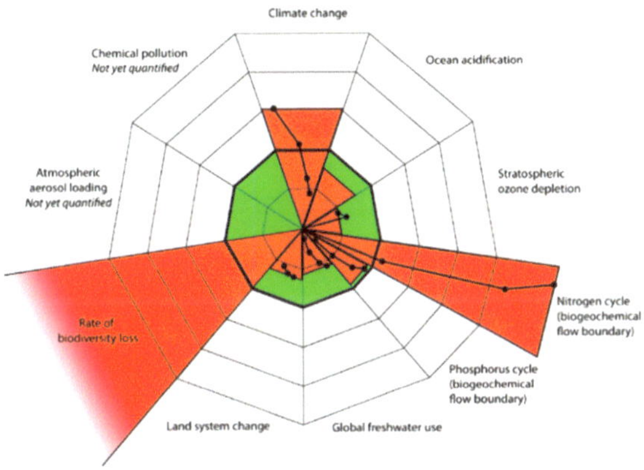

Figure 10. Planetary boundaries stated by Rockstrom et al. (2009) [2].

On the other hand, based on the order of magnitude of the environmental impacts from the house, particularly on global warming, the amount of CO_2 equivalent produced by the concrete structure made with 50% slag concrete is 379.3 kg equivalent CO_2 (Table 6), which is approx. 27% smaller than the equivalent CO_2 that might be released by traditional concrete (100% OPC). This percentage is slightly higher than the average obtained in Colombia (3 to 25%) with other waste valorization strategies in the cement industry [30].

Table 6. Equivalent CO_2 released by different concrete elements from the designed sustainable house in Santiago de Cali (Colombia).

Element	kg CO_2/yd^3	kg CO_2/m^3	Concrete (m^3)	kgCO_2/House
Beams 50% slag concrete (aerial)	12	15.8	7.88	124.5
Beams 50% slag concrete (foundation)	12	15.8	9.00	142.2
Columns 50% slag concrete	15.3	20.1	5.60	112.6
		Total		379.3

Therefore, considering the current use of fly ash and blast furnace slag in Colombia, which are 3.5 and 72.3 kg per ton of cement, a gradual and safe transition plan for the cement industry should be designed to achieve the sustainable development goals in 2050, particularly regarding climate change. In order to do so, it is important to indicate that the growing annual CO_2 emissions from Colombia in 2021 were 91.7 million tons, which is 2.5 times smaller than the decreasing CO_2 annual emissions from Spain, which is a country belonging to the European Union with a similar population size (Figure 11). Moreover, with 5.87 million CO_2 tons in 2021, the Colombian cement industry had lower emissions than the oil, gas and coal sectors (Figure 12).

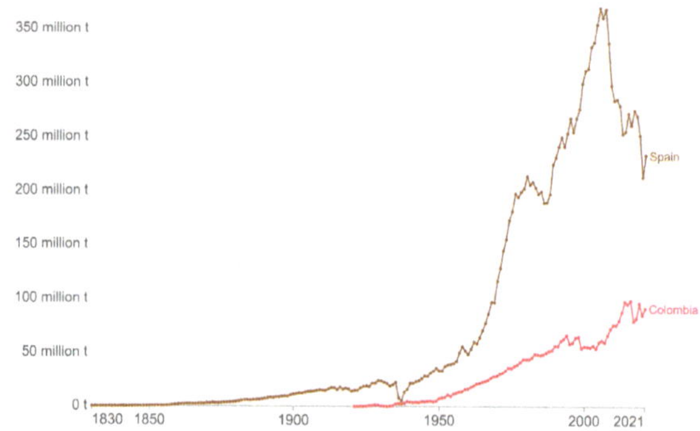

Figure 11. Annual CO_2 emissions from fossil fuels and industry of Colombia and Spain taken from Our World in Data based on the Global Carbon Project (2022) [31].

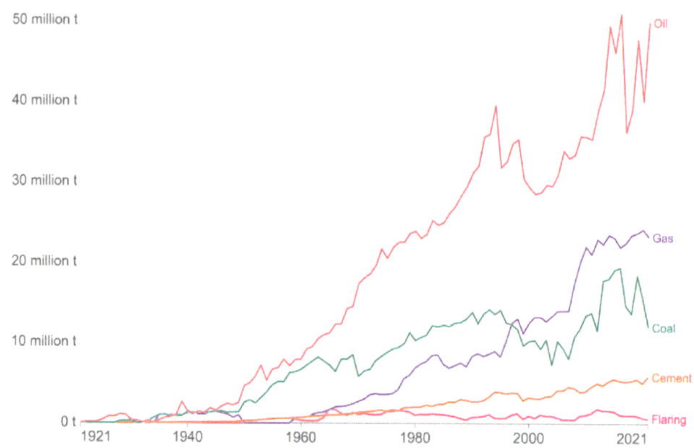

Figure 12. CO_2 emissions by fuels and industry of Colombia taken from Our World in Data based on the Global Carbon Project (2022) [31].

5. Conclusions

First, the performed LCA and LCC using the software BEES allows estimating the environmental and economic impacts of using fly ash and blast furnace slag as replacements of Portland cement in concrete. In this case study, 40% fly ash concrete or 50% blast furnace slag concrete would be recommended for reducing acidification or global warming potential, respectively. Second, considering the net zero carbon program signed by Santiago de Cali and the planetary boundaries, the best option for the sustainable house is the 50% blast furnace slag concrete.

Third, the article describes the environmental and economic differences from valorizing wastes potentially coming from thermoelectric power plants and steel plants in the construction sector under the concept of industrial symbiosis. Future research includes durability differences between concrete using blast furnace slag and fly ash. Also, the social impacts of eco-concretes are taken into account for future works. Finally, considering the current and projected consumption of cement in Colombia, detailed roadmaps to decarbonize the construction sector of Santiago de Cali and other major cities such as Bogotá, Medellin and Barranquilla are required. These roadmaps should work on decarbonizing construction materials as performed with FA and BFS, reducing the needed quantity of materials for new developments and stimulating the rehabilitation of existing buildings and infrastructure. The last strategies can be implemented through public policies that stimulate the use of sustainable building materials as used with the CE model from Santiago de Cali, but they also limit the embedded CO_2 per area in new developments, reduce taxes for building and infrastructure rehabilitation, stimulate research and development in the construction sector, and support standardization committees such as on sustainability life cycle assessment, which is urgently required at the national level to develop Colombian databases.

Author Contributions: Conceptualization and methodology, A.M.-R. and N.D.B.; validation, A.M.-R.; formal analysis, A.M.-R.; investigation and resources, A.M.-R. and N.D.B.; data curation, A.M.-R.; writing—original draft preparation, A.M.-R.; writing—review and editing, A.M.-R. and N.D.B.; visualization, A.M.-R. All authors have read and agreed to the published version of the manuscript.

Funding: The authors would like to thank the financial support given by the Global Minds Fund for the Short Teaching Stay of Anibal Maury Ramirez at Ghent University during May 2023.

Data Availability Statement: Not applicable.

Conflicts of Interest: The authors declare no conflict of interest.

References

1. La Economía Circular y su Papel en el Cambio Climático. Available online: https://www.elespectador.com/actualidad/la-economia-circular-y-su-papel-en-el-cambio-climatico/ (accessed on 6 April 2023).
2. Rockström, J.; Steffen, W.; Noone, K.; Persson, Å.; Chapin, F.S., III; Lambin, E.; Lenton, T.M.; Scheffer, M.; Folke, C.; Schellnhuber, H.J.; et al. Planetary boundaries: Exploring the safe operating space for humanity. *Ecol. Soc.* **2009**, *14*, 32. [CrossRef]
3. Raworth, K. *A Safe and Just Space for Humanity: Can We Live within the Doughnut? Oxfam Discussion Papers*; Oxfam International: Oxford, UK, 2012; ISBN 978-1-78077-059-8.
4. Osuna Motta, I.; Villa Franco, J.F.; Maury-Ramírez, A.; Valencia Ojeda, L.A.; Giraldo Yepes, C.M.; Colonia Restrepo, A.F.; Galarza, S.; Carmona Ceballos, J.I.; Rincón Laverde, E.J.; Gutiérrez Robledo, A.; et al. Manual de Construcción Sostenible de Santiago de Cali. Alcaldía de Santiago de Cali, 2022 (Under City Hall Revision for Approval). Available online: https://www.concejodecali.gov.co/documentos/?genPag=3 (accessed on 26 May 2023).
5. Economía Circular—DANE. Available online: http://www.andi.com.co/Uploads/economia-circular-1-reporte.pdf (accessed on 6 April 2023).
6. Maury-Ramírez, A.; Illera-Perozo, D.; Mesa, J.A. Circular economy in the construction sector: A case study of Santiago de Cali (Colombia). *Sustainability* **2022**, *14*, 1923. [CrossRef]
7. Medina Medina, A.F.; Torres Rojas, D.F.; Meza Girón, G.; Villota Grisales, R.A. Diseño de un Sistema Para Generar Purificación del Aire y Auto-Limpieza en las Superficies del Túnel de la Avenida Colombia (Cali). Ph.D. Thesis, Pontificia Universidad Javeriana Cali, Santiago de Cali, Colombia, May 2016.
8. Naranjo, A.; Colonia, A.; Mesa, J.; Maury-Ramírez, A. Evaluation of Semi-Intensive Green Roofs with Drainage Layers Made Out of Recycled and Reused Materials. *Coatings* **2020**, *10*, 525. [CrossRef]
9. Naranjo, A.; Colonia, A.; Mesa, J.; Maury, H.; Maury-Ramírez, A. State-of-the-art green roofs: Technical performance and certifications for sustainable construction. *Coatings* **2020**, *10*, 69. [CrossRef]
10. Bravo-German, A.M.; Bravo-Gómez, I.D.; Mesa, J.A.; Maury-Ramírez, A. Mechanical Properties of Concrete Using Recycled Aggregates Obtained from Old Paving Stones. *Sustainability* **2021**, *13*, 3044. [CrossRef]
11. Hurtado Portocarrero, I.A. Evaluación del Efecto del Reemplazo Parcial del Cemento Portland por Ceniza de la Industria del Papel en Morteros y Muretes. Master's Thesis, Pontificia Universidad Javeriana Cali, Santiago de Cali, Colombia, September 2019.
12. Shobeiri, V.; Bennett, B.; Xie, T.; Visintin, P. Mix design optimization of concrete containing fly ash and slag for global warming potential and cost reduction. *Case Stud. Constr. Mater.* **2023**, *18*, e01832. [CrossRef]
13. Das, P.; Sankar Cheela, V.R.; Mistri, A.; Chakraborty, S.; Dubey, S.; Barai, S.V. Performance assessment and life cycle analysis of concrete containing ferrochrome slag and fly ash as replacement materials—A circular approach. *Constr. Build. Mater.* **2022**, *347*, 128609. [CrossRef]
14. Chen, X.; Wang, H.; Najm, H.; Venkiteela, G.; Hencken, J. Evaluating engineering properties and environmental impact of pervious concrete with fly ash and slag. *J. Clean. Prod.* **2019**, *237*, 117714. [CrossRef]
15. van den Heede, P.; de Belie, N. Environmental impact and life cycle assessment (LCA) of traditional and 'green' concretes: Literature review and theoretical calculations. *Cem. Concr. Compos.* **2012**, *34*, 431. [CrossRef]
16. Kara, S.; Herrmann, C.; Hauschild, M. Operationalization of life cycle engineering, Resources. *Conserv. Recycl.* **2023**, *190*, 106836. [CrossRef]
17. de Pedro, J.P.Q.; Lagao, J.A.T.; Ongpeng, J.M.C. Life Cycle Assessment of Concrete Using Copper Slag as a Partial Cement Substitute in Reinforced Concrete Buildings. *Buildings* **2023**, *13*, 746. [CrossRef]

18. Orozco, C.R.; Babel, S.; Tangtermsirikul, S.; Sugiyama, T. Understanding the environmental, economic, and social impact of fly ash utilization on early-age high-strength mass concrete using life cycle analysis. *Mater. Today Proc.* 2023, in press. [CrossRef]
19. De Lassio, J.; França, J.; Espirito Santo, K.; Haddad, A. Case Study: LCA Methodology Applied to Materials Management in a Brazilian Residential Construction Site. *J. Eng.* **2016**, *2016*, 8513293. [CrossRef]
20. Global Consensus on Sustainability in the Built Environment—GLOBE. Decarbonising Global Construction. Available online: 221109-GlobePolicyAdviceDocument-revised-new-1.pdf (accessed on 6 April 2023).
21. Lombardi, D.R.; Laybourn, P. Redefining Industrial Symbiosis, Crossing Academic–Practitioner Boundaries. *J. Ind. Ecol.* **2012**, *16*, 28. [CrossRef]
22. Cagno, E.; Negri, A.M.; Neri, A.; Giambone, M. One Framework to Rule Them All: An Integrated, Multi-level and Scalable Performance Measurement Framework of Sustainability, Circular Economy and Industrial Symbiosis. *Sustain. Prod. Consum.* **2023**, *35*, 55. [CrossRef]
23. Mesa, J.A.; Fúquene-Retamoso, C.; Maury-Ramírez, A. Life cycle assessment on construction and demolition waste: A systematic literature review. *Sustainability* **2021**, *13*, 7676. [CrossRef]
24. Ruiz-Izquierdo, J.; Valencia-Chavez, D.; Salas-Nuñez, J.L.; Maldonado, J.F.; Yama, C.E. *Edificación Sostenible, Proyecto Final Curso Sostenibilidad en la Construcción*; Pontificia Universidad Javeriana Cali: Santiago de Cali, Colombia, 2022.
25. National Institute of Standards—NIST. Building for Environmental and Economic Sustainability (BEES) Online 2.1 Technical Manual. Available online: https://ws680.nist.gov/BEES2/Home/UserManual/ (accessed on 6 April 2023).
26. De Belie, N.; Gruyaert, E.; Soutsos, M. (Eds.) *Properties of Fresh and Hardened Concrete Containing Supplementary Cementitious Materials, State-of-the-Art Report of the RILEM Technical Committee 238-SCM, Working Group 4*; Springer: Berlin/Heidelberg, Germany, 2018. [CrossRef]
27. Villagrán-Zaccardi, Y.; Pareja, R.; Rojas, L.; Irassar, E.; Torres-Acosta, A.; Tobón, J.; Vanderley, J. Overview of cement and concrete production in Latin America and the Caribbean with a focus on the goals of reaching carbon neutrality. *RILEM Tech. Lett.* **2022**, *7*, 30. [CrossRef]
28. American Standard of Testing Materials—ASTM. *Standard Practice for Measuring Life-Cycle Costs of Buildings and Building Systems*; ASTM: Conshohocken, PA, USA, 2017.
29. World Green Building Council—WGBC. Policy Pathway Santiago de Cali, Colombia. Available online: https://worldgbc.org/signatory/santiago-de-cali-colombia/ (accessed on 6 April 2023).
30. Tecnalia, Diagnóstico de Eficiencia en el Uso de Materiales y Cierre de Ciclos en los Sectores Manufacturero y de Construcción en Colombia: Contraste Frente a Experiencias Internacionales. Available online: https://bibliotecadigital.ccb.org.co/bitstream/handle/11520/21034/Diagn%c3%b3stico%20Tecnalia.pdf?sequence=1&isAllowed=y (accessed on 6 April 2023).
31. Ritchie, H.; Roser, M.; Rosado, P. CO_2 and Greenhouse Gas Emissions. Our World in Data. 2023. Available online: https://ourworldindata.org/co2/country/colombia#citation (accessed on 6 April 2023).

sustainability

Article

Mechanical Performance of Mortars with Partial Replacement of Cement by Aluminum Dross: Inactivation and Particle Size

Daniel Parra-Molina [1], Manuel Alejandro Rojas-Manzano [1], Adriana Gómez-Gómez [1], Mario Fernando Muñoz-Vélez [1] and Aníbal Maury-Ramírez [1,2,*]

1 Department of Civil and Industrial Engineering, Faculty of Engineering and Sciences, Pontificia Universidad Javeriana Cali, Calle 18# 118–250, Santiago de Cali 760031, Colombia; danielparra17@javerianacali.edu.co (D.P.-M.); alejandro.rojas@javerianacali.edu.co (M.A.R.-M.); adrianagomez@javerianacali.edu.co (A.G.-G.); mario.munoz@javerianacali.edu.co (M.F.M.-V.)

2 Architecture Department, Faculty of Design Sciences, University of Antwerp, Mutsaardstraat 29, 2000 Antwerp, Belgium

* Correspondence: anibal.maury-ramirez@uantwerpen.be

Abstract: Although the use of primary aluminum dross as cement replacement has shown promising results in mortars and concretes, there is a knowledge gap between the effect of the secondary dross inactivation process and particle sizes on the mechanical properties and consistency. So, by using X-ray diffraction, laser granulometry, and scanning electron microscopy, this article describes first the inactivation process applied to a secondary aluminum dross. Second, this manuscript presents the fresh and hardened properties of mortar mixes containing 5, 10, and 20% inactivated secondary aluminum dross with three different particle sizes (i.e., fine, intermediate, and coarse). Mortar flow test results indicate that compressive and flexural strengths of mixes containing up to 20% fine and intermediate aluminum dross as cement replacement were satisfactory, respectively. These results have the potential to reduce the environmental and health impacts caused by cement production and secondary aluminum dross disposal, respectively. Moreover, the durability aspects of the mortar mixes, as well as the effectivity of the investigated inactivation process, are identified as future research topics.

Keywords: aluminum dross; eco-mortar; dross inactivation; dross particle size; mechanical properties; industrial symbiosis; circular economy; sustainable building materials; sustainable construction

Citation: Parra-Molina, D.; Rojas-Manzano, M.A.; Gómez-Gómez, A.; Muñoz-Vélez, M.F.; Maury-Ramírez, A. Mechanical Performance of Mortars with Partial Replacement of Cement by Aluminum Dross: Inactivation and Particle Size. *Sustainability* **2023**, *15*, 14224. https://doi.org/10.3390/su151914224

Academic Editor: Estefanía Cuenca Asensio

Received: 4 September 2023
Revised: 17 September 2023
Accepted: 21 September 2023
Published: 26 September 2023

1. Introduction

Although there are significant environmental impacts at local and global scales, the construction sector is strongly connected to economic growth and social development by promoting a large number of jobs (direct and indirect) and energizing other subsectors of the economy [1]. The last is particularly important in developing countries such as Colombia, which was ranked third (13 Mt), after Brazil (56.6 Mt) and Mexico (40.0 Mt), based on cement production in Latin America and the Caribbean [2]. So, in order to minimize the environmental impacts while increasing social development and economic growth, the City Hall of Santiago de Cali is evaluating the implementation of a circular economy model for the construction sector, which understands circular economy (CE) as a production and consumption system that promotes efficiency in the use of materials, water, and energy, taking into account the resilience of ecosystems and the circular use of material flows through the implementation of technological innovations, alliances, and collaborations between stakeholders (e.g., raw material producers, building companies, users, and final disposal actors) and the promotion of business models that respond to the fundamentals of sustainable development [3]. Among other strategies, the CE model stimulates the use of eco-concretes and eco-mortars that replace cement with by-products from local industries [4].

Aluminum dross is a by-product of aluminum production that can be classified based on the aluminum content in white (15 to 70% Al), black (12 to 18% Al), and salt cake (3 to 5% Al). Most frequently, white aluminum dross is reprocessed due to its high quality. On the other hand, black dross and aluminum salt cake, which are normally landfilled, have been investigated in building materials as additive, calcium aluminate cement, cement, and fine aggregate replacements [5] (Figure 1).

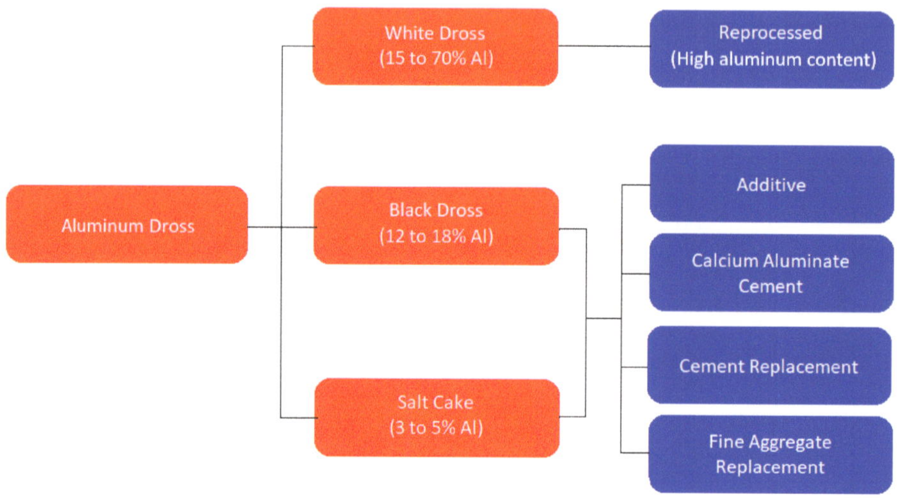

Figure 1. Some aluminum dross uses in building materials [5].

Pereira et al. (2000) investigated the use of salt cake, containing oxides of aluminum, as a replacement for cement and sand in mortars. They reported satisfactory cement replacement around 10% and sand replacement from 30% to 50% [6]. As supplementary cementitious materials, Elinwa and Mbadike (2011) produced concrete by replacing various percentages of cement with aluminum waste. From their results, it was found that the compressive and flexural strengths of concrete with replacement levels of 10 to 15% are comparable to those of control concrete. Another result was that concrete-containing aluminum waste retards the concrete setting times, which is desirable for hot-weather concreting [7]. Similarly, Reddy & Neeraja (2016) studied the use of aluminum dross as a cement replacement in concrete. They reported that for up to 15% replacement of cement by secondary aluminum dross, the performance is comparable with conventional concrete [8]. In the same way, Mailar et al. (2016) reported superior mechanical and durability properties with a concrete by replacing 20% of the cement with aluminum dross. They also noticed a delay in the setting process [9]. Aditionally, Panditharadhya et al. (2018) investigated the partial replacement of Portland cement with aluminum dross (5%, 10%, 15%, and 20%) and its combined use with a pozzolan in concrete. They reported that by increasing the percentage of dross, the mixture required more water to keep a normal consistency. They also found that as the percentage of dross replacement increased, the initial setting time increased, but the final setting time decreased. Regarding the mechanical properties, they observed that as the percentage of cement substitution increased, the compressive, tensile, and flexural strengths decreased, an aspect that was associated with the increase of voids in the mixture by the addition of dross [10]. A detailed compilation of these results is presented in Table 1.

Table 1. Results from mortars and concretes using aluminum dross and salt cake as supplementary cementitious materials.

Reference	Material with Aluminum By-Products as Cement Phase	Compressive Strength	Flexural Strength	Consistency Indicated by the Setting Time
[6]	Mortar with 10% aluminum salt cake	↔	↔	N.A.
[7]	Concrete with 10 to 15% aluminum dross	↔	↔	↑
[8]	Concrete with 15% aluminum dross	↔	↔	N.A.
[9]	Concrete with 20% aluminum dross	↑	↑	↑
[10]	Concrete with 5%, 10%, 15% and 20% aluminum dross	↓	↓	↓

↑: Increase, ↓: Decrease, ↔: Not affected.

Although the above-mentioned aluminum dross and salt cake valorization experiences are significantly important for sustainable development, as they reduce Portland cement consumption and mitigate landfill-associated problems, the relationship between the dross characteristics and the mechanical performance of the investigated building materials is not completely clear. In this sense, it is quite important to understand the aluminum dross inactivation process, which will certainly impact the dross properties and, therefore, the mechanical performance of building materials using this by-product. For example, Dai and Apelian (2017) used primary (rich in aluminum and magnesium) and secondary (complex impurities mixture) aluminum dross in cement mortars and studied the effect of particle size, weight fraction, and dross origin on the mechanical properties of mortar specimens. Obtaining only promising results with the primary dross, they concluded that these factors have important effects on the uniformity of the microstructure of mortar specimens [11]. Therefore, this research aims to fill the knowledge gap in the relationship between secondary aluminum dross inactivation and particle size and the mechanical performance of mortars containing 5, 10, and 20% aluminum dross as cement replacement.

2. Materials and Methods

The proposed methodology for this study consisted of five phases, which are (1) Preliminaries, (2) By-product Characterization, (3) Mix Design, (4) Mortars and Properties, and (5) Results Analysis (Figure 2). It is noteworthy that the Colombian norms—NTC standards—are an adaptation of the American Society for Testing and Materials (ASTM) standards (Appendix A) and were used in this project. A brief description of each phase is presented as follows:

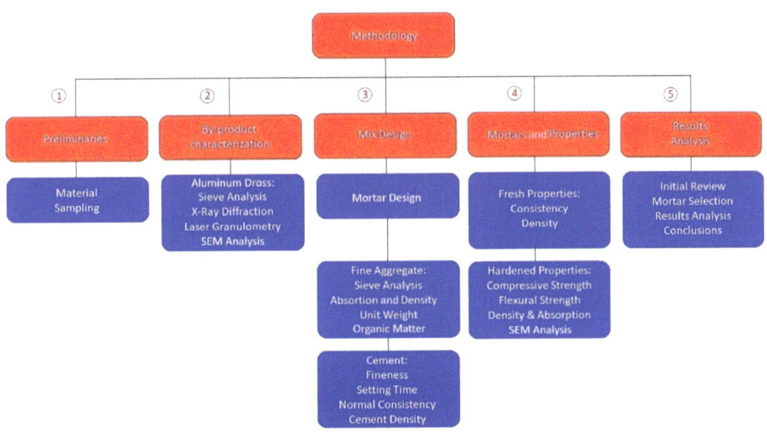

Figure 2. Methodology, which includes five phases: (1) Preliminaries, (2) By-product Characterization, (3) Mix Design, (4) Mortars and Properties, and (5) Results Analysis. Adapted from [12].

The first phase included all the preliminary activities to obtain the materials for mortar manufacturing. In this case, river sand from the Cauca River (fine aggregate), high early strength cement (Argos), and aluminum dross—a by-product obtained from a local aluminum industry—were selected. The chemical properties of the cement are presented in Appendix B, which is based on a quality control report done by the producer during the project.

The second phase aimed to exclusively characterize the aluminum dross by means of sieve analysis, X-ray diffraction (Rigaku Multiflex with Search/Match analysis), SEM—Scanning Electron Microscopy (Japanese Manufacturer of Scientific Instruments JEOL Model JSM-6490)—and laser granulometry (Malverm Instruments Model Mastersizer 2000). The X-ray diffraction measurements were obtained with Brag-Brentano geometry and Cu kα. The angular range was $2\theta = 6°–110°$ in steps of $0.02°$, and the counting time was 8 s/step. SEM analysis at 100×, 1000×, and 5000× on gold-coated samples was employed to examine the morphology. The particle-size measurements were conducted using the laser diffraction method with wet dispersion in water. The characterization of aluminum dross was done before and after inactivation and also after grinding to obtain the fine ($\leq$75 μm), intermediate (75–150 μm), and coarse particle sizes (150–300 μm).

The aim of the third phase was the mortar mix design using characterization of the fine aggregate and cement. The mix design methodology was developed by Sánchez de Guzmán [13] and requires definition of a goal of compressive strength and consistency, which in this case were 28 MPa and plasticity over 110% in the mortar test flow. Additionally, this methodology requires the characterization of all components, which in this case was done through sieve analyses (NTC 77), absorption and density (NTC 237 and 176), unit weight (NTC 92), and organic matter (NTC 127) for the fine aggregate characterization. Similarly, fineness (NTC 33), setting time (NTC 118), normal consistency (NTC 110), and density (NTC 221) were used for the cement [14–21].

In the fourth phase, after the mix design was obtained for the reference mortar, mixes with cement replacements of 5, 10, and 20% (on a weight basis) using aluminum dross with different particle sizes (fine, intermediate, and coarse) were manufactured. Investigated mixes were then evaluated using three replicates regarding their fresh and hardened properties using the consistency (NTC 5784), density (NTC 1926), compressive and flexural strengths (NTC 220 and NTC 120), SEM analyses, and density and absorption tests [22–25]. Thus, mix designations, descriptions, and sample numbers are presented in Table 2.

Table 2. Mortar mix designation, aluminum dross content, and particle size.

Mix Designation	Aluminum Dross Content (%)	Aluminum Dross Particle Size (μm)
REF	0	0
5F	5	Fine ($\leq$75 μm)
5I	5	Intermediate (75–150 μm)
5C	5	Coarse (150–300 μm)
10F	10	Fine ($\leq$75 μm)
10I	10	Intermediate (75–150 μm)
10C	10	Coarse (150–300 μm)
20F	20	Fine ($\leq$75 μm)
20I	20	Intermediate (75–150 μm)
20C	20	Coarse (150–300 μm)

Lastly, the fifth phase consisted of analyzing the results for selecting the best-performance mortar mixtures in terms of the obtained mechanical properties and consistency. In this case, mix design and application criteria such as NTC–2017 were used to perform the analysis [26].

3. Results Analysis

3.1. Preliminary Activities

The aim of this phase was to obtain high-quality raw materials for the mortars, especially regarding the aluminum dross. Twenty-five kilograms of this by-product was obtained from a local industry that mainly produces aluminum profiles by primary and secondary processes that generate white dross, black dross, and salt cake. The dross used in this study is that which passes through a 2 mm sieve and is collected from different batches following the Orozco-Erazo et al. (2022) procedure [27].

At the university laboratory, the aluminum dross inactivation was done by water immersion for 5 days, followed by drying in an oven at 115 to 120 °C for 24 h. In this process, which is represented by equation (1), water reacts with aluminum nitride (AlN) to form ammonia (NH_3) and aluminum oxide (Al_2O_3), as indicated by Shinzato & Hypolito (2005) [28]. Finally, the aluminum dross was exposed to air for 7 days. The whole process was also called dross washing due to the use of water as the main inactivator.

$$AlN + H_2O \rightarrow NH_3 + Al(OH)_3 \tag{1}$$

Afterwards, the dross was exposed to abrasion at the Los Angeles machine, which was operated at 1000 rpm with 12 iron balls for 30 min. Then, a separation process with sieves #50 (300 μm opening size), #100 (150 μm opening size), #200 (75 μm opening size), and a bottom sieve was performed to classify the different particle sizes required. More details from the aluminum dross collection and inactivation processes can be observed in Figure 3.

(**a**) Aluminum dross collection at the local aluminum industry.

(**b**) Abrasion of aluminum dross using Los Angeles machine.

(**c**) Dross separation using sieves #50 (300 μm), #100 (150 μm), #200 (75 μm), and bottom sieve.

Figure 3. Aluminum dross collection and inactivation processes.

3.2. By-Product Characterization

3.2.1. Aluminum Dross Inactivation

The washing process or inactivation of the aluminum dross caused chemical and physical changes that were detected by XRD, SEM, and laser granulometry. First, there was a reduction in aluminum nitride (AlN) and formation of aluminum hydroxide (Al $(OH)_3$). This can be observed when comparing the XRD analyses after (a) and before (b) washing in Figure 4. There was a reduction in and appearance of peaks 2 and 6, respectively.

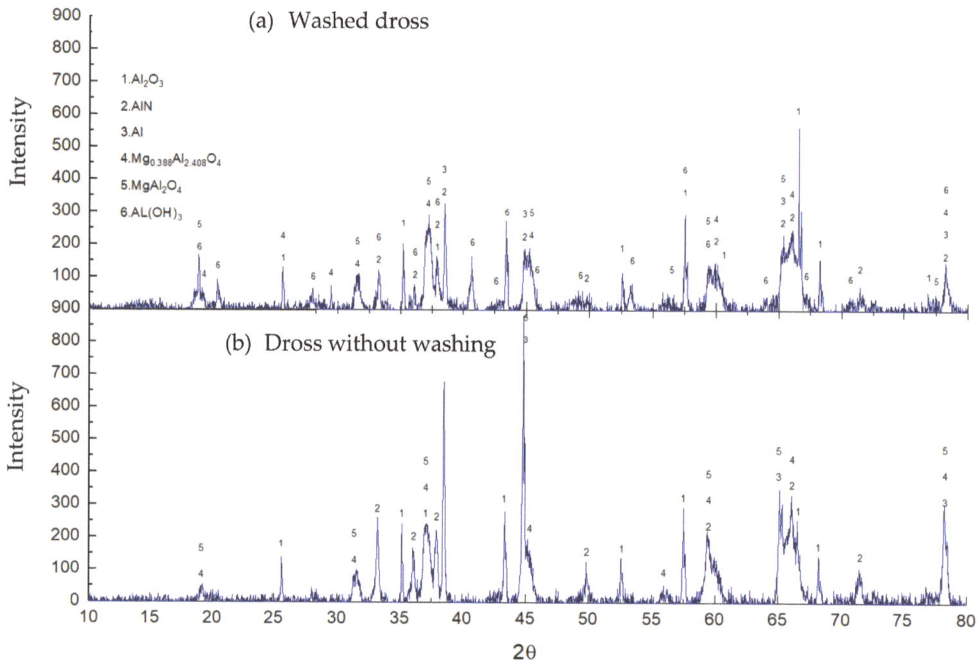

Figure 4. XRD analyses from aluminum dross washed (**a**) and without washing (**b**).

Second, there was agglomeration of the dross particles, which increased the particle size range from 20–800 μm to 100–1000 μm (Figure 5). Also, some morphological changes can be observed in Figure 6 with the SEM analyses at 100×, 1000×, and 5000× from aluminum dross washed (a) and without washing (b). The appearance of larger particles with softer surfaces (fiber type) after the aluminum dross inactivation is noticeable. Similarly, the increase in particle size due to dross inactivation can be observed in the higher specific surface area, uniformity coefficient, and average particle diameter that were calculated from the laser granulometry results and compiled in Table 3.

3.2.2. Aluminum Dross Grinding

Although the aluminum dross grinding did not cause chemical changes, there were significant ones in the physical properties, mainly particle size. Laser granulometry results indicate a particle size between 100 and 600 μm, with an average particle diameter of 234.9 μm for the coarse fraction of the aluminum dross. Similarly, a particle size between 30 and 300 μm, with an average particle diameter of 80.0 μm for the intermediate fraction. Also, a particle size between 1 and 100 μm, with an average particle diameter of 22.9 μm, was determined for the fine fraction (Figure 7).

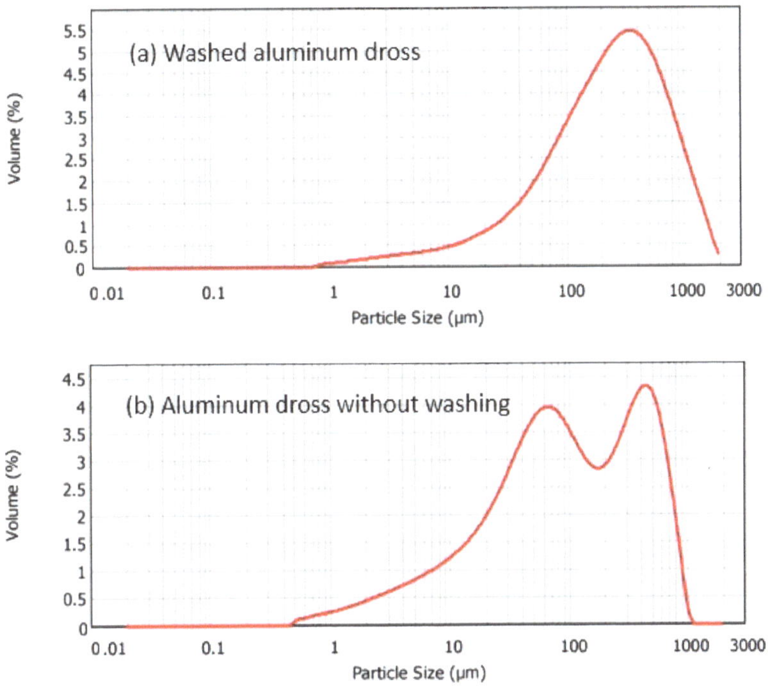

Figure 5. Particle-size distribution obtained by laser granulometry from aluminum dross washed (**a**) and without washing (**b**).

(**a**) Washed aluminum dross.

(**b**) Aluminum dross without washing.

Figure 6. SEM analyses at 100×, 1000×, and 5000× from aluminum dross washed (**a**) and without washing (**b**).

Table 3. Specific surface, uniformity coefficient, and average particle diameter of aluminum dross at different conditions.

Dross Condition	Specific Surface (m^2/g)	Uniformity Coefficient	Average Particle Diameter (μm)
Washed dross	0.121	1.03	243.105
Dross without washing	0.316	1.71	92.432
Coarse dross (washed)	0.329	0.48	234.895
Intermediate dross (washed)	0.500	0.624	80.022
Fine dross (washed)	0.829	0.809	22.912

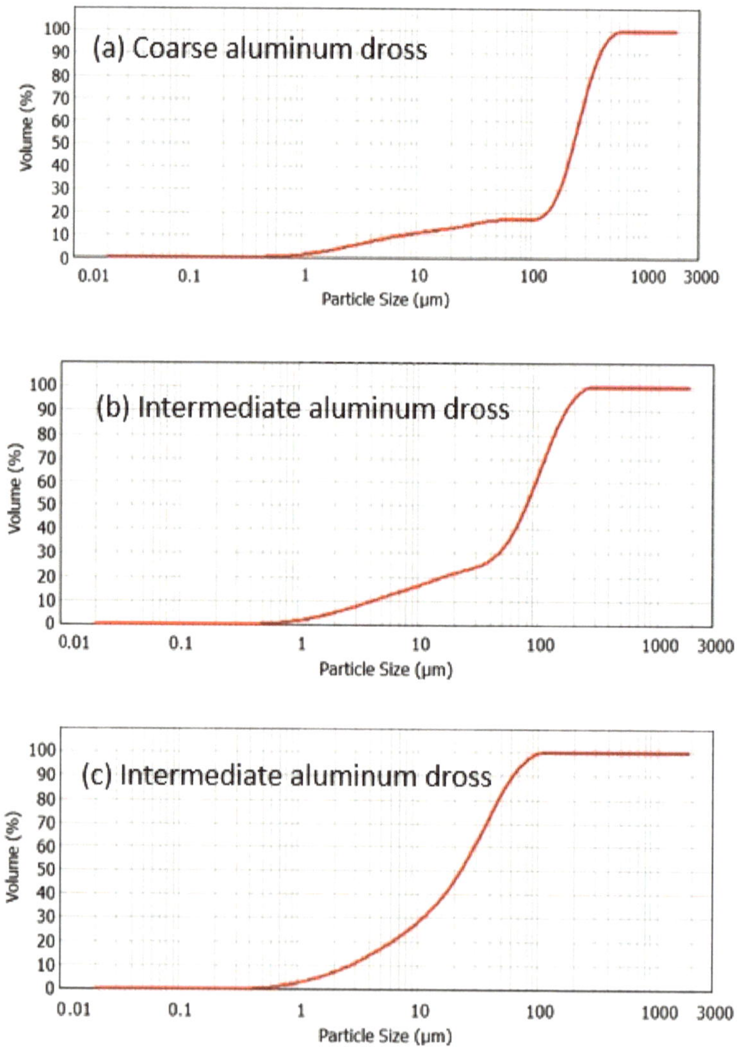

Figure 7. Particle-size distribution obtained by laser granulometry from coarse (**a**), intermediate (**b**), and fine (**c**) washed aluminum dross.

Additionally, a summary of the specific surface areas, uniformity coefficients, and average particle diameters of the three fractions is presented in Table 3. Considering that the aluminum dross will be part of the cementitious phase in this research, it is important to compare these parameters with those obtained for the cement. In this context, with a specific surface of 0.342 m^2/g (3420 cm^2/g), cement is much more similar to coarse dross. Considering its uniform coefficient of 0.701, cement is much more similar to intermediate and fine dross. Also, in relation to its average particle diameter, fine dross almost doubles (22.9 μm) that of the used cement (14.36 μm).

3.3. Mix Design

Following the corresponding standards, first, the fine aggregate—which is river sand from the Cauca River (Colombia)—was physically characterized by calculating the sand density at saturated and surface-dried conditions, absorption, natural humidity, fineness modulus and nominal max. particle size. These two last parameters were calculated from the sieve analysis and granulometric curve presented in Figure 8. The parameters used in the reference mortar mix design are presented in Table 4.

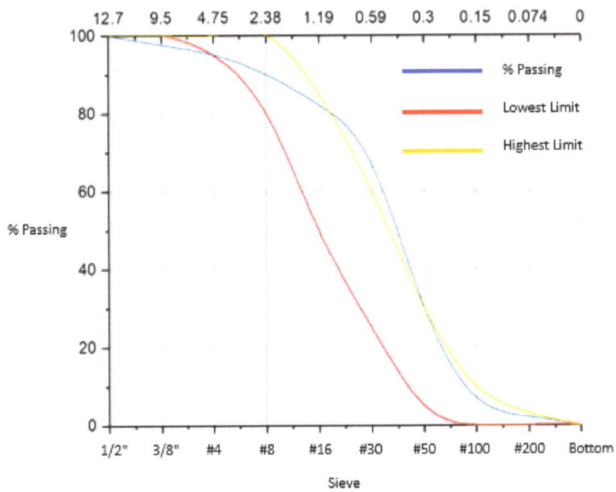

Figure 8. Granulometry analysis of fine aggregate.

Table 4. Physical parameters of the fine aggregate and cement used in the mortar mix design.

Parameter	Value
Sand density [a]	2.68 g/cm^3
Sand absorption	2.56%
Sand natural humidity	8.50%
Sand fineness modulus	2.31
Sand nominal max. particle size	4.75 mm (Sieve #4)
Initial setting time (cement)	53 min
Final setting time (cement)	195 min (3.25 h)
Cement density	2.88 g/cm^3

[a] Fine aggregate at saturated and surface-dried conditions.

Second, the reference mortar composition was determined by the mix design method proposed by Sanchez Guzman (2011) [13], which requires defining the compressive strength (28 MPa) and consistency (plastic) of the mortar. So, using the physical characterization of the fine aggregate and cement, the following mix composition was defined for one cubic meter of reference mortar (Table 5). Then, mixes with cement replacements of 5, 10, and 20% aluminum dross (fine, intermediate, and coarse) were performed to this mix design.

Table 5. Components for one cubic meter of reference mortar with a compressive strength of 28 MPa and plastic consistency.

Component	Mass (kg)	Volume (L)
Fine aggregate	1255.0 [a]	467.94
Cement	550.0	190.97
Water	340.0	340.0

[a] Fine aggregate at saturated and surface-dried conditions.

3.4. Fresh and Hardened Properties of Mortars

Mortar flow results show that mortar mixes with aluminum dross content up to 20% did not substantially affect their design consistencies, which were over 110% in all cases. However, there was a slight consistency decrease when increasing the aluminum dross content in the mortar mixes. These results are in agreement with those reported in [10], where there was a decrease in the consistency, indicated by a decrease in the final setting time.

On the other hand, the mechanical properties indicated by the compressive strength at 7, 28, and 56 days of the investigated mortar mixes are presented in Figure 9. Results show that mortar mixes containing up to 20% fine aluminum dross satisfy the required 28 MPa compressive strength after 28 and 56 days. On the contrary, samples using 20% aluminum dross with coarse and intermediate particle sizes did not satisfy the compressive strength required in the mix design. Although there was an increase in the mechanical performance after 7, 28, and 56 days of casting, this was not enough to achieve the required performance in these samples. This trend was also reported by Panditharadhya, Mulangi, and Shankar (2019), who studied concrete with 5%, 10%, 15%, and 20% aluminum dross [10].

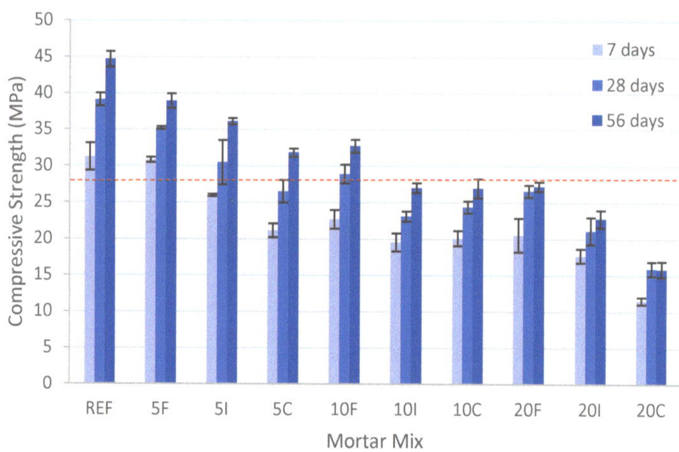

Figure 9. Compressive strength at 7, 28, and 56 days from the different mortar mixes partially replacing cement (0, 5, 10, and 20%) with coarse (C), intermediate (I), and fine (F) aluminum dross.

Similar to the compressive strength, the flexural strength decreased with the aluminum dross content used in the mortar samples (Figure 10). In this case, mortar mixes containing up to 20% aluminum dross with fine and intermediate particle sizes have acceptable mechanical performances. For example, these satisfactorily comply with the minimum mechanical resistance required for paving stones (4.2 MPa according to the NTC 2017 standard). Similar flexural strengths were obtained recently by one of the authors using recycled aggregates obtained from old paving stones in Cauca (Colombia) [29].

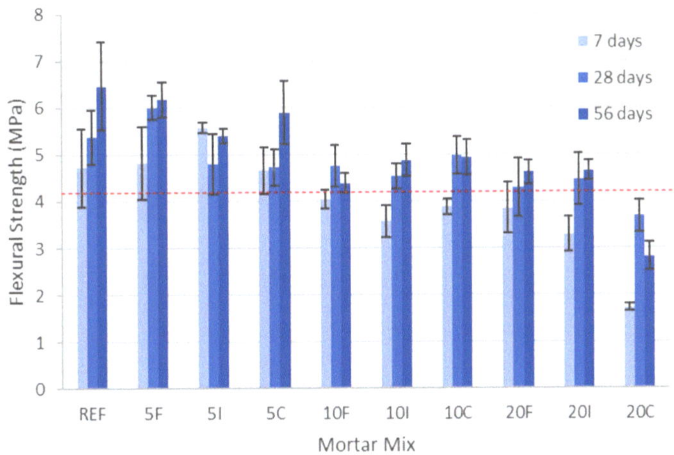

Figure 10. Flexural strength at 7, 28, and 56 days from the different mortar mixes partially replacing cement (0, 5, 10, and 20%) with coarse (C), intermediate (I), and fine (F) aluminum dross.

Moreover, when analyzing the correlation between the compressive and flexural strengths (Figure 11), it is observed that the compressive strength is between five and seven times the flexural strength, which is in agreement with other reported results. Similarly, the decreasing trend in the flexural and compressive strengths when increasing the amount of aluminum dross from 0% (REF) to 20% can be explained by the combined effect of the mortar porosity and dross particle sizes. Indeed, although not reflected in the density and absorption tests, there was a slight porosity increase observed in the SEM images at 100×, 2000×, and 5000× of mortar samples (Figure 12). This porosity was produced by the residual gases (e.g., NH_3) released during the curing process by the aluminum dross. This can be confirmed through the XRD analyses (Figure 4), where there was a reduction in peaks 2 and 6, but still, there was not complete elimination of the aluminum nitride. However, compressive and flexural strengths reported here with the secondary aluminum dross are significantly higher than those reported by Dai and Apelian (2017) when evaluating mortars with 20% secondary dross. They reported a decrease in the mortar flexural strengths from 86 to 100% for fine and coarse particle sizes, respectively [11].

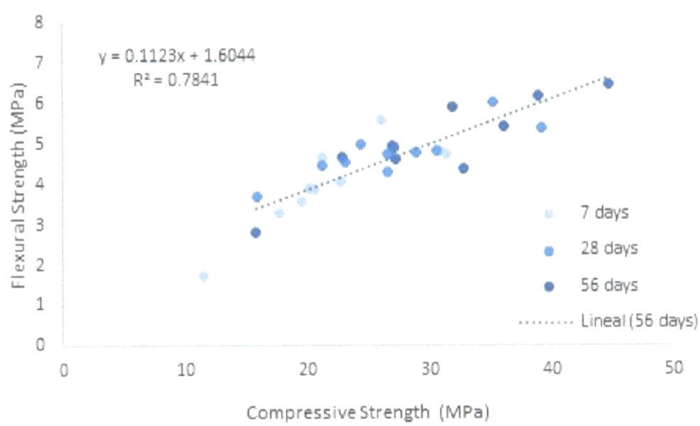

Figure 11. Relationship between the flexural and compressive strengths of mortar mixtures using aluminum dross as cement replacement (0, 5, 10, and 20%) with different particle sizes (F: Fine, I: Intermediate, C: Coarse).

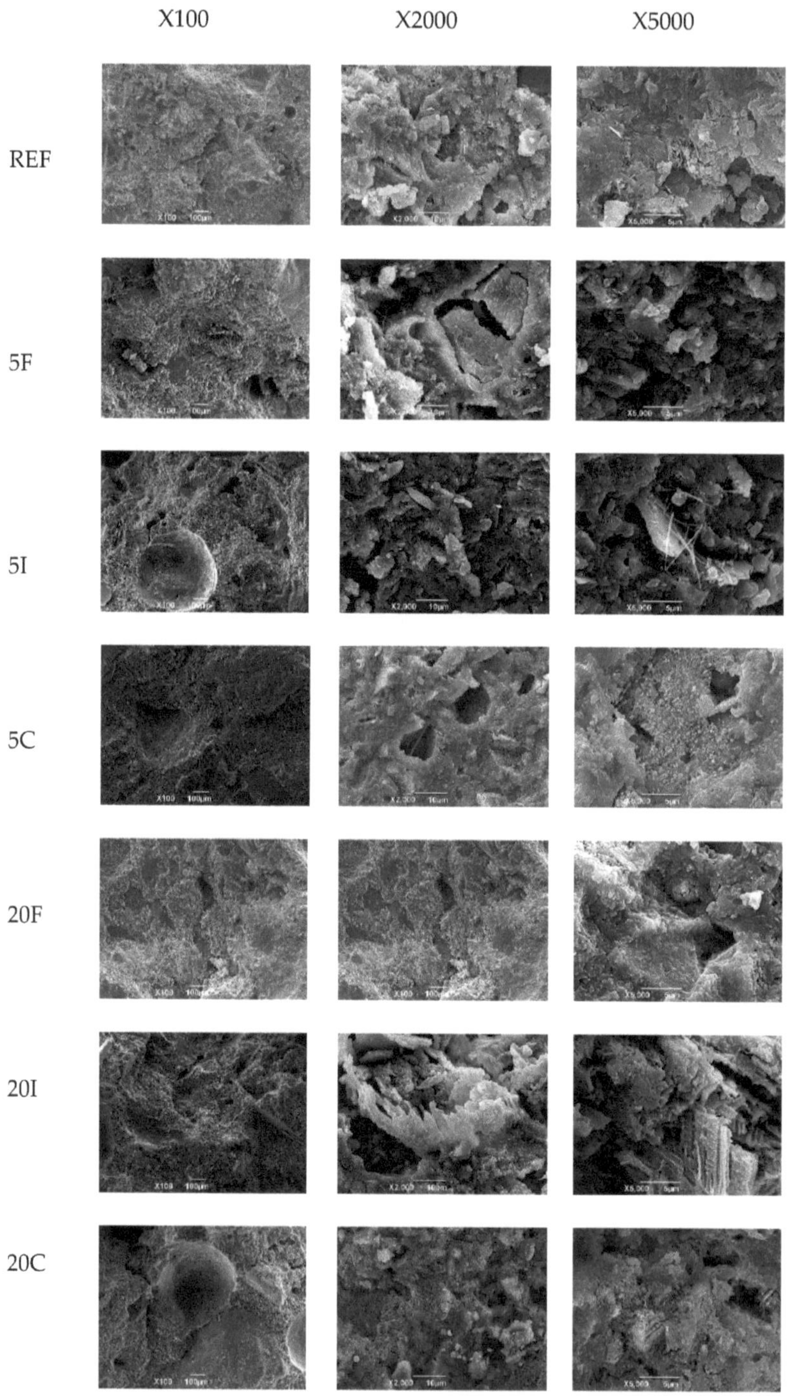

Figure 12. SEM images at 100×, 2000×, and 5000× of mortar mixes with partial replacement of cement with aluminum dross.

Finally, it must be pointed out that prior to real-scale application of the investigated mortars, their durability aspects and environmental impacts should be investigated to include a complete life-cycle approach to evaluation. Also, optimization of the dross inactivation process is identified as a key factor in significantly contributing to the net zero carbon program signed by Santiago de Cali, which is aimed at reducing the current and projected consumption of cement in Colombia, the third-largest CO_2 producer from cement in the Latin America region, after Brazil and Mexico (Figure 13) [30,31]. So, simultaneous research is currently being performed by the authors on the design of an industrial-scale production process and a financial feasibility analysis for a specific case study. The findings validated the potential of the proposed recovery process and provided insights into determining the market price for the resulting product [32].

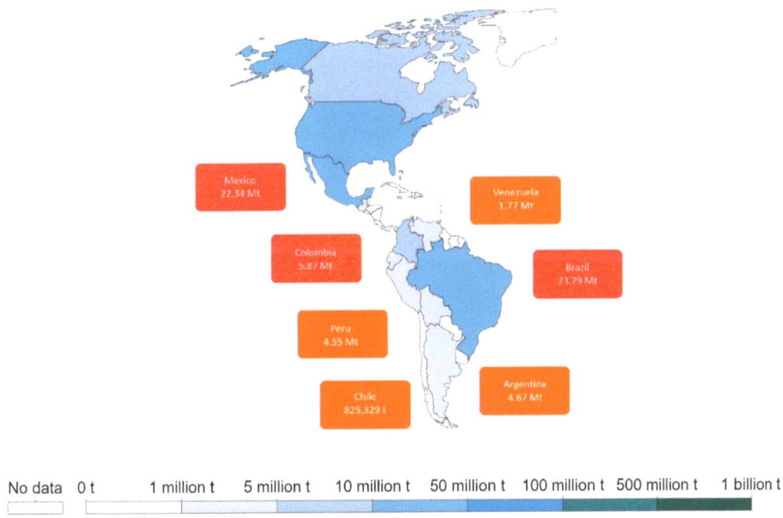

Figure 13. CO_2 emissions from cement in the Latin America region modified from Our World Data based on the Global Carbon Budget (2022) [30].

4. Conclusions

This article described first the inactivation of aluminum dross and second the evaluation of this aluminum dross with three different particle sizes as supplementary cementitious materials in mortar mixes designed for a compressive strength of 28 MPa and a plastic consistency of 110% in the mortar flow test. Mortar flow test results and compressive and flexural strengths of mortar mixes containing up to 20% fine and intermediate aluminum dross as cement replacement were satisfactory. However, mortar mixes containing 20% coarse aluminum dross did not satisfy the design applications. This might be attributed to the combined effect of a higher mortar porosity and large dross particle size. Mortar porosity was increased by the gases released by the aluminum dross during the curing process. Although aluminum nitride reacted with water to form ammonia and aluminum oxide during the aluminum dross inactivation process, there was still non-reacted AlN releasing gaseous NH_3 during the curing process of this mortar mix.

Although more research should be performed prior to the mortar's real-scale application, this article reports the importance of industrial symbiosis between the construction and aluminum industries from Santiago de Cali (Colombia) in reducing carbon dioxide emissions due to the lower consumption of cement. Also, this reports the use of aluminum dross as a supplementary cementitious material in mortars to reduce the potential impacts on the environment and public health caused by aluminum dross disposal.

Author Contributions: Conceptualization, M.A.R.-M., A.G.-G. and A.M.-R.; methodology, M.A.R.-M. and A.G.-G.; validation, D.P.-M., M.A.R.-M. and A.G.-G.; formal analysis, D.P.-M., M.A.R.-M. and A.G.-G.; investigation, D.P.-M., M.A.R.-M. and A.G.-G.; resources, M.F.M.-V.; data curation, A.M.-R.; writing—original draft preparation, A.M.-R.; writing—review and editing, A.M.-R.; visualization, D.P.-M. and A.M.-R.; supervision, M.F.M.-V.; project administration, M.F.M.-V.; funding acquisition, M.F.M.-V. All authors have read and agreed to the published version of the manuscript.

Funding: The authors would like to thank the financial support given by the internal call of Pontificia Universidad Javeriana Cali through the grant 2168-2021.

Institutional Review Board Statement: Not applicable.

Informed Consent Statement: Not applicable.

Data Availability Statement: Not applicable.

Acknowledgments: The authors would like to thank Alumina S.A. for providing the aluminum dross generated during the aluminum smelting processes. They also acknowledge the Servicio Geológico Colombiano sede Cali for allowing them to use their laboratories and equipment.

Conflicts of Interest: The authors declare no conflict of interest.

Appendix A

This section presents the equivalence between the NTC and ASTM Standards used in this research project (Table A1).

Table A1. Equivalence between NTC and ASTM standards.

NTC Standard	ASTM Standard
NTC–77 Método de ensayo para el análisis por tamizado de los agregados finos y gruesos	ASTM C136/C136M-19 Standard Test Method for Sieve Analysis of Fine and Coarse Aggregates
NTC–237 Método de ensayo para determinar la densidad relativa (gravedad especifica) y la absorción del agregado fino	ASTM C128-22 Standard Test Method for Relative Density (Specific Gravity) and Absorption of Fine Aggregate
NTC–176 Método de ensayo para determinar la densidad relativa (gravedad específica) y la absorción del agregado grueso	ASTM C127-15 Standard Test Method for Relative Density (Specific Gravity) and Absorption of Coarse Aggregate
NTC–92 Método de ensayo para la determinación de la densidad volumétrica (masa unitaria) y vacíos en agregados	ASTM C29/C29M-17a Standard Test Method for Bulk Density ("Unit Weight") and Voids in Aggregate
NTC–33 Método de ensayo para determinar la finura del cemento hidráulico por medio del aparato Blaine de permeabilidad al aire	ASTM C204-23 Standard Test Methods for Fineness of Hydraulic Cement by Air-Permeability Apparatus
NTC–118 Método de ensayo para determinar el tiempo de fraguado del cemento hidráulico mediante aguja de Vicat	ASTM C191-21 Standard Test Methods for Time of Setting of Hydraulic Cement by Vicat Needle
NTC–110 Cantidad de agua requerida para la consistencia normal de una pasta de cemento hidráulico	ASTM C187-23 Standard Test Method for Amount of Water Required for Normal Consistency of Hydraulic Cement Paste
NTC–221 Método de ensayo para determinar la densidad del cemento hidráulico	ASTM C188-17 Standard Test Method for Density of Hydraulic Cement
NTC–5784 Método de ensayo para determinar la fluidez de morteros de cemento hidráulico	ASTM C1437-20 Standard Test Method for Flow of Hydraulic Cement Mortar
NTC–1926 Método de ensayo para determinar la densidad (masa unitaria), el rendimiento y el contenido de aire por gravimetría del concreto	ASTM C138/C138M-23 Standard Test Method for Density (Unit Weight), Yield, and Air Content (Gravimetric) of Concrete
NTC–220 Determinación de la resistencia de morteros de cemento hidráulico a la compresión, usando cubos de 50 mm o 2 pulgadas de lado	ASTM C109/C109M-21 Standard Test Method for Compressive Strength of Hydraulic Cement Mortars (Using 2-in. or [50 mm] Cube Specimens)
NTC–120 Método de ensayo para determinar la resistencia a la flexión de morteros de cemento hidráulico	ASTM C348-21 Standard Test Method for Flexural Strength of Hydraulic-Cement Mortars
NTC–2017 Adoquines de concreto para pavimento	ASTM C1319-23 Standard Specification for Concrete Grid Paving Units

Appendix B

This section describes the chemical properties of the cement used in this project. This information comes from a quality control report performed by the cement producer from 1 to 30 September 2021 in the manufacturing plant located in Yumbo (Colombia). Table A2 presents the main cement phases and compounds statistically analyzed from thirty samples of cement. With 68%, C_3S (alite) and C_2S (belite) represent the major phases responsible for the mechanical performance. Similarly, the C_3A (aluminate) phase, which is lower than the maximum allowed (15%), is responsible for the early strength obtained. Finally, it is observed that the chemical compounds are among the limits established for cement durability. This is particularly important regarding MgO (periclase) and alkali compounds (K_2O and N_2O).

Table A2. Main cement phases and compounds of the cement used in this project.

Abbreviation	Average	SD	CV	Max	Min	Range
C_3S	53.91	1.07	1.98	56.65	52.25	4.40
C_2S	14.44	0.86	5.96	16.39	12.95	3.45
C_4AF	11.83	0.21	1.77	12.25	11.42	0.83
C_3A	4.85	0.29	6.06	5.54	4.39	1.15
SiO_2	20.85	0.32	1.56	21.55	20.33	1.22
Al_2O_3	4.68	0.12	2.57	4.94	4.44	0.50
Fe_2O_3	4.26	0.02	0.41	4.29	4.23	0.07
CaO	61.77	0.47	0.75	62.72	60.71	2.00
MgO	1.38	0.03	2.23	1.43	1.33	0.11
Na_2O	0.12	0.00	3.40	0.13	0.12	0.02
K_2O	0.20	0.02	10.61	0.24	0.16	0.08
SO_3	2.93	0.13	4.43	3.23	2.66	0.56
Free lime	0.97	0.18	18.65	1.49	0.67	0.82
Alkaline equivalent	0.26	0.02	6.39	0.29	0.23	0.07

SD: Standard Deviation, CV: Variation Coefficient, Max: Maximum datum, Min: Minimum datum.

References

1. Maury-Ramírez, A.; Illera-Perozo, D.; Mesa, J.A. Circular Economy in the Construction Sector: A Case Study of Santiago de Cali (Colombia). *Sustainability* **2022**, *14*, 1923. [CrossRef]
2. Villagrán-Zaccardi, Y.; Pareja, R.; Rojas, L.; Irassar, E.; Torres-Acosta, A.; Tobón, J.; Vanderley, J. Overview of cement and concrete production in Latin America and the Caribbean with a focus on the goals of reaching carbon neutrality. *RILEM Tech. Lett.* **2022**, *7*, 30. [CrossRef]
3. Economía Circular—DANE. Available online: http://www.andi.com.co/Uploads/economia-circular-1-reporte.pdf (accessed on 6 April 2023).
4. Maury-Ramírez, A.; De Belie, N. Environmental and Economic Assessment of Eco-Concrete for Residential Buildings: A Case Study of Santiago de Cali (Colombia). *Sustainability* **2023**, *15*, 12032. [CrossRef]
5. Lemos-Micolta, E.D.; Chilito-Bolaños, L.C.; Maya-Soto, J.C.; Gómez-Gómez, A.; Rojas-Manzano, M.A. Uso de la Escoria de Aluminio en el Concreto–Revision del Estado del Arte. In Proceedings of the IX Congreso Internacional y 23a Reunión Técnica (AATH 2020), La Plata, Argentina, 2–6 November 2020.
6. Pereira, D.A.; de Aguiar, B.; Castro, F.; Almeida, M.F.; Labrincha, J.A. Mechanical behaviour of Portland cement mortars with incorporation of Al-containing salt slags. *Cem. Concr. Res.* **2000**, *30*, 1131–1138. [CrossRef]
7. Elinwa, A.U.; Mbadike, E. The use of aluminium waste for concrete production. *J. Asian Architect. Build Eng.* **2011**, *10*, 217–220. [CrossRef]
8. Reddy, M.S.; Neeraja, D. Mechanical and durability aspects of concrete incorporating secondary aluminium slag. *Resour. Effic. Technol.* **2016**, *2*, 225–232. [CrossRef]
9. Mailar, G.; Raghavendra, N.S.; Sreedhara, B.M.; Manu, D.S.; Hiremath, P.; Jayakesh, K. Investigation of concrete produced using recycled aluminium dross for hot weather concreting conditions. *Resour. Effic. Technol.* **2016**, *2*, 68–80. [CrossRef]
10. Panditharadhya, B.J.; Sampath, V.; Mulangi, R.H.; Shankar, A.U.R. Mechanical properties of pavement quality concrete with secondary aluminium dross as partial replacement for ordinary portland cement. *IOP Conf. Ser. Mater. Sci. Eng.* **2018**, *431*, 032011. [CrossRef]
11. Dai, C.; Apelian, D. Fabrication and characterization of aluminum dross-containing mortar composites: Upcycling of a waste product. *J. Sustain. Metall.* **2017**, *3*, 230–238. [CrossRef]
12. Parra-Molina, D. Estudio del Comportamiento Mecánico de Morteros con Sustitución Parcial de Cemento por Escoria de Aluminio y Efecto de la Reducción del Tamaño de Partícula. Master's Thesis, Universidad Javeriana Cali, Santiago de Cali, Colombia, 23 June 2023.
13. Sánchez De Guzmán, D. *Tecnología del Concreto y del Mortero*; Bhandar Editores: Bogotá, Colombia, 2001.

14. *NTC-77;* Método de Ensayo para el Análisis por Tamizado de los Agregados Finos y Gruesos. ICONTEC: Bogotá, Colombia, 2018.
15. *NTC-237;* Método de Ensayo para Determinar la Densidad Relativa (Gravedad Especifica) y la Absorción del Agregado Fino. ICONTEC: Bogotá, Colombia, 2020.
16. *NTC-176;* Método de Ensayo para Determinar la Densidad Relativa (Gravedad Específica) y la Absorción del Agregado Grueso. ICONTEC: Bogotá, Colombia, 2019.
17. *NTC-92;* Método de Ensayo para la Determinación de la Densidad Volumétrica (Masa Unitaria) y Vacíos en Agregados. ICONTEC: Bogotá, Colombia, 1995.
18. *NTC-33;* Método de Ensayo para Determinar la Finura del Cemento Hidráulico por Medio del Aparato Blaine de Permeabilidad al Aire. ICONTEC: Bogotá, Colombia, 2019.
19. *NTC-118;* Método de Ensayo para Determinar el Tiempo de Fraguado del Cemento Hidráulico Mediante Aguja de Vicat. ICONTEC: Bogotá, Colombia, 2020.
20. *NTC-110;* Cantidad de Agua Requerida para la Consistencia Normal de una Pasta de Cemento Hidráulico. ICONTEC: Bogotá, Colombia, 2019.
21. *NTC-221;* Método de Ensayo para Determinar la Densidad del Cemento Hidráulico. ICONTEC: Bogotá, Colombia, 2019.
22. *NTC-5784;* Método de Ensayo para Determinar la Fluidez de Morteros de Cemento Hidráulico. ICONTEC: Bogotá, Colombia, 2021.
23. *NTC-1926;* Método de Ensayo para Determinar la Densidad (Masa Unitaria), el Rendimiento y el Contenido de Aire por Gravimetría del Concreto. ICONTEC: Bogotá, Colombia, 2013.
24. *NTC-220;* Determinación de la Resistencia de Morteros de Cemento Hidráulico a la Compresión, Usando Cubos de 50 mm o 2 Pulgadas de Lado. ICONTEC: Bogotá, Colombia, 2021.
25. *NTC-120;* Método de Ensayo para Determinar la Resistencia a la Flexión de Morteros de Cemento Hidráulico. ICONTEC: Bogotá, Colombia, 2022.
26. *NTC-2017;* Adoquines de Concreto para Pavimento. ICONTEC: Bogotá, Colombia, 2018.
27. Orozco-Erazo, D.A.; Vega-Báez, J.A. Estudio del Efecto del Tamaño de Partícula de Escoria de Aluminio Calcinada Utilizada como Reemplazo Parcial del Cemento en las Propiedades Mecánicas de Morteros. Bachelor's Thesis, Pontificia Universidad Javeriana Cali, Santiago de Cali, Colombia, September 2022.
28. Shinzato, M.C.; Hypolito, R. Solid waste from aluminum recycling process: Characterization and reuse of its economically valuable constituents. *Waste Manag.* **2005**, *25*, 37–46. [CrossRef] [PubMed]
29. Bravo-German, A.M.; Bravo-Gómez, I.D.; Mesa, J.A.; Maury-Ramírez, A. Mechanical Properties of Concrete Using Recycled Aggregates Obtained from Old Paving Stones. *Sustainability* **2021**, *13*, 3044. [CrossRef]
30. Ritchie, H.; Roser, M.; Rosado, P. CO_2 and Greenhouse Gas Emissions. Our World in Data. 2023. Available online: https://ourworldindata.org/grapher/annual-co2-cement?tab=map (accessed on 15 September 2023).
31. Global Consensus on Sustainability in the Built Environment—GLOBE. Decarbonising Global Construction. Available online: https://www.rilem.net/globe (accessed on 20 September 2023).
32. Muñoz-Velez, M.F.; Salazar-Serna, K.; Escobar-Torres, D.; Rojas-Manzano, M.A.; Gómez-Gómez, A.; Maury-Ramírez, A. Circular Economy: Adding value to the post-industrial waste through the transformation of aluminum dross for cement-matrix applications. *Sustainability* **2023**, *15*, 13952. [CrossRef]

sustainability

MDPI

Article

Circular Economy: Adding Value to the Post-Industrial Waste through the Transformation of Aluminum Dross for Cement Matrix Applications

Mario Fernando Muñoz-Vélez [1], Kathleen Salazar-Serna [1], Daniela Escobar-Torres [1], Manuel Alejandro Rojas-Manzano [1], Adriana Gómez-Gómez [1] and Aníbal Maury-Ramírez [1,2,*]

1 Department of Civil and Industrial Engineering, Faculty of Engineering and Sciences, Pontificia Universidad Javeriana Cali, Calle 18 # 118-250, Santiago de Cali 760031, Colombia; mario.munoz@javerianacali.edu.co (M.F.M.-V.); kathleen.salazar@javerianacali.edu.co (K.S.-S.); danielaet@javerianacali.edu.co (D.E.-T.); alejandro.rojas@javerianacali.edu.co (M.A.R.-M.); adrianagomez@javerianacali.edu.co (A.G.-G.)

2 Architecture Department, Faculty of Design Sciences, University of Antwerp, Mutsaardstraat 29, 2000 Antwerp, Belgium

* Correspondence: anibal.maury@javerianacali.edu.co

Abstract: In light of globalization and escalating environmental concerns, society is increasingly confronted with the challenge of implementing the concept of a circular economy, which promotes the recycle of waste materials and offers a promising solution. Aluminum dross, a byproduct of the aluminum production process, poses environmental issues when not properly managed. Therefore, this study examined the technical and financial feasibility of implementing an industrial process for the recovery and transformation of aluminum dross into raw materials for use in cementitious materials. From a technical perspective, two processes were evaluated: washing and the grindability of the material. An X-ray diffraction analysis allowed to verify an approximately 88% reduction in AlN (a compound that produces ammonia gases when reacting with water) after washing the material. The most efficient grinding process was achieved using an impact mill. The financial feasibility study was carried out through cash flow forecasting, which revealed that a minimum selling price of USD 0.12 per kilogram of processed material could generate a return rate of 9.7% over a five-year period. These results present opportunities for the metal and construction industries to develop products with low CO_2 emissions by reintegrating aluminum dross into a productive cycle. Moreover, this work serves as a valuable reference for policymakers and environmental authorities seeking to formulate new legislation or incentives that encourage companies to invest in environmentally focused projects.

Keywords: aluminum dross; waste management; circular economy; feasibility study; cement–matrix composites; industrial symbiosis

Citation: Muñoz-Vélez, M.F.; Salazar-Serna, K.; Escobar-Torres, D.; Rojas-Manzano, M.A.; Gómez-Gómez, A.; Maury-Ramírez, A. Circular Economy: Adding Value to the Post-Industrial Waste through the Transformation of Aluminum Dross for Cement Matrix Applications. *Sustainability* **2023**, *15*, 13952. https://doi.org/10.3390/su151813952

Academic Editor: Hosam Saleh

Received: 17 August 2023
Revised: 10 September 2023
Accepted: 16 September 2023
Published: 20 September 2023

1. Introduction

Aluminum is one of the most commonly used materials in manufacturing, construction, transportation, and packaging, among other industries, and it has a significant impact on the world economy. Global aluminum production grew by 139% in the last decade, reaching its peak in 2023 (International Aluminium Institute, 2023) [1], with countries such as China, India, and Russia leading the production. While the increased use of aluminum brings dynamism to international markets, it also leads to negative environmental impacts. For example, during the production process, waste such as white aluminum dross (AD) is generated. This waste is classified as hazardous due to the heavy metals that can leach from landfills into water sources. Additionally, one of its components, aluminum nitride (AlN), undergoes a transformation into ammonia gas (NH_3) when exposed to water. This transformation not only affects the environment but also poses risks to human health due to its irritating and corrosive nature. Considering the aforementioned context, it is necessary to seek alternatives that strike a balance between economic growth, social welfare, and environmental issues. This entails developing profitable and sustainable strategies in the long term, such as the recycling of production waste within circular economy models (Moreno, 2018) [2].

Dross is an inevitable residue generated during the aluminum smelting process as a result of the reaction between the liquid aluminum surface and the atmosphere. It is categorized into three types based on the production technique: white, black, and saline dross (David & Kopac, 2013) [3]. White dross is produced during the production of primary and secondary aluminum in extrusion plants, rolling mills, and foundries. Its name stems from its light color. On the other hand, black and saline dross are produced in recycling processes and are characterized by their dark color. Black dross typically has a granular shape and contains more salts and gases compared to white dross, while saline dross has fewer metal residues and more impurities.

The structure and composition of dross cannot be predicted precisely, as various factors such as temperature, alloy composition, and atmospheric conditions influence the oxidation process. However, dross generally consists of aluminum oxides, residual metallic aluminum, nitride, carbide, aluminum sulfide, salts, and some alloying elements (Dai, 2012) [4]. The chemical composition of AD (aluminum dross) varies depends on the alloying elements present in the casting process. However, it is typically composed mainly of aluminum oxide (Al_2O_3), spinel ($MgAl_2O_4$), and sodium chloride (NaCl) in higher proportions. Other reported compounds include silicon dioxide (SiO_2), sodium superoxide (Na_2O), ferric oxide (Fe_2O_3), aluminum metal, diaoyudaoite ($NaAl_{11}O_{17}$), aluminum oxide nitride (Al_5O_6N), hibonite ($CaAl_{12}O_{19}$), fluorite (CaF_2), and calcite ($CaCO_3$). Additionally, AD may contain traces of silicon (Si), aluminum carbide (Al_4C_3), magnesium fluoride (MgF_2), sodium tetrachloroaluminate ($NaAlCl_4$), potassium tetrachloroaluminate ($KAlCl_4$), magnesium oxide (MgO), parascandolaite ($KMgF_3$), and elpasolite (K_2NaAlF_6). Other minor crystalline phases identified through X-ray diffraction analysis (XRD) include cryolite (Na_3AlF_6) and potassium chloride (KCl) (Mahinroosta & Allahverdi, 2018; Srivastava & Meshram, 2023) [5,6].

There are opportunities for the reincorporation of AD into various production chains due to its Al_2O_3 content, which can be utilized as a mineral addition in the production of construction materials, among other applications. Several studies in the literature have reported the use of AD as a source of Al_2O_3. For instance, it has been employed in the manufacturing of calcium aluminate cement (Ünlü & Drouet, 2002) [7] and refractories (Ibarra Castro et al., 2009) [8] as well as in the production and synthesis of different composite materials (Ewais et al., 2009; Huang et al., 2014; Kim et al., 2009; Murayama et al., 2009; Yoshimura et al., 2008) [9–13]. AD has also been used as a substitute for cementitious material (Pereira et al., 2000) [14] and in the production of cellular concrete (Araújo & Tenório, 2005; Hwang & Song, 1997) [15,16], aluminous cement manufacturing (Ewais et al., 2009) [9], partial replacement of fine aggregate in mortars (Borhan & Janna, 2016; Llanos & Rodríguez, 2011; Pereira et al., 2000; Puertas et al., 1999) [14,17–19], concrete block production (Shinzato & Hypolito, 2005) [20], and conventional concrete manufacturing (Elinwa & Mbadike, 2011; Mailar et al., 2016; Ozerkan et al., 2014; Reddy & Neeraja, 2016) [21–24].

The results of these studies indicate the potential to explore circular economy alternatives with this hazardous waste, thereby contributing to the reduction of the 2.5 million tons of dross that are annually disposed of in landfills worldwide. These landfills generate undesirable heat, liquid leachate with heavy metals, and approximately 40 million m^3 of toxic, flammable, and foul-smelling gases such as ammonia, phosphine, hydrogen sulfide, and methane (David & Kopac, 2013) [3]. However, it should be noted that while the physical, chemical, and mechanical properties of certain applications for the reuse of dross have been studied, no technical and financial feasibility studies for its industrial-scale production have been conducted so far. Thus, this project aims to contribute with a proposal for the application of white dross, facilitating its reincorporation into a closed production cycle. To achieve this, the physicochemical characterization of the material was conducted to identify its potential uses. Subsequently, an industrial-scale production process was designed for a specific application case in a Colombian aluminum manufacturer. Finally, a financial feasibility analysis was carried out to assess its implementation.

2. Materials and Methods

2.1. Materials

The present study utilized AD obtained from Alumina S.A, a company located in Santiago de Cali, Colombia. This AD is categorized as white dross, originating from primary and secondary foundries. It is noteworthy to mention that this waste is classified as hazardous due to the potential leaching of heavy metals into groundwater from landfills, and the emission of hydrogen gases (H_2), NH_3, and methane (CH_4) upon interaction with aqueous substances, which are hazardous to both health and the environment (Dai, 2012; Siddique Pasley, 2003; Tang et al., 2022) [4,25,26].

2.2. Methods

In the present study, the methodology depicted in Figure 1 was followed. This diagram illustrates a sequential approach used in the research process. Initially, various applications of AD were explored through a comprehensive literature review, and the Disney method, which involves four thinking styles sequentially employed to analyze problems and generate ideas, was applied. This approach led to the identification of five feasible alternatives. Subsequently, the analytic hierarchy process (AHP) method was employed to determine the most favorable alternative. Following the selection of the preferred alternative, treatments and laboratory tests were conducted to obtain transformation parameters for AD. Finally, the industrial-scale process design and a corresponding financial analysis were carried out to assess the technical and financial viability of AD recovery and transformation in the case study.

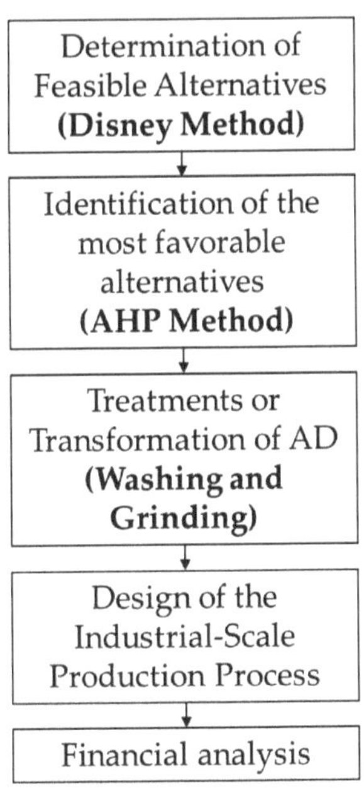

Figure 1. Flowchart of the developed methodology.

2.2.1. Identification of the Most Favorable Alternatives

A multi-criteria selection process was conducted using the AHP method based on the identified feasible alternatives (Saaty, 2008) [27]. Experts from various fields participated in defining the evaluation criteria and their respective weights of importance, as presented in Table 1.

Table 1. Selection criteria application of the AHP method.

Criterion	Weight	Score	Scale
C1: Associated implementation cost	10%	0	High cost
		1	Average cost
		2	Low cost
C2: Technical feasibility of implementation	16%	0	High level of implementation difficulty
		1	Medium difficulty level
		2	Low difficulty level
C3: Availability of information on the alternative	6%	0	No information
		1	Information is limited
		2	High amount of information available
C4: Compliance with design constraints established by stakeholders	41%	0	Affects or worsens properties
		1	Does not affect or improve properties
		2	Properties are maintained or improved
C5: Environmental impact	28%	0	High environmental impact
		1	Average environmental impact
		2	Low environmental impact

The research paper evaluated the identified alternatives based on five criteria. Criterion C1 assessed the costs of implementing AD in the alternative. Criterion C2 evaluated the feasibility of implementing the alternatives in terms of equipment, materials, and workspace. Criterion C3 was based on information from the literature review. Criterion C4 evaluated if the alternative could affect the final product's properties and performance, while criterion C5 assessed energy consumption and waste generated by the alternatives.

Then, scores on a scale of 0 to 2 for each criterion were assigned. Criteria C1, C2, and C5 were scored based on high cost, high difficulty, or high environmental impact, receiving a score of zero. Criteria C3 and C4 were evaluated based on high information availability and compliance with specifications, receiving a score of two. After scoring each alternative, a pair comparison matrix was constructed, and the priority vector's consistency was verified. The alternative with the highest total hierarchical analysis score was then selected.

2.2.2. Transformation of Aluminum Dross (Laboratory Scale)

After identifying the most promising alternative, the required chemical treatments and physical transformations were carried out to prepare the AD for industrial use. In this regard, a washing process was initially conducted on the AD; the material was distributed on a surface, and approximately 20 L of water were added daily for five days. Subsequently, the washed aluminum dross (WAD) was dried in an oven at a temperature of 120 °C for 24 h. Chemical analyses were conducted using XRD and EDS techniques to evaluate the composition of the AD before and after washing. The XRD analysis was performed using a Bruker model D8 Advance series 1 with CuK-α radiation (λ = 1.5406 Å), employing a step size of 0.02° for each 3 s. In addition, scanning electron microscopy (SEM) was employed to examine the morphology of the material, utilizing a JOEL model JSM-6490 with a gold sample coating. The effect of washing on the material's morphology was also assessed using SEM in the same electron microscope.

Furthermore, to meet the particle size requirements outlined in ASTM C618-19 (Standard Specification for Coal Fly Ash and Raw or Calcined Natural Pozzolan for Use in Concrete) [28], it was necessary to evaluate the grinding process for WAD to ensure it met the specified size criteria. Specifically, at least 66% of the material should pass through a sieve size of 325 (particle size < 45 μm). To prepare the AD for reincorporation, it was important to analyze its grindability in different equipment and methods to identify the most efficient particle reduction mechanisms.

Two grinding methods were examined: ball mill grinding and impact grinding. Ball mill grinding was conducted using a mill with a capacity of 500 g, operating at a speed of 90 rpm. Various grinding times were tested, including 15, 30, 45, 60, 90, 120, and 150 min, in order to assess the effect of grinding duration on the material. In contrast, impact grinding was performed using a ring mill with a capacity of 100 g. The grinding times evaluated were 2, 4, 6, 8, and 10 min. For durations exceeding 2 min, the intervals of grinding were followed by 10 min of rest, as impact grinding tends to cause significant wear on the mill.

2.2.3. Development of an Industrial Process for Transforming Aluminum Dross

Based on the findings, a production process was proposed, including the required machinery and personnel. To determine the plant layout, the systematic layout planning (SLP) method was applied (Suhardini et al., 2017) [29]. This involved defining the macro and micro location, creating a relationship matrix to establish distances between working areas, and using a diagram of relationships to determine the appropriate order for the plant's different areas. Finally, the Guerchet method was used to calculate the areas of occupation for each equipment and propose a general distribution for the plant.

Additionally, to determine the operational parameters, various equipment options were evaluated, and a choice was made based on the minimum installed capacity needed for a scenario where a Colombian aluminum manufacturer applies the circular economy principle, as the aim is to encourage waste-generating companies to implement such processes and develop new business ventures. Production times and workforce requirements were established by analyzing process activities and equipment capabilities using a logistics process simulation software (SIMIO version 9.1).

2.2.4. Assessing the Financial Viability of Utilizing Aluminum Dross in the Chosen Application

After defining and validating the requirements for implementing the AD recovery production process, the necessary investments and operating costs were estimated. Using this information, a 5-year projection of the project's income statement and cash flow was created. Financial evaluation indicators such as net present value (NPV), internal rate of return (IRR), and benefit/cost ratio (B/C) were then calculated (Ross et al., 2010) [30]. As there is currently no active market for AD that is suitable for reincorporation into other industrial applications, the sale price that meets the minimum viability conditions for a project (NPV = 0) was determined. This sale price is important for potential investors and allows for the development of a market for AD as recovered waste material in the construction sector by industrial symbiosis.

3. Results and Discussion

This section presents the outcomes of the analysis and selection of the application alternative and proposes a transformation process for the reuse of AD. The feasibility of implementing this process has been validated through a case study conducted in a Colombian aluminum manufacturing company.

3.1. Exploring Potential Applications of Aluminum Dross and Selecting an Optimal Use

The utilization of dross as an engineering product or as a component in an engineered product system is a subject of interest due to various reasons. Firstly, recycling Al_2O_3 from dross can offer an alternative source to many primary materials. Secondly, if the dross can be directed towards a valuable product, aluminum smelters can profit by charging an entry fee for handling and processing of residual dross, which can be utilized for producing new or existing products, thereby improving some of its physical, mechanical, or chemical characteristics (Crespo, 2015) [31].

To achieve the above objectives, an analysis of possible uses of WAD was conducted, employing the ideation technique known as the Disney method. The method involved exploring various alternatives through the dreamy, realistic, and critical phases and ultimately evaluating the possible risks (Appendix A).

After applying the Disney method, it was determined that five alternatives successfully passed through each phase of the analysis. The selected alternatives for further investigation were identified as follows: NMP refractories without additions, preparation of gamma alumina (γ-Al_2O_3) via a hydrometallurgical process, partial replacement of Portland cement for concrete production, replacement of Portland cement in the production of mortars, and production of polypropylene compound and aluminum dross. These options are marked with an asterisk (*) in Table A1.

Table 2 presents the matrix used to determine the final decision vector, which represents the best alternative considering the priority vectors of the criteria, the priority vectors of the alternatives for each criterion, and the assigned importance values for each criterion. The analysis resulted in the identification of the two most favorable alternatives for the utilization of AD: A4 (27.8%) and A3 (24.5%). These alternatives involve the partial replacement of Portland cement for the production of mortars and concrete, respectively. Based on these findings, a circular economy approach was implemented, focusing on the recovery of AD for its incorporation into cementitious matrices such as mortars and concrete. Appendix B shows the procedure that was carried out to obtain Table 2.

Table 2. Final decision vector determination matrix (best alternative).

Alternatives	Vector of Priorities by Criterion					Final Vector	Best Alternative
	C1	C2	C3	C4	C5		
A1	5.6%	35.8%	9.1%	7.7%	28.1%	17.4%	
A2	13.0%	6.5%	9.1%	23.1%	5.1%	13.8%	
A3	34.2%	15.5%	27.3%	23.1%	28.1%	24.5%	A4 A3
A4	34.2%	35.8%	27.3%	23.1%	28.1%	27.8%	
A5	13.0%	6.5%	27.3%	23.1%	10.8%	16.4%	
Vector Criteria	9.9%	16.1%	6.2%	41.6%	26.2%		

3.2. Circular Economy Proposal for the Recovery of Aluminum Dross

Pilot processing tests for use in cementitious matrices: Pilot processing tests for the utilization of AD in cementitious matrices were conducted to validate its suitability for such applications. It is important to note that AD primarily consists of aluminum oxide (15–30% Al_2O_3), but it may also contain approximately 8.0% AlN, a compound that produces ammonia gases upon contact with water (Attia et al., 2018; López-Delgado et al., 2020) [32,33]. The presence of AlN in cement-based mixtures poses risks to both worker safety and the formation of voids within the cementitious matrix due to the trapped gas (Dai, 2012; Lemos Micolta et al., 2020; Mahinroosta & Allahverdi, 2018) [4,5,34].

To address this issue, a "washing" or inactivation process was implemented to reduce the percentage of aluminum nitrides in the AD before its incorporation into cementitious matrices. This washing process aimed to minimize the risks associated with ammonia gas generation and ensure the integrity and performance of the cement-based materials. The AD underwent a series of processing steps to prepare it for use in cementitious matrices. Initially, the collected AD was subjected to sieving using mesh #10 (Figure 2a–c). This sieving process helped to remove any larger particles or impurities present in the material.

Figure 2. (**a**) AD at the collection site, (**b**) initial sieve, (**c**) AD, (**d**) washing process, and (**e**) drying process.

For the subsequent washing process, the AD was dispersed and evenly distributed on a suitable surface. Approximately 20 L of water per day were added to the AD over a period of 5 days (Figure 2d). After the washing process, the wet AD was dried in an oven at a temperature of 120 °C for 24 h (Figure 2e). This drying step aimed to remove excess moisture from the material.

In order to confirm the removal of AlN from AD and determine the effect of the washing process on chemical and morphological characteristics, dross evaluation was performed before and after washing using XRD and SEM. The results of the chemical characterization conducted by X-ray diffraction (XRD) are presented in Table 3. It was observed that the percentage of AlN decreased from 18.2% in AD to 2.2% in WAD, confirming the successful removal of aluminum nitrides through the washing process. Additionally, a significant increase in the presence of gibbsite was observed, which can be attributed to the interaction between aluminum phases (Al) and water (H_2O) used during the washing process (Singh, 1982) [35]. Furthermore, compounds such as aluminum oxide, iron oxides, and quartz showed an increase in their percentage, while the percentages of aluminum and spinel decreased. These findings provide valuable insights into the changes in the chemical composition of the AD following the washing process.

Table 3. XRD results with percentage values of the presence of chemical compounds in AD and WAD.

Compound	Chemical Formula	Content (%)	
		AD	**WAD**
Gibbsite	$Al(OH)_3$	3.7	27.5
Aluminum Nitrides	AlN	18.2	2.2
Aluminum Oxide	Al_2O_3	14.8	17.1
Aluminum	Al	6.3	3.86
Diaoyudaoite	$NaAl_{11}O_{17}$	0	0.81
Iron Oxides	Fe_2O_3	0.3	2.4
Quartz	SiO_2	1.5	5.2
Spinel	$MgAl_2O_4$	34	25.5

On the other hand, Figure 3 displays the surface morphology of the AD particles, revealing an apparent presence of aluminum oxide, spinel, and aluminum nitrides. Additionally, the micrographs of WAD demonstrated the presence of other crystal structures derived from aluminum oxide, such as corundum. These findings are consistent with previous studies (Braulio et al., 2011; Chaplianko & Nikichanov, 2021; Zuo et al., 2021) [36–38]. Notably, the micrographs indicated a change in the material's morphology following the washing process, with an increased presence of structures exhibiting a crystalline morphology.

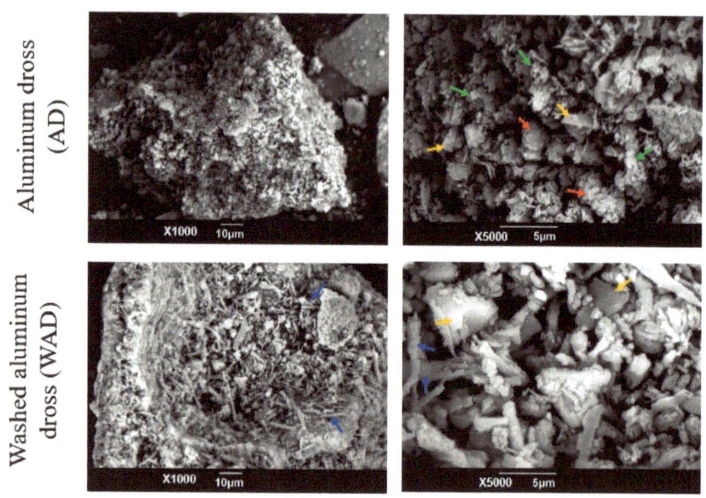

Figure 3. SEM of AD and WAD. Aluminum oxide (orange arrow), spinel (red arrow), aluminum nitride (green arrow), and corundum (blue arrow).

The results of the millability test conducted in the ball mill are presented in Figure 4. The findings indicate that even with the maximum grinding time evaluated (150 min), only 18% of the particles achieved the desired size of 45 μm. The influence of time on the particle size change was not clearly observed. In a study conducted by Ramezani and Neitzert in 2012 (Ramezani & Neitzert, 2012) [39] on the grinding process of aluminum powder using a planetary ball mill, it was observed that prolonged grinding time led to morphological changes in aluminum particles, resulting in their transformation into flakes and a substantial increase in size from approximately 32 μm to over 1400 μm within just one hour. These findings provide insights into the behavior observed in Figure 4, which could be attributed to a lamination process involving the soft phases present in the WAD, such as aluminum (Al).

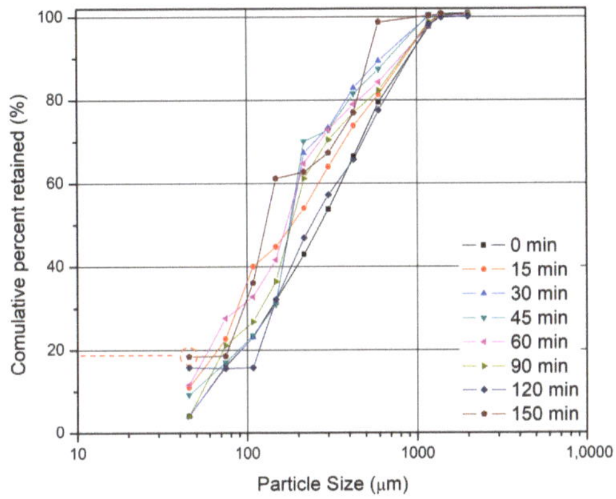

Figure 4. Granulometric distribution curves of WAD for different process times (ball mill).

Figure 5a illustrates the granulometric distribution curves of the WAD material processed in the impact mill. This grinding method yielded a particle size distribution where up to 56.21% of the particles were smaller than 45 μm, achieved within a processing time of 10 min. Although the obtained results do not fully meet the particle size requirements set by ASTM C618-19, a projection based on the kinetic analysis of grinding proposed by Leyva Ramírez et al. (Leyva Ramírez et al., 2009) [40] demonstrated that a particle size distribution with 66% of material below 55 μm (maximum retained in sieve 325 of 34%) could be achieved within a time frame of 12 min (Figure 5b). While no specific studies were found regarding the vibrational impact mill grinding process for aluminum dross or pure aluminum, ref. (Karakas & Kanca, 2020) [41] conducted a study on the grindability of alpha iron oxide using this technique. They observed a reduction in particle size by approximately 76.6%. This suggests that the ring mill, which imparts more energy to the system compared to a ball mill (Plescia & Tempesta, 2017) [42], has the potential to break down flakes and achieve smaller particle sizes despite the deformation phenomenon mentioned earlier.

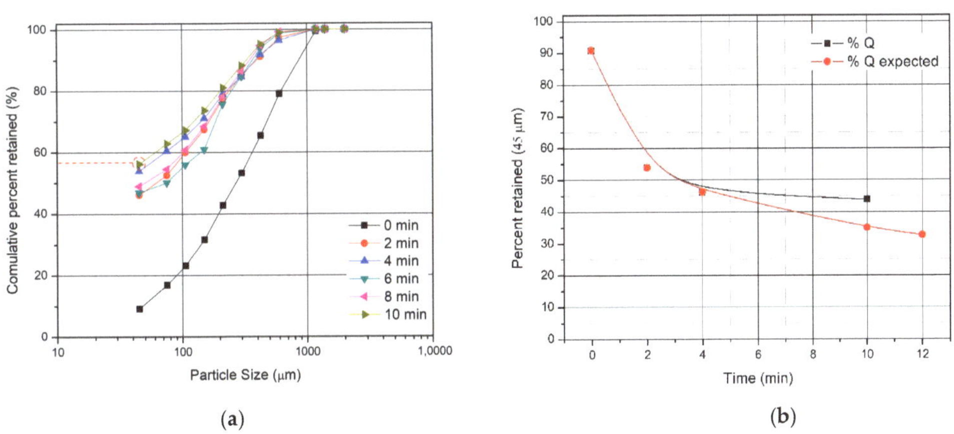

Figure 5. (**a**) Granulometric distribution curves of aluminum dross for different process times (**left**). Impact mill (rings); (**b**) grinding kinetics (**right**).

3.3. Development of an Industrial Process for Transforming Aluminum Dross

Proposal to implement at a production process for the recovery of aluminum dross on an industrial scale: The process design and operational scheme were developed based on the results obtained from the laboratory pilot test and the analysis of available equipment options in the market. In order to achieve this, the necessary machinery for dross transformation was identified, the plant layout was designed, and the required workforce for operation was defined. To validate the design, data from a specific case study in a Colombian aluminum production company were utilized. The objective is to encourage the implementation of circular economy processes by the companies responsible for generating the waste while also enabling the development of new business opportunities. Additionally, the analysis presented in this proposal may be of interest to potential investors seeking to establish the production process as an independent business model.

Production process flow and equipment: Figure 6 presents the flowchart with the activities associated with the AD recovery process. This dross transformation allows the waste to become raw material to be used as a partial replacement for Portland cement.

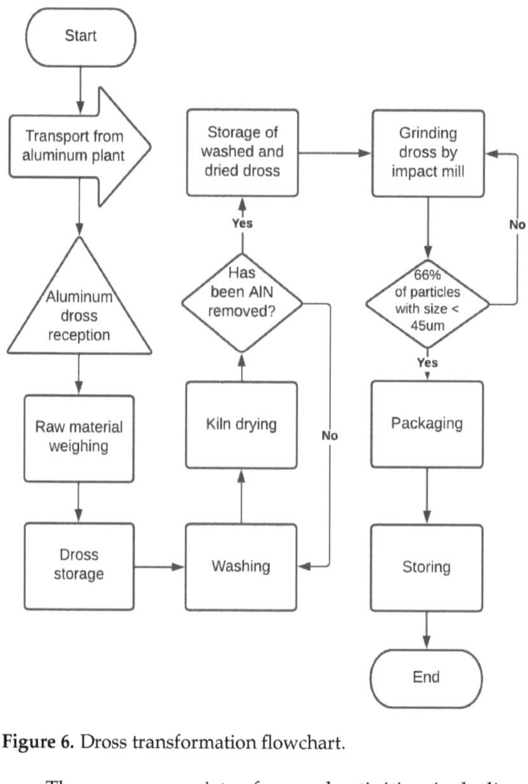

Figure 6. Dross transformation flowchart.

The process consists of several activities, including transportation, raw material reception, six operations, three storage stages, and two operational decisions. Depending on the location of the recovery plant, the material may need to be transported internally or externally from the aluminum production plant. If road transport is required, a dry bulk truck equipped with a rear valve for material extraction is necessary. In the analyzed case, 75 tons of AD are transported internally on a monthly basis from the main plant to the recovery plant.

Upon arrival at the reception area, the truck undergoes inspection and weighing. The material is then transferred through a hose to a conical bottom storage silo with automatic dosing and cumulative weighing capabilities, which are desirable features for storing this type of material. The silo is connected to a pool with a washing wheel where the dross-washing process takes place. An endless screw transports the fine dross from the bottom of the tank to a drain wheel, which recovers, drains, and discharges the material into another endless screw connected to a tilting rotary furnace model FARB-2, where the dross is dried.

The dry dross is stored in another silo designated for the product in the process, which will supply the ALPA MZ500 impact mill. This mill is used to reduce the particle size of the dross to meet ASTM C618-19 standards. The vibratory mill rapidly rotates the material, subjecting it to powerful impacts and frictional forces, resulting in the production of ultra-fine and uniform powder. This stage is crucial, as it ensures that the particle size of the dross is suitable for replacing cement in the manufacturing of mortars.

The milled dross is stored in a final silo for the finished products and then undergoes the packaging and storage activities. The transformed dross is packed in semi-extendable Kraft paper bags with filling valves. The paper used weighs between 82 and 95 g/m^2 and is designed with anti-tear technology to minimize damage during transport and handling. Micro perforations are incorporated into the bags to eliminate air accumulation during the filling process. Although the bags have a maximum capacity of 50 kg, it is recommended to pack them in 30 kg bags to ensure better ergonomic conditions for workers during loading (Hernández & Nieves, 2015) [43]. These bags are then arranged on pallets measuring 1 m x 1.2 m and stored in the designated storage area. Table 4 presents a comprehensive list of the equipment chosen for the various stages of the dross recovery process, including their respective installed capacities and acquisition prices.

Table 4. Equipment required in the process.

Process	Machine	Speed (kg/h)	Capacity (kg/mth)
Raw material inspection and weighing	Floor scale	N/A	80,000
Raw material storage	Storage silo	N/A	30,000
Drying	Tilting rotary furnace FARB-2	667	118,667
In-process product storage	Storage silo	N/A	30,000
Milling	Vibratory mill MZ 500	532	94,667
Finished product storage	Storage silo	N/A	30,000
Packing	Manual packing	421	75,000

The production capacities per hour for each activity were estimated based on the installed capacities of the equipment and the process configuration, as illustrated in Table 5.

Table 5. Capacity by process activities.

Activity	Capacity
Washing	686 kg/h
Drying	667 kg/h
Milling	479 kg/h
Packed	421 kg/h
Storage (total capacity)	30,000 kg

With the specified processing capacities, the equipment utilization for the washing tanks, rotary furnace, vibratory mill, and packaging ranges between 40% and 60%, resulting in a processing rate of 3125 kg per day.

Location and Plant Layout: Considering that the waste-generating plant is situated in the Valle del Cauca department, potential locations in proximity to this area were reviewed. Utilizing Logware software version 5.0, the optimal coordinates for the recovery plant were determined. The chosen location for the plant is within the ACOPI industrial zone in the city of Yumbo.

Subsequently, the space requirements for establishing the plant were assessed. Based on the process requirements, eleven specific areas were defined: truck entrance, raw material reception and storage area, washing area, drying zone, storage area for washed and dry dross, grinding area, storage area for milled dross, packing area, storage area for packed dross, office areas (management/customer service), and washrooms/dressing rooms. These areas were further analyzed using the relationship matrix presented in Figure 7.

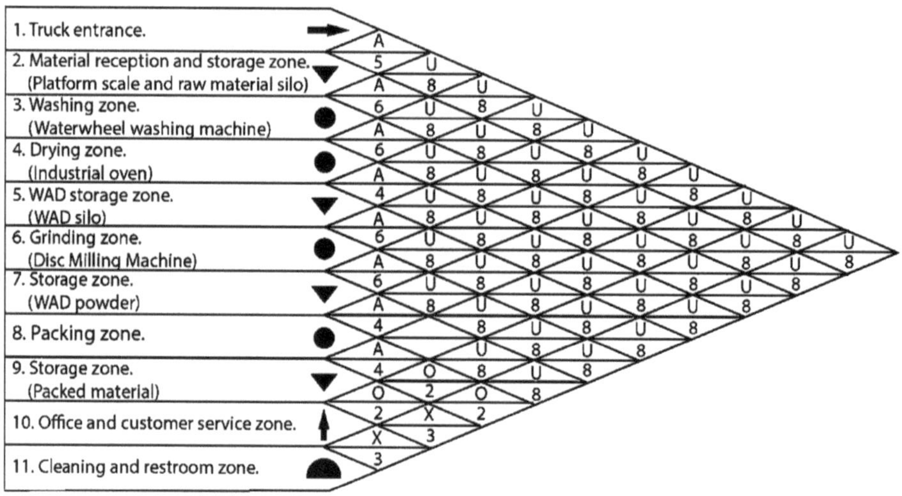

Figure 7. Relationship matrix.

The relationship diagram shown in Figure 8 was established to determine the appropriate distances between different zones within the plant. The first space is used for a transport activity which is represented by a horizontal arrow. Triangles represent storage activities, circles refer to operations, vertical arrow denotes administrative areas and the half oval refer to service areas.

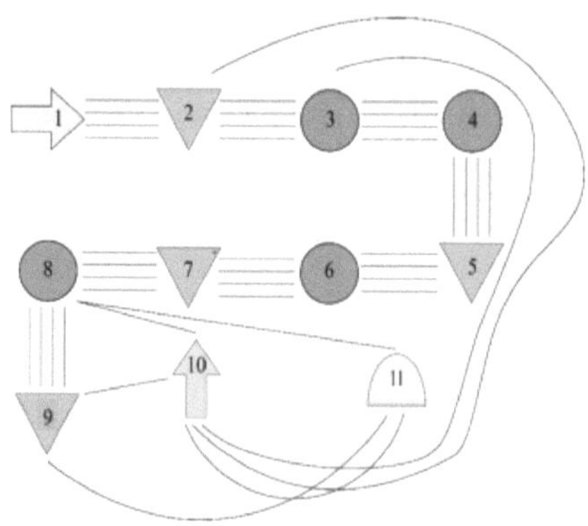

Figure 8. Relationship diagram.

This diagram served as the foundation for the general distribution of the plant, which can be seen in Figure 9.

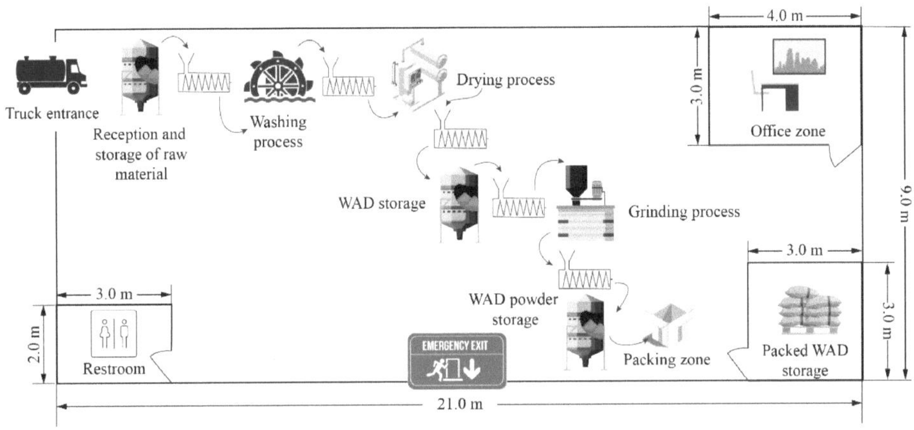

Figure 9. General plant layout.

To calculate the required areas, the Guerchet method was applied, considering the equipment dimensions and the space requirements for work and storage areas. The finished product storage area considered the size and capacity of pallets for stacking the packaged material. Based on these considerations, it was determined that the plant requires an area of 283 m^2. Figure 10 presents the theoretical plant distribution in equivalent surface units (ESU), with a conversion base of four used for the dimension conversions.

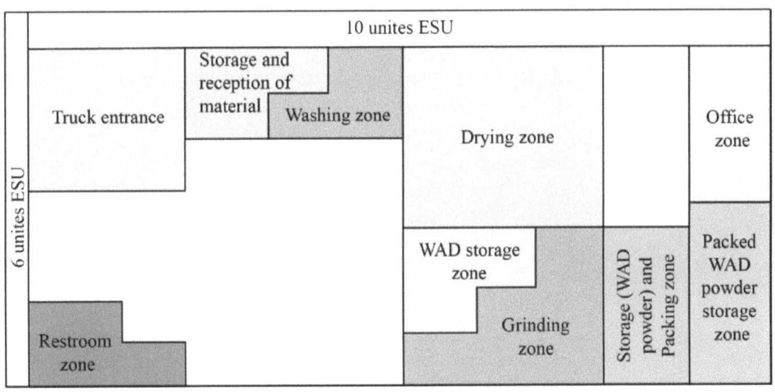

Figure 10. Theoretical plant layout.

Workforce Requirements: Based on the technical specifications of the equipment and the installed capacity of the process, parameters such as processing time and speed were established. By utilizing SIMIO software, the process was simulated using the waste generation amount of 75 tons per month as per the application case. The simulation revealed that the process requires 12 h to complete. However, considering preparation times, cleaning, active breaks, and administrative activities, this translates to two work shifts of 8 h each. Each shift requires two operatives to execute all process activities. Therefore, a total of four operatives needs to be hired. Additionally, the hiring of a plant manager is necessary to oversee process control and staff inspection. It is important to note that the workforce analysis focused solely on the personnel required for production and did not consider commercial or administrative activities.

3.4. Financial Analysis for the Application Case

Once the resources required for the AD transformation were determined, the total investment value for equipment and adaptations was calculated at USD 43,000 (Meneses-Núñez et al., 2022) [44]. Subsequently, a financial projection of the income statement was conducted, serving as the basis for projecting cash flow over a 5-year period, as presented in Table 6. The cash flows were projected with a constant inflation rate of 3.6%, based on the projections from the Colombian central bank at the time the study. Additionally, a tax rate of 35% was considered in accordance with the tax regulations for 2022.

Table 6. Projected cash flow.

Project Cash Flow	0	1	2	3	4	5
Operational Profit		7.527	10.848	11.393	11.957	12.542
Operating taxes		2.635	3.797	3.987	4.185	4.390
Net Operating Income		4.893	7.051	7.405	7.772	8.152
+ Depreciation or amortization		4.283	4.283	4.283	4.283	4.283
Gross Cash Flow		9.176	11.335	11.689	12.055	12.435
− Fixed Asset Investments	42.835					
Project Free Cash Flow	42.835	9.176	11.335	11.689	12.055	12.435

Among the most significant direct costs, workforce expenses account for 45% of the total production cost, followed by energy costs representing 19.5% of the total. In this particular application case, no cost was assigned to the raw material, as the waste generator itself will be responsible for transforming it into a reusable material.

Regarding the total investments required, it was determined that 50% of the financing would be obtained from financial institutions at an annual effective rate of 8.7%. This rate was based on the financing rates available to the company in the application case from the financial sector. With this capital structure in place, a weighted average cost of capital (WACC) of 9.7% annually was calculated.

Using these data, the sales value per kilogram of recovered dross was determined to satisfy the expression NPV = 0, forming the basis for projecting the cash flow income. It was found that the sale price must be at least USD 0.12 per kilogram for the project to be feasible. This price corresponds to having an internal rate of return (IRR) equal to or greater than the WACC and a benefit-to-cost ratio of 1. According to financial theory, this implies that the project's revenues compensate for investments and operating costs and generate returns that are at least equal to the cost of capital.

These findings not only demonstrate the feasibility of the specific project in the Colombian case but also contribute to the broader field of research by showcasing a circular economy application that goes beyond technical validation. Analyzing the financial implications of implementing such a project highlights the potential for developing new markets with a stronger environmental focus. Calculation of the minimum market price serves as an initial step that reveals opportunities for the expansion of environmentally-oriented markets. In fact, the City Hall of Santiago de Cali (Colombia) is evaluating a circular economy model for the construction sector, which stimulates the use of sustainable building materials containing by-products coming from different local industrial sectors [45]. Although initial evaluation has been oriented towards the environmental and economic waste valorization coming from the steel production and coal combustion factories [46], the local aluminum industry has a great potential to generate supplementary cementitious materials as reported here.

4. Conclusions

The chemical characterization through XRD revealed that the washing process resulted in a significant reduction in the AlN content, decreasing from 18.2% in AD to 2.3% in WAD. This finding holds considerable importance as it is well known that AlN, when exposed to water, produces ammonia gases, which can be harmful to the health of operators. Furthermore, when used in cementitious materials, the accumulation of these gases during mixing with water can lead to void formation, potentially compromising the mechanical properties of the materials.

The impact mill yielded superior results, achieving up to 56.21% of particles smaller than 45 μm in just 10 min of processing. Kinetic analysis projected that 66% of material with particle sizes below 55 μm can be obtained in 12 min. This is crucial to satisfy the particle size requirements stipulated by ASTM C618-19 for the use of WAD in cementitious matrices.

With the chosen equipment's capacity, an estimated 80,000 kg of aluminum dross can be recycled monthly. A manufacturing plant of 283 m^2 can be allocated to accommodate various operation stages, ensuring ample space for the 11 defined areas involved in the process. These capacity and facility requirements are pivotal considerations for the practical implementation of AD recycling. To execute the proposed process in the case study, an investment of USD 43,000 is required. Cash flow analysis revealed that a minimum selling price of USD 0.12/kg of processed material can generate a return rate of at least 9.7% over a five-year period. These findings hold significance for aluminum factories, cement producers, investment institutions, policymakers, and environmental authorities, as they illustrate the potential for cultivating a new market for this material, with both economic and environmental implications at the forefront.

Author Contributions: Conceptualization, M.F.M.-V., M.A.R.-M. and A.G.-G.; methodology, M.F.M.-V., K.S.-S., D.E.-T., M.A.R.-M. and A.G.-G.; formal analysis, M.F.M.-V., K.S.-S., and D.E.-T.; investigation, M.F.M.-V., K.S.-S. and D.E.-T.; data curation, M.F.M.-V. and D.E.-T.; writing—original draft preparation, M.F.M.-V., K.S-S., D.E-T. and A.M-R.; writing—review and editing, M.F.M.-V., K.S.-S., D.E.-T. and A.M.-R.; supervision, M.F.M.-V. and K.S.-S.; project administration, M.F.M.-V.; funding acquisition, M.F.M.-V., M.A.R.-M. and A.G.-G. All authors have read and agreed to the published version of the manuscript.

Funding: This research was funded by Pontificia Universidad Javeriana, Sede Cali, through the internal call, grant number 2168-2021.

Institutional Review Board Statement: Not applicable.

Informed Consent Statement: Not applicable.

Data Availability Statement: The data presented in this study are available on request from the corresponding author.

Acknowledgments: The authors would like to thank Alumina S.A. for providing the aluminum dross generated during the aluminum smelting processes. They also acknowledge the Servicio Geológico Colombiano sede Santiago de Cali for allowing them to use their laboratories and equipment.

Conflicts of Interest: The authors declare no conflict of interest.

Abbreviations

Al	Aluminum
Al_4C_3	Aluminum carbide
AD	Aluminum dross
AlN	Aluminum nitride
Al_2O_3	Aluminum oxide
Al_5O_6N	Aluminum oxide nitride
NH_3	Ammonia
ASTM	American Society for Testing and Materials
AHP	Analytic hierarchy process
B/C	Benefit/cost ratio
$CaCO_3$	Calcite
CaO	Calcium oxide

ACOPI	Colombian Association of Micro, Small, and Medium Enterprises
CR	Consistency ratio
CuK-α	Copper k-α
Na_3AlF_6	Cryolite
$NaAl_{11}O_{17}$	Diaoyudaoite
K_2NaAlF_6	Elpasolite
EDS	Energy-dispersive X-ray spectroscopy
ESU	Equivalent surface units
Fe_2O_3	Ferric oxide
CaF_2	Fluorite
γ-Al_2O_3	Gamma alumina
$CaAl_{12}O_{19}$	Hibonite
H_2	Hydrogen gases
IRR	Internal rate of return
MgF_2	Magnesium fluoride
MgO	Magnesium oxide
CH_4	Methane
NPV	Net present value
NMP	Non-metallic products
$KMgF_3$	Parascandolaite
KCl	Potassium chloride
$KAlCl_4$	Potassium tetrachloroaluminate
PV	Priority vector
SEM	Scanning electron microscopy
Si	Silicon
SiO_2	Silicon dioxide
NaCl	Sodium chloride
Na_2O	Sodium superoxide
$NaAlCl_4$	Sodium tetrachloroaluminate
$MgAl_2O_4$	Spinel
SLP	Systematic layout planning
WAD	Washed aluminum dross
λ	Wavelength
WACC	Weighted average cost of capital
XRD	X-ray diffraction analysis

Appendix A

Table A1. Disney Method: Recovering Alternatives for Aluminum Dross.

Dreamer (Why Not?)	Realistic (How?)	Critical (What is Wrong?)
Guarantee Creativity	**Ensures Feasibility**	**Prevents Possible Risks**
Expanded clay aggregates (Bajare et al., 2012) [47]	Removal of impurities by heat treatment, preparation of clay pastes and non-metallic products (NMP) samples, drying, and finally synthesis in an oven at 1170–1210 °C.	• Risk of not having the necessary implements for the process; • Risk of not achieving the removal of impurities; • Risk of increased costs.
*NMP refractories without additions (Ramaswamy et al., 2019) [48]	Washing the NMP at 200 °C to remove salts, calcination at 100 °C, and NMP compaction and calcination at 1500 °C. Thermal shock tests at 660 °C.	• Risk of not having the necessary implements for the process.; • Risk of not achieving salt removal; • Risk of increased costs.

Table A1. *Cont.*

Dreamer (Why Not?)	Realistic (How?)	Critical (What is Wrong?)
Guarantee Creativity	**Ensures Feasibility**	**Prevents Possible Risks**
Preparation of gamma alumina (γ-Al_2O_3) by pyrometallurgical process (Mahinroosta & Allahverdi, 2018) [5]	The NMP is fed to the plasma flame, and argon is used as the carrier gas. The particle size to be obtained is 8 µm.	• Sophisticated and expensive process; • Risk due to the difficulty of implementation.
*Preparation of gamma alumina (γ-Al_2O_3) by hydrometallurgical process (Shen et al., 2021). [49]	These processes consist of three steps: alkaline or acid solution of NMP, precipitation of the filter liquid, and calcination of the precipitate.	• Risk of not having the necessary implements for the process; • Risk of not achieving the removal of impurities; • Risk of increased costs.
Replacement of aluminum powder as a foaming agent for synthesizing light cellular concrete (Liu et al., 2017). [50]	Grinding and sieving to achieve a particle size of 45 µm.	• Risk of not having the necessary implements for the process; • Risk of increased costs.
*Partial replacement of Portland cement for concrete production (Elinwa & Mbadike, 2011; Javali et al., 2017; Mailar et al., 2016; Ozerkan et al., 2014; Reddy & Neeraja, 2016) [21–24,51]	The dross should be ground, sieved using a 90 µm sieve, and a shutdown process should be carried out.	• Risk of not having the necessary implements for the process; • Risk of not achieving contact with cement companies.
*Replacement of Portland cement in mortar production (Dai & Apelian, 2017; Pereira et al., 2000) [14,52]	The dross should be washed in distilled water, dried on a heating plate, and then added to the mortar mixture.	• Risk of not having the necessary implements for the process; • Risk of not achieving contact with cement companies.
Production of ceramics based on magnesium titanate and aluminum (Ewais & Besisa, 2018) [53]	Grinding and sieving to achieve a particle size less than 90 µm; powders must be mixed using a mill, impurities are removed with boiling water and then with cold water, and the synthesized materials are obtained by cooking at 1300 °C.	• Risk of not having the necessary implements for the process; • Risk of increased costs.
*Production of polypropylene compound and aluminum dross (Adeosun et al., 2012; Samat et al., 2017) [54,55]	Lumps of dross should be crushed and sieved into particles of size from 53 µm to 150 µm.	• Risk of not having the necessary implements for the process; • Risk of increased costs; • Risk of not achieving contact with polypropylene companies.
Production of silicate-based glass (Mahinroosta & Allahverdi, 2018) [5].	The NMP-washing process must be carried out, and the residue or mineral glass of low silicon content must be melted in a CaO–Al_2O_3 system.	• Risk of not having the necessary implements for the process; • Risk of increased costs; • Risk due to the difficulty of implementation.

Appendix B

The scores assigned to the five selected alternatives in the AHP are presented in Table A2. It is noteworthy that Alternative 1 had the highest associated cost due to the high energy consumption required in the sintering process of refractories. In terms of "ease of implementation", Alternatives 1 and 4 were classified as having a low level of difficulty in implementation, as the necessary equipment and spaces for the development of these applications, such as mills, muffles, and sieves, are available for this study. In contrast, in the "environmental impact" criterion, Alternatives 1, 3, and 4 were evaluated as having a low impact. This is because the substitution of raw materials like cement with post-industrial waste like AD leads to a dual benefit of reducing the use of materials with high carbon footprint and utilizing a waste that could cause environmental harm if disposed of improperly.

Table A2. Evaluation of AHP criteria for each alternative.

	Alternatives	Criteria				
		C1: Cost	**C2: Ease of Implementation**	**C3: Availability of Information**	**C4: Specification Compliance**	**C5: Environmental Impact**
1	NMP refractories without additions	0	2	1	1	2
2	Preparation of gamma alumina (γ-Al_2O_3) by hydrometallurgical process	1	0	1	2	0
3	Partial replacement of Portland cement for concrete production	2	1	2	2	2
4	Replacement of Portland cement in mortar production	2	2	2	2	2
5	Production of polypropylene composite and AD	1	0	2	2	1

For the purpose of establishing the relationships between the alternatives and criteria, a visual tool (Table A3) was employed to ensure the proportionality and transitivity of the paired comparisons assigned to the AHP methodology, thereby affirming the consistency of the results. The table shows the proportional relationship between the reference and the comparison, be it an alternative or criterion. Moving columns to the right indicates a directly proportional relationship between the number of movements (e.g., C2 = 3C3), whereas moving columns to the left signifies an inversely proportional relationship (e.g., A1 = 1/5 A3).

Table A3. Visual tool to establish comparisons between criteria and alternatives.

		More Preference than Reference						**←**		**Reference**		**Less Preference than Reference**						
		1/9	**1/8**	**1/7**	**1/6**	**1/5**	**1/4**	**1/3**	**1/2**	**1**	**2**	**3**	**4**	**5**	**6**	**7**	**8**	**9**
Criteria comparison								**C4**	**C5**	**C2**	**C1**	**C3**						
Comparison of alternatives by criterion	C1					A3 A4		A2 A5		A1								
	C2									A1 A4		A3		A2 A5				
	C3							A3 A4 A5		A1 A2								
	C4							A2 A3 A4 A5		A1								
	C5									A1 A3 A4		A5		A2				

After establishing the relationships between criteria and alternatives using the visual tool, the assigned values were transferred to the AHP comparison matrix (Table A4). This allowed the determination of the priority vector (PV), which represents the relative importance of the criteria when compared to each other and the alternatives when compared within each criterion. Additionally, the consistency ratio (CR) was calculated for each of the 5 × 5 paired matrices using the AHP. It is worth noting that the consistency ratios for all cases were found to be less than 10%, indicating that the comparisons adhere to the principles of transitivity and proportionality.

Table A4 shows the relative importance of the selection criteria when compared to each other. It is evident that C4 (compliance with specifications) holds the highest significance, with a priority of 41.6%, followed by C5 (environmental impact) with 26.2%. Additionally, when comparing the alternatives within each criterion, certain patterns emerge. For C1 (cost), Alternatives A3 and A4 have the lowest cost priority, both at 34%. In terms of C2 (reliability of implementation), A1 and A4 exhibit higher priority (36%). Regarding C3 (availability of information), A3, A4, and A5 indicate a higher level of information availability. In terms of C4 (compliance with specifications), all alternatives require compliance at a priority of 27% except for A1. Lastly, A1, A3, and A4 are noted for their higher environmental impact (C5), with a priority of 28%.

Table A4. Matrix of comparison by pairs according to criteria and alternatives for each criterion.

Criteria	C1	C2	C3	C4	C5	PV	CR
C1	**1**	1/2	2	1/4	1/3	9.9%	
C2	2	**1**	3	1/3	1/2	16.1%	
C3	1	1/3	**1**	1/5	1/4	6.2%	1.4%
C4	4	3	5	**1**	2	41.6%	
C5	3	2	4	1/2	**1**	26.2%	

C3	A1	A2	A3	A4	A5	PV	CR
A1	**1**	1	1/3	1/3	1/3	9.1%	
A2	1	**1**	1/3	1/3	1/3	9.1%	
A3	3	3	**1**	1	1	27.3%	0.0%
A4	3	3	1	**1**	1	27.3%	
A5	3	3	1	1	**1**	27.3%	

C1	A1	A2	A3	A4	A5	PV	CR
A1	**1**	1/3	1/5	1/5	1/3	5.6%	
A2	3	**1**	1/3	1/3	1	13.0%	
A3	5	3	**1**	1	3	34.2%	1.2%
A4	5	3	1	**1**	3	34.2%	
A5	3	1	1/3	1/3	**1**	13.0%	

C4	A1	A2	A3	A4	A5	PV	CR
A1	**1**	1/3	1/3	1/3	1/3	7.7%	
A2	3	**1**	1	1	1	23.1%	
A3	3	1	**1**	1	1	23.1%	0.0%
A4	3	1	1	**1**	1	23.1%	
A5	3	1	1	1	**1**	23.1%	

C2	A1	A2	A3	A4	A5	PV	CR
A1	**1**	5	3	1	5	35.8%	
A2	1/5	**1**	1/3	1/5	1	6.5%	
A3	1/3	3	**1**	1/3	3	15.5%	1.2%
A4	1	5	3	**1**	5	35.8%	
A5	1/5	1	1/3	1/5	**1**	6.5%	

C5	A1	A2	A3	A4	A5	PV	CR
A1	**1**	5	1	1	3	28.1%	
A2	1/5	**1**	1/5	1/5	1/3	5.1%	
A3	1	5	**1**	1	3	28.1%	0.9%
A4	1	5	1	**1**	3	28.1%	
A5	1/3	3	1/3	1/3	**1**	10.8%	

References

1. International Aluminium Institute Primary Aluminium Production. Available online: https://international-aluminium.org/statistics/primary-aluminium-production/ (accessed on 9 May 2023).
2. Moreno, A. *Economía Circular: Crecimiento Inteligente, Sostenible e Integrador*; Universidad de ciencias Aplicadas y Ambientales: Bogotá, Colombia, 2018.
3. David, E.; Kopac, J. Aluminum Recovery as a Product with High Added Value Using Aluminum Hazardous Waste. *J. Hazard. Mater.* **2013**, *261*, 316–324. [CrossRef] [PubMed]

4. Dai, C. *Development of Aluminium Dross-Based Material for Engineering Applications*; Worcester Polytechnic Institute: Worcester, MA, USA, 2012.
5. Mahinroosta, M.; Allahverdi, A. Hazardous Aluminum Dross Characterization and Recycling Strategies: A Critical Review. *J. Environ. Manag.* **2018**, *223*, 452–468. [CrossRef] [PubMed]
6. Srivastava, A.; Meshram, A. On Trending Technologies of Aluminium Dross Recycling: A Review. *Process Saf. Environ. Prot.* **2023**, *171*, 38–54. [CrossRef]
7. Ünlü, N.; Drouet, M.G. Comparison of Salt-Free Aluminum Dross Treatment Processes. *Resour. Conserv. Recycl.* **2002**, *36*, 61–72. [CrossRef]
8. Ibarra Castro, M.N.; Almanza Robles, J.M.; Cortés Hernández, D.A.; Escobedo Bocardo, J.C.; Torres Torres, J. Development of Mullite/Zirconia Composites from a Mixture of Aluminum Dross and Zircon. *Ceram. Int.* **2009**, *35*, 921–924. [CrossRef]
9. Ewais, E.M.M.; Khalil, N.M.; Amin, M.S.; Ahmed, Y.M.Z.; Barakat, M.A. Utilization of Aluminum Sludge and Aluminum Slag (Dross) for the Manufacture of Calcium Aluminate Cement. *Ceram. Int.* **2009**, *35*, 3381–3388. [CrossRef]
10. Huang, J.; Fang, M.; Huang, Z.; Liu, Y.; Yang, J.; Huang, S.; Xu, Y.; Chen, K.; Yi, S.; Zhang, S. Preparation, Microstructure, and Mechanical Properties of Spinel-Corundum-Sialon Composite Materials from Waste Fly Ash and Aluminum Dross. *Adv. Mater. Sci. Eng.* **2014**, *2014*, 789867. [CrossRef]
11. Yoshimura, H.N.; Abreu, A.P.; Molisani, A.L.; de Camargo, A.C.; Portela, J.C.S.; Narita, N.E. Evaluation of Aluminum Dross Waste as Raw Material for Refractories. *Ceram. Int.* **2008**, *34*, 581–591. [CrossRef]
12. Kim, J.; Biswas, K.; Jhon, K.-W.; Jeong, S.-Y.; Ahn, W.-S. Synthesis of AlPO4-5 and CrAPO-5 Using Aluminum Dross. *J. Hazard. Mater.* **2009**, *169*, 919–925. [CrossRef]
13. Murayama, N.; Arimura, K.; Okajima, N.; Shibata, J. Effect of Structure-Directing Agent on AlPO4-n Synthesis from Aluminum Dross. *Int. J. Miner. Process.* **2009**, *93*, 110–114. [CrossRef]
14. Pereira, D.; de Aguiar, B.; Castro, F.; Almeida, M.; Labrincha, J. Mechanical Behaviour of Portland Cement Mortars with Incorporation of Al-Containing Salt Slags. *Cem. Concr. Res.* **2000**, *30*, 1131–1138. [CrossRef]
15. Hwang, J.-Y.; Song, X.M. Replacing Al Powder with Al Slag or Recycled Foil in Cellular Concrete. *JOM* **1997**, *49*, 29–30. [CrossRef]
16. de Araújo, E.G.; Tenório, J.A.S. Cellular Concrete with Addition of Aluminum Recycled Foil Powders. *Mater. Sci. Forum* **2005**, *498–499*, 198–204. [CrossRef]
17. Puertas, F.; Blanco-Varela, M.T.; Vazquez, T. Behaviour of Cement Mortars Containing an Industrial Waste from Aluminium Refining. *Cem. Concr. Res.* **1999**, *29*, 1673–1680. [CrossRef]
18. Llanos, K.; Rodríguez, J. *Estudio Del Aprovechamiento de La Escoria de Aluminio de La Empresa Fundición Agram*; Pontificia Universidad Javeriana Cali: Cali, Colombia, 2011.
19. Borhan, T.M.; Janna, H. Thermal Properties of Cement Mortar Containing Waste Aluminium Fine Aggregate. *J. Kerbala Univ.* **2016**, *14*, 193–200.
20. Shinzato, M.C.; Hypolito, R. Solid Waste from Aluminum Recycling Process: Characterization and Reuse of Its Economically Valuable Constituents. *Waste Manag.* **2005**, *25*, 37–46. [CrossRef]
21. Elinwa, A.U.; Mbadike, E. The Use of Aluminum Waste for Concrete Production. *J. Asian Arch. Build. Eng.* **2011**, *10*, 217–220. [CrossRef]
22. Ozerkan, N.; Maki, O.; Anayeh, M.; Tangen, S.; Abdullah, A. The Effect of Aluminium Dross on Mechanical and Corrosion Properties of Concrete. *Int. J. Innov. Res. Sci. Eng. Technol.* **2014**, *3*, 9912–9922.
23. Mailar, G.; Sreedhara, B.M.; Manu, D.S.; Hiremath, P.; Jayakesh, K. Investigation of Concrete Produced Using Recycled Aluminium Dross for Hot Weather Concreting Conditions. *Resour.-Effic. Technol.* **2016**, *2*, 68–80. [CrossRef]
24. Reddy, M.S.; Neeraja, D. Mechanical and Durability Aspects of Concrete Incorporating Secondary Aluminium Slag. *Resour.-Effic. Technol.* **2016**, *2*, 225–232. [CrossRef]
25. Siddique Pasley, R. The Physical and Chemical Reclamation and Recycling of Elements from Black Aluminium Furnace Residues. Ph.D. Thesis, Brunei University, Seri Begawan, Brunei, 2003.
26. Tang, J.; Liu, G.; Qi, T.; Zhou, Q.; Peng, Z.; Li, X.; Yan, H.; Hao, H. Two-Stage Process for the Safe Utilization of Secondary Aluminum Dross in Combination with the Bayer Process. *Hydrometallurgy* **2022**, *209*, 105836. [CrossRef]
27. Saaty, T.L. Relative Measurement and Its Generalization in Decision Making Why Pairwise Comparisons Are Central in Mathematics for the Measurement of Intangible Factors the Analytic Hierarchy/Network Process. *Rev. Real Acad. Ciencias Exactas Fis. Nat. Ser. A Mat.* **2008**, *102*, 251–318. [CrossRef]
28. *ASTM C618-19*; Standard Specification for Coal Fly Ash and Raw or Calcined Natural Pozzolan for Use in Concrete. ASTM: West Conshohocken, PA, USA, 2022.
29. Suhardini, D.; Septiani, W.; Fauziah, S. Design and Simulation Plant Layout Using Systematic Layout Planning. *IOP Conf. Ser. Mater. Sci. Eng.* **2017**, *277*, 12051. [CrossRef]
30. Ross, S.; Westerfield, R.; Jordan, B. *Fundamentals of Corporate Finance*, 9th ed.; Mc Graw Hill: New York, NY, USA, 2010; ISBN 978-0-07-338239-5.
31. Crespo, R. Incorporación de Escorias de Aluminio En La Fabricación de Productos de Arcilla Cocida. Ph.D. Thesis, Universidad politécnica de Madrid, Madrid, Spain, 2015.
32. Attia, N.; Hassan, K.M.; Hassan, M.I. *Environmental Impacts of Aluminum Dross after Metal Extraction BT—Light Metals 2018*; Martin, O., Ed.; Springer International Publishing: Cham, Switzerland, 2018; pp. 1155–1161.

33. López-Delgado, A.; Robla, J.I.; Padilla, I.; López-Andrés, S.; Romero, M. Zero-Waste Process for the Transformation of a Hazardous Aluminum Waste into a Raw Material to Obtain Zeolites. *J. Clean. Prod.* **2020**, *255*, 120178. [CrossRef]
34. Lemos Micolta, E.D.; Chilito Bolaños, L.C.; Maya Soto, J.C.; Gómez Gómez, A.; Rojas Manzano, M.A. Uso de La Escoria de Aluminio En El Concreto—Revisión Del Estado Del Arte. In Proceedings of the IX Congreso Internacional y 23a Reunión Técnica, Virtual, 4–6 November 2020; pp. 125–132.
35. Singh, S.S. The Formation and Coexistence of Gibbsite, Boehmite, Alumina and Alunite at Room Temperature. *Can. J. Soil Sci.* **1982**, *62*, 327–332. [CrossRef]
36. Braulio, M.A.L.; Rigaud, M.; Buhr, A.; Parr, C.; Pandolfelli, V.C. Spinel-Containing Alumina-Based Refractory Castables. *Ceram. Int.* **2011**, *37*, 1705–1724. [CrossRef]
37. Chaplianko, S.V.; Nikichanov, V.V. Influence of Reactive Alumina Type on the Properties of Corundumspinel Castables (Review). *Sci. Res. Refract. Tech. Ceram.* **2021**, *121*, 103–112. [CrossRef]
38. Zuo, Z.; Lv, H.; Li, R.; Liu, F.; Zhao, H. A New Approach to Recover the Valuable Elements in Black Aluminum Dross. *Resour. Conserv. Recycl.* **2021**, *174*, 105768. [CrossRef]
39. Ramezani, M.; Neitzert, T. Mechanical Milling of Aluminum Powder Using Planetary Ball Milling Process. *J. Achiev. Mater. Manuf. Eng.* **2012**, *55*, 790–798.
40. Leyva Ramírez, E.; de La Fuente Fernández, M.; Leyva González, O.S.; Sánchez Cruz, A.; Ferreiro Guerrero, Y. Estudio de La Cinética de Molienda de La Mena de Cromita Del Yacimiento Albertina. *Tecnol. Química* **2009**, *29*, 55–63.
41. Karakas, O.; Kanca, E. An Investigation on Optimum Grinding System and Conditions for Steel Plant ARP By-Product α-Fe2O3 for Pigment Industry. *Eng. Sci. Technol. Int. J.* **2020**, *23*, 1266–1272. [CrossRef]
42. Plescia, P.; Tempesta, E. Analysis of Friction Coefficients in a Vibrating Cup Mill (Ring Mill) during Grinding. *Tribol. Int.* **2017**, *114*, 458–468. [CrossRef]
43. Hernández, M.Á.; Nieves, C.E. *Perfil Logístico Del Sector Cemento En Colombia*; Repositorio Institucional EdocUR: Bogotá, Colombia, 2015.
44. Meneses-Núñez, L.S.; Escobar, D.; Ibarra, S.J.; Núñez-Navia, J.C.; Muñoz-Velez, M.; Salazar-Serna, K. *Diseño de un Proceso Productivo que Permita la Recuperación de Escoria de Aluminio Para su uso Como Materia Prima en Morteros. Capstone Project for Industrial Engineering Degree*; Pontificia Universidad Javeriana Cali: Cali, Colombia, 2022.
45. Maury-Ramírez, A.; Illera-Perozo, D.; Mesa, J.A. Circular Economy in the Construction Sector: A Case Study of Santiago de Cali (Colombia). *Sustainability* **2022**, *14*, 1923. [CrossRef]
46. Maury-Ramírez, A.; De Belie, N. Environmental and Economic Assessment of Eco-Concrete for Residential Buildings: A Case Study of Santiago de Cali (Colombia). *Sustainability* **2023**, *15*, 12032. [CrossRef]
47. Bajare, D.; Korjakins, A.; Kazjonovs, J.; Rozenstrauha, I. Pore structure of lightweight clay aggregate incorporate with non-metallic products coming from aluminium scrap recycling industry. *J. Eur. Ceram. Soc.* **2012**, *32*, 141–148. [CrossRef]
48. Ramaswamy, P.; Gomes, S.A.; Ravichander, N.P. Utilization of aluminum dross: Refractories from industrial waste. *IOP Conf. Ser. Mater. Sci. Eng.* **2019**, *577*, 012101. [CrossRef]
49. Shen, H.; Liu, B.; Ekberg, C.; Zhang, S. Harmless disposal and resource utilization for secondary aluminum dross: A review. *Sci. Total Environ.* **2021**, *760*, 143968. [CrossRef] [PubMed]
50. Liu, Y.; Leong, B.S.; Hu, Z.-T.; Yang, E.-H. Autoclaved aerated concrete incorporating waste aluminum dust as foaming agent. *Constr. Build. Mater.* **2017**, *148*, 140–147. [CrossRef]
51. Javali, S.; Chandrashekar, A.R.; Naganna, S.R.; Manu, D.S.; Hiremath, P.; Preethi, H.G.; Vinod Kumar, N. Eco-concrete for sustainability: Utilizing aluminium dross and iron slag as partial replacement materials. *Clean Technol. Environ. Policy* **2017**, *19*, 2291–2304. [CrossRef]
52. Dai, C.; Apelian, D. Fabrication and characterization of aluminum dross-containing mortar composites: Upcycling of a waste product. *J. Sustain. Metall.* **2017**, *3*, 230–238. [CrossRef]
53. Ewais, E.M.M.; Besisa, N.H.A. Tailoring of magnesium aluminum titanate based ceramics from aluminum dross. *Mater. Des.* **2018**, *141*, 110–119. [CrossRef]
54. Adeosun, S.O.; Usman, M.A.; Ayoola, W.A.; Sekunowo, I.O. Evaluation of the mechanical properties of polypropylene-aluminum-dross composite. *Int. Sch. Res. Not.* **2012**, *2012*, 1–6. [CrossRef]
55. Samat, N.; Sabaruddin, F.A.; Meor Yusoff, M.S.; Dayang Habibah, A.I.H. Evaluation of waste from aluminum industry as filler in polypropylene composites. *JOM* **2017**, *69*, 790–795. [CrossRef]

sustainability MDPI

Article

3D Printing of Hybrid Cements Based on High Contents of Powders from Concrete, Ceramic and Brick Waste Chemically Activated with Sodium Sulphate (Na_2SO_4)

Rafael Robayo-Salazar, Fabio Martínez, Armando Vargas and Ruby Mejía de Gutiérrez *

Composite Materials Group (GMC-CENM), Universidad del Valle, Cali 760033, Colombia;
rafael.robayo@correounivalle.edu.co (R.R.-S.); fabio.martinez@correounivalle.edu.co (F.M.);
vargas.armando@correounivalle.edu.co (A.V.)
* Correspondence: ruby.mejia@correounivalle.edu.co

Abstract: This article evaluates the synthesis, characterization and 3D printing of hybrid cements based on high (70%) contents of powders from concrete waste (CoW), ceramic waste (CeW) and red clay brick waste (RCBW) from construction and demolition waste. For the synthesis of the hybrid cements, 30% (by weight) of ordinary Portland cement (OPC) was added. Sodium sulphate (Na_2SO_4) (4%) was used as a chemical activator. The effect of the liquid/solid ratio on the properties in the fresh state of the mixes was studied by means of minislump, flowability index, and buildability tests. The compressive strength was evaluated at 3, 7, 28 and 90 days of curing at room temperature (≈25 °C), obtaining strengths of up to 30.7 MPa (CoW), 37.0 MPa (CeW) and 33.2 MPa (RCBW) with an L/S ratio of 0.30. The results obtained allowed selecting the CoW 0.30, CeW 0.33 and RCBW 0.38 mixes as optimal for carrying out 3D printing tests on a laboratory scale, successfully printing elements with good print quality, adequate buildability, and compressive strength (CoW 0.30 = 18.2 MPa, CeW 0.33 = 27.7 MPa and RCBW 0.38 = 21.7 MPa) higher than the structural limit (≥17.5 MPa) established for concrete by Colombian Regulations for Earthquake Resistant Construction (NSR-10).

Keywords: additive manufacturing; 3D printing; construction and demolition waste; sodium sulphate; alkali-activated materials; geopolymers

Citation: Robayo-Salazar, R.; Martínez, F.; Vargas, A.; Mejía de Gutiérrez, R. 3D Printing of Hybrid Cements Based on High Contents of Powders from Concrete, Ceramic and Brick Waste Chemically Activated with Sodium Sulphate (Na_2SO_4). *Sustainability* **2023**, *15*, 9900. https://doi.org/10.3390/su15139900

Academic Editor: José Ignacio Alvarez

Received: 1 June 2023
Revised: 16 June 2023
Accepted: 20 June 2023
Published: 21 June 2023

1. Introduction

The worldwide 3D printing construction market was valued at USD 11 million in 2021 and expected to grow to USD 48 million in 2030, according to Grand View Research [1]. In fact, the implementation of additive manufacturing technology in the construction sector has brought into play a new market and given rise to multiple advantages for this industry compared to conventional construction methods. Among such advantages are a higher construction speed [2], reduced labour costs [3], greater energy efficiency [4], lower consumption of materials [5], decreased waste generation and the possibility of producing elements with complex geometries almost impossible to obtain using conventional methods [6]. As highlighted in [7], factors necessary to position 3D printing as a sustainable construction method include using non-conventional cementitious materials [8], including alkali-activated cements, geopolymers and hybrid cements [9–14].

Synthesis of these non-conventional cementitious materials is based on chemical activation of a material rich in aluminosilicates (precursor) through the use of alkaline activators (hydroxide type (ROH, $R(OH)_2$), weak acid salts (R_2CO_3), strong acid salts (Na_2SO_4, $CaSO_4{\cdot}2H_2O$), and siliceous salts $R_2O(n)SiO_2$, where R is an alkaline ion of the Na, K or Li type [15]. This process gives rise to materials with physical, mechanical and durable properties similar or even superior to traditional cementitious materials such as ordinary Portland cement (OPC) [16]. Another advantage of these types of non-conventional cementitious material is their potential to reduce the carbon footprint, making it possible to call them environmentally friendly cements [17,18].

In particular, hybrid cements are able to use a small amount (≤30%) of OPC that promotes a gain of strength at room temperature (≈25 °C) and that can be chemically activated with smaller amounts (2–6% by weight) of Na_2SO_4 (sodium sulphate) [19]. Na_2SO_4 has a lower economic and energy cost than traditional alkaline activators (NaOH (sodium hydroxide) and Na_2SiO_3 (waterglass)) [20]. The chemical activation mechanism of these hybrid cements (OPC ≤ 30%) via incorporation of Na_2SO_4 has been described by other authors [21]. The role of the SO_4^{2-} ion in these non-conventional binders consists of (1) accelerating the hydration process of the alite (C_3S) phase present in the clinker; (2) the formation of ettringite from the reaction with the celite phase (C_3A); and (3) the formation of NaOH as a by-product of the reaction between Na_2SO_4 and Portlandite ($Ca(OH)_2$) generated during the hydration of calcium silicates (C_3S (alite) and C_2S (belite)) present in the clinker. An additional hypothesis derived from these reactions is that the $Ca(OH)_2$ and NaOH formed can alkaline-activate the reactive phase of the precursor (aluminosilicate) and form (N,C)-A-S-H type hybrid gels [21].

Precursors that can be used for the synthesis of alkali-activated materials include a wide range of pozzolanic additions (Supplementary Cementitious Materials (SCM)) of natural or artificial origin, and industrial by-products with high aluminosilicate contents, including natural pozzolans, fly ash, steel slag and thermally activated clays (metakaolin), among others. Construction and demolition waste (CDW) is made up mostly of concrete waste (CoW), ceramic waste (CeW) and red clay brick waste (RCBW). All of these are aluminosilicate in nature, so they feature a certain degree of reactivity to equally be used as precursors. In this regard, in previous studies [22,23], it was shown that the fine fractions (powders) of CDW can be used through chemical activation processes in the synthesis of alkali-activated materials, geopolymers and/or hybrid cements.

According to Raza et al. [11], the use of alkali-activated materials in 3D printing was introduced in 2016 by Xia and Sanjayan [14], and from that moment this research topic quickly became an innovative trend for research groups around the world. The application of CDW-based alkali-activated materials however in the field of 3D printing has hardly been explored at all. As highlighted in [7], regardless of its nature, the cementitious material suitable for 3D printing must have an adequate extrusion capacity (mouldable and extrudable material), be fluid, be buildable, with an adequate setting time (open time), have dimensional stability (low shrinkage), and achieve a certain level of mechanical strength to be used in structural applications. In this context, Şahin et al. [24] studied the rheological properties for 3D printing of geopolymers based on hollow brick (HB), red clay brick (RCB), roof tile (RT) and glass (G), activated with combinations of sodium hydroxide (NaOH), calcium hydroxide ($Ca(OH)_2$) and sodium silicate (Na_2SiO_3). The mix activated with 6.25 M NaOH and 10% $Ca(OH)_2$ exhibited the best rheological and mechanical properties and was selected for laboratory-scale 3D printing tests. Based on that study [24] and using the same geopolymeric cement, Ilcan et al. [25] demonstrated the possibility of incorporating a fine aggregate of recycled concrete (aggregate-to-binder ratio of 0.38) in the production of low and high viscosity mortars, successfully applying the aggregate in 3D printing without affecting the rheological and mechanical properties of the mortar mixes. Demiral et al. [26] subsequently evaluated the effect of anisotropy (dependence on the direction of 3D printing) on compressive strength in three directions (perpendicular, parallel and lateral) and flexural strength in two directions (perpendicular and lateral), in the geopolymeric mortars produced in the abovementioned study [25]. They further evaluated the adhesion between layers through direct and indirect traction tests. The authors conclude that interlayer adhesion influences the anisotropic behaviour of 3D printed elements. They state however that 3D-printed specimens tested in the direction perpendicular to the printing direction showed similar performance to mould-casted specimens, indicating that interlayer adhesion had little influence in the perpendicular loading direction.

Despite the recent advances, the use of low economic, low energy cost alternative activators such as sodium sulphate (Na_2SO_4) in the synthesis of hybrid cements based on high CDW contents and their application in 3D printing has not yet been reported. This article aims to synthesize and characterize hybrid cements based on high contents (70% by weight) of concrete waste (CoW), ceramic waste (CeW), and red clay brick waste (RCBW), derived from the fine fraction (powder) of construction and demolition waste (CDW), and ordinary Portland cement (OPC) (30% by weight). Na_2SO_4 was used for the chemical activation of the hybrid cements. The effect of the liquid/solid (L/S) ratio on the properties of the fresh state (mini slump, flowability index, workability, and open time) and hardened state (compressive strength) of the mixtures was evaluated, and the optimal ranges of these properties were determined for their application in 3D printing. The optimal mixtures were used in laboratory-scale printing tests, demonstrating their potential application in additive manufacturing processes. These are the first reported results of 3D printing for this type of hybrid cement based on CDW powders.

2. Materials and Methods

2.1. Raw Materials

Concrete waste (CoW), ceramic waste (CeW) and red clay brick waste (RCBW) from construction and demolition activities (CDW) were used to produce the mixes. These residues were finely ground using a ball mill. The particle size was estimated by laser granulometry using a Mastersizer-2000 equipment (Malvern Panalytical, Madrid, Spain). For the synthesis of hybrid cements, ordinary Portland cement (OPC) was used. An Ultrapyc 3000 helium pycnometer (Anton Paar, Graz, Austria) was used to determine the density of the raw materials. The chemical composition was determined by X-ray fluorescence (XRF) using a MagiX-Pro PW-2440 spectrometer (Phillips, Eindhoven, The Netherlands). Industrial grade sodium sulphate (Na_2SO_4) was used for chemical activation of hybrid cements produced.

2.2. Production of Mixes and Characterization

A total of 10 mixes (hybrid cements) (Table 1) were designed based on a 70% precursor content (CoW, CeW or RCBW) and the addition of 30% (by weight) of OPC. In order to evaluate the effect of the liquid/solid (L/S) ratio on the fresh and hardened properties of the mixes, this design variable was modified between 0.30–0.38. For the calculation of the L/S ratio, liquids correspond to the mixing water and solids correspond to the sum of the waste and the OPC (precursor). The Na_2SO_4 content was 4% by weight with respect to the precursor (waste + OPC). The determination of this optimal content (4% by weight) of chemical activator (Na_2SO_4) was based on a previous study [19].

Table 1. Design of mixes and proportioning of raw materials.

Mix	L/S Ratio	Proportion (g)			
		Waste	OPC	Na_2SO_4	Water
CoW 0.30	0.30	70	30	4	30
CoW 0.33	0.33	70	30	4	33
CoW 0.36	0.36	70	30	4	36
CeW 0.30	0.30	70	30	4	30
CeW 0.33	0.33	70	30	4	33
CeW 0.36	0.36	70	30	4	36
RCBW 0.30	0.30	70	30	4	30
RCBW 0.33	0.33	70	30	4	33
RCBW 0.36	0.36	70	30	4	36
RCBW 0.38	0.38	70	30	4	38

The mixes were produced in a Hobart mixer with a mixing time of 5 min. Initially, the waste (precursor) was dry homogenized with the addition of OPC. Subsequently, the chemical activator, previously dissolved in the mixing water, was added to the mix.

In order to correlate the rheological behaviour of hybrid cements with their 3D printing capacity, the mixes were characterized in the fresh state by adapting minislump, flow rate and buildability tests. The minislump (Figure 1a) was determined as the settlement shown by the mix due to its own weight after removing the conical mold according to ASTM C230 standard [27].

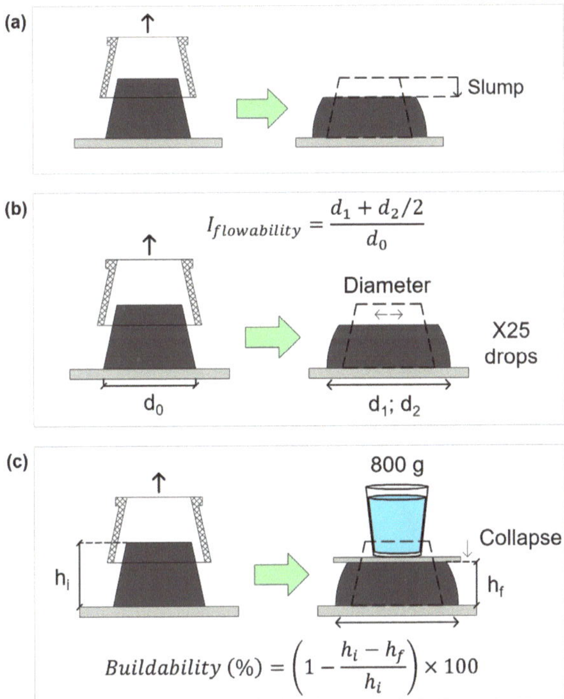

Figure 1. Characterization tests in the fresh state of the mixes: (**a**) minislump, (**b**) flowability index and (**c**) buildability of the mixes.

Flowability index (Figure 1b) was determined according to the procedure established in ASTM C230 [27], taking into account the average diameter reached by the mix after being subjected to 25 drops from the flow table. Buildability (Figure 1c) was determined from the collapse caused by an 800 g weight placed on the mix immediately after carrying out the minislump test. This weight (800 g) is equivalent to the fresh weight of the mix used to fill the conical mould.

The setting time (initial and final) of the mixes was determined according to the procedure described in the ASTM C191 standard (method B) [28] using a Vicat apparatus. Additionally, the effect of setting time on ultrasonic pulse velocity was evaluated using a Pundit PL-200 unit (Proceq, Schwerzenbach, Swiss) with P-type wave transducers of 54 kHz frequency, a pulse voltage of 200 V, and a sensor gain of 500×. For the measurement of ultrasonic pulse velocity in the fresh state, an acrylic cubic mould with a side of 75 mm and a wall thickness of 1.3 mm was used. Additionally, the effect of mixing time on the loss of workability of the mixes was established through the minislump and flowability tests. This evaluation was carried out up to a maximum mixing time of 90 min. Together, these tests allowed us to study the open time of the mixes for 3D printing.

The compressive strength of the hybrid cements was evaluated in an INSTRON 3369 (Instron, Norwood, MA, USA) universal testing machine with a 50 kN capacity, using a testing speed of 1 mm/min. Conventionally moulded 20 mm cubes were tested to calculate the average strength of the mixes at 3, 7, 28 and 90 days of curing at room temperature (25 °C) (relative humidity (RH) ≈ 80%). Each value of strength corresponds to the average of three test samples.

2.3. Additive Manufacturing (3D Printing) and Tests

The additive manufacturing process carried out is summarized in Figure 2, starting with the computer-aided design (CAD) of a solid part exported in .STL format, followed by the printing parameterization process through the free software Ultimaker Cura 5.0 and generating a file in .gcode format. Finally, the execution of the printing process was carried out using a Creality Ender-3 printer (Creality, Shenzhen, China), to which a Ceramic 3D Printer Kit (Eazao) was adapted. The optimum printing speed was 7 mm/s. The nozzle used corresponds to a circular geometry of 8 mm in diameter. The parameterization of the 3D printing process included a layer height of 6 mm, with the layer height/width ratio being 0.75 (6 mm/8 mm). This optimum ratio (0.75) was determined following preliminary printing tests.

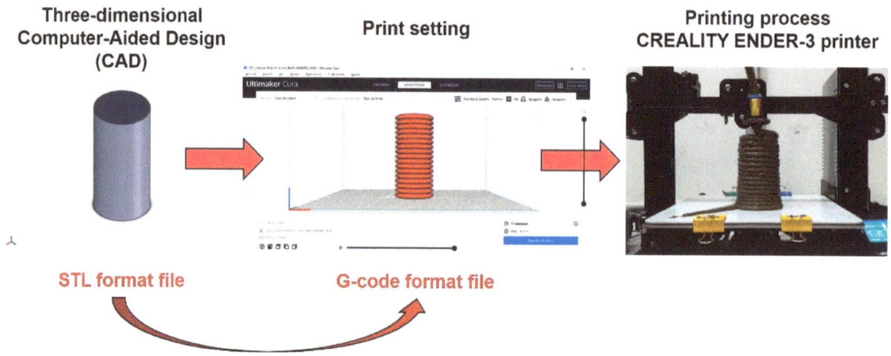

Figure 2. Graphic summary of the methodology followed for the additive manufacturing process (3D printing).

To evaluate the printability of the mixes, hollow (without filling) cylindrical specimens of 50.8 mm in diameter × 101.6 mm in height (17 layers) were printed (Figure 2). At the end of the printing tests, the actual heights of the 3D specimens were verified with the help of a metric rule to validate their buildability.

Solid beam-type specimens of 45 mm × 30 mm × 140 mm (width × height × length) were printed to evaluate the mechanical strength (compressive and flexural) of the 3D printed mixes. A total of 3 solid beams were produced for each mix. A concentric filling pattern (from outside to inside) was used, considering 100% filling. The specimens were removed from the impression base (plate) 24 h after their production and were subjected to a curing process in a controlled environment (RH ≈ 80% and 25 °C) until the corresponding test age.

The beams were flexural tested (3 points) after 7 days and the compressive strength was determined at 7 and 28 days with the halves of the flexural test beams (Figure 3), according to the procedure described in the UNE-EN 1015 standard [29]. The direction of application of the flexural and compressive loads was perpendicular to the direction of printing (Figure 3). Additionally, the density, absorption and porosity at 28 days were determined according to the ASTM C642 standard [30] from beams of the same type.

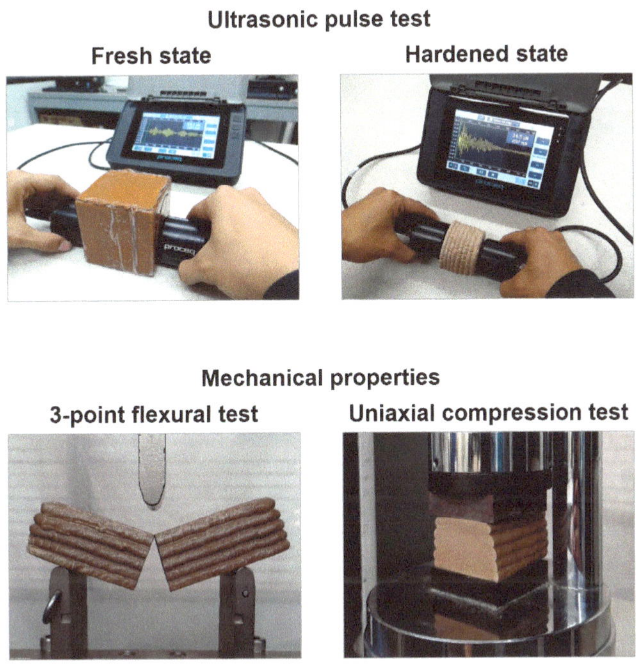

Figure 3. Characterization tests in the hardened state of the mixes: 3D printed specimens.

The ultrasonic pulse velocity was determinate at 28 days on solid 3D-printed cylinders 50.8 mm in diameter and 50.8 mm in height according to the procedure established in ASTM C597 [31]. These results were compared with that obtained in specimens made using the conventional casting process (mould-casted). A Pundit 200 instrument was used with P-wave transducers of 54 kHz frequency, a pulse voltage of 100 V and a sensor gain of 1×. Before the measurements, a calibration of the wave transmission time was carried out with the calibration pattern of the equipment. The specimens were tested in a dry condition (ambiently dried). The measurements were made on the lower and upper faces of the specimens (direction perpendicular to the printing direction) (Figure 3). The ultrasonic pulse velocity reported for each mix corresponds to the average of three measurements.

The macroscopic observation of the interface zone between layers was carried out through the inspection of a cross section of the 3D printing specimens in a stereomicroscope. The microstructural analysis was performed on this same area by means of scanning electron microscopy (SEM), using a JEOL JSM-6490LV microscope (Jeol, Tokio, Japan) with an accelerating voltage of 20 kV. An Oxford Instruments Link-Isis X-ray spectrometer was coupled to the microscope (EDS).

3. Results and Discussion

3.1. Materials Characterization

The results of the chemical composition demonstrate the aluminosilicate nature (SiO_2 + Al_2O_3) of the CoW, CeW and RCBW, representing 44.4, 75.4 and 77.4% of their total composition, respectively (Table 2).

The densities of the CoW, CeW and RCBW were 2.68, 2.71 and 2.75 g/cm^3, respectively. The OPC meanwhile reported a density of 3.00 g/cm^3. The average particle size of the CoW, CeW and RCBW was 24.6, 25.8 and 23.6 µm, respectively (Figure 4). The average particle size of the OPC was 22.5 µm.

Table 2. Chemical composition of raw materials (XRF).

Material	SiO_2	Al_2O_3	CaO	Fe_2O_3	MgO	Na_2O	K_2O	Others	LOI [1]
CoW	36.1	8.3	28.7	6.8	1.9	0.6	0.6	1.1	15.9
CeW	59.3	16.1	9.8	5.5	0.8	0.5	1.6	2.3	4.1
RCBW	59.0	18.4	5.4	7.8	2.4	1.1	1.5	1.5	2.9
OPC	19.4	4.1	55.7	4.7	1.7	0.3	0.3	4.6	9.2

[1] Loss on ignition (LOI)

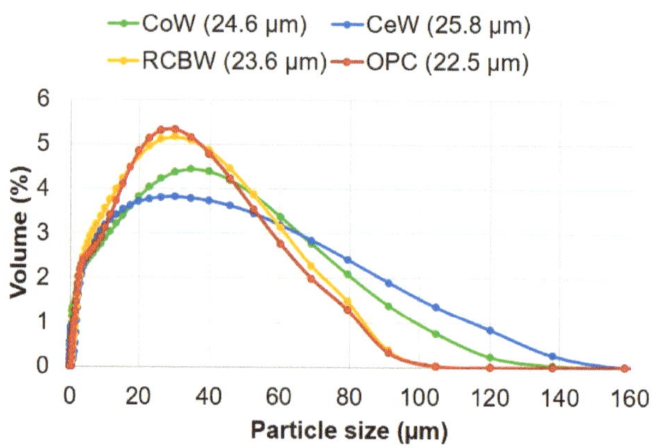

Figure 4. Particle size distribution (laser granulometry) of the materials.

3.2. Fresh Properties

3.2.1. Minislump, Flow Rate and Buildability

The effect of the L/S ratio on the properties in the fresh state of the CoW, CeW and RCBW mixes, included in Table 1, can be seen in Figure 5 (minislump), Figure 6 (flowability index) and Figure 7 (buildability). In general, it is observed that the higher the L/S ratio, the higher the workability (minislump and flowability) of the mixes and as a consequence the lower the buildability, a behaviour that has been reported elsewhere [32]. According to Tay et al. [33], a high water content reduces the internal frictions between the cement particles and resulting in greater flowability. Additionally, it is evident that waste type exerts some control over the rheology of the mixes, suggesting that optimization of each mix design must consider the properties in the fresh state that the type of waste fosters and the effects of these on the 3D printing process. In this regard, the CoW mixes tend to be the most flowability (lowest water demand), followed by the CeW mixes and then the RCBW mixes; the latter demand a higher L/S ratio (0.38) to achieve the level of workability required by the 3D printing process.

In the case of the minislump (Figure 5) and flowability (Figure 6), the CeW 0.30 and RCBW 0.30 and 0.33 mixes have a very dry consistency and fall below the optimal printing region. In relation to the above, the mixes must have an acceptable extrusion capacity, which is affected by a very dry consistency. In contrast, the CoW 0.36 mix had a very fluid consistency that places it above the optimal printing area. Regarding buildability (Figure 7), the very fluid mixes reported a low shape retention capacity (buildability < 80%), which affects the ability to support the weight of the subsequent layers without collapsing and this behaviour is not adequate for the 3D printing process. In contrast, the very dry mixes presented a high buildability (close to 100%), but at the same time a low extrusion capacity (equally unsuitable for 3D printing). In conclusion, it was necessary to find a balance between flowability and buildability in selecting the optimal mixes. Considering this, only the mixes with minislump between 10–20 mm, flowability index between 2.0–2.4 and buildability greater than 80% could be used in the 3D printing process. These correspond to mixes CoW 0.30; CeW 0.33 and RCBW 0.38. The results of the printing tests of these mixes are included in Section 3.4.1.

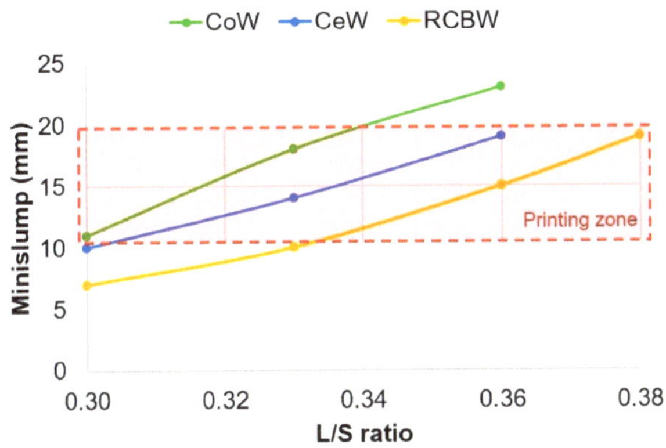

Figure 5. Minislump of the mixes.

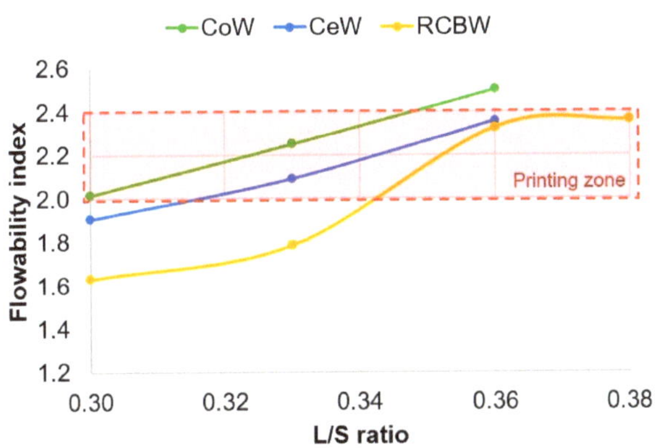

Figure 6. Flowability index of the mixes.

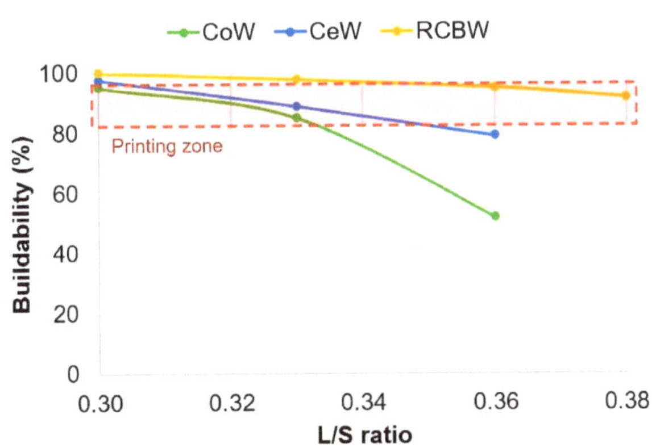

Figure 7. Buildability of the mixes.

3.2.2. Open Time

Open time is defined as the time interval in which the mix is able to be printed before its properties in the fresh state are altered [34]. 3D printing actually requires a sufficient setting time to maintain the consistency of the mix during the extrusion, pumping and deposition process, and thus avoid possible blockages in the pipe and/or nozzle of the printer. However, at the same time a mix with adequate buildability is required; a property that is promoted with short setting times that ensure the necessary strength for the lower layers to support the weight of the upper layers. The open time adjustment must also take into account that a very short setting time could affect adhesion between layers and therefore the mechanical strength of the printed element [2]. Given the above, the effect of L/S ratio on the setting time of the mixes is presented in Figure 8.

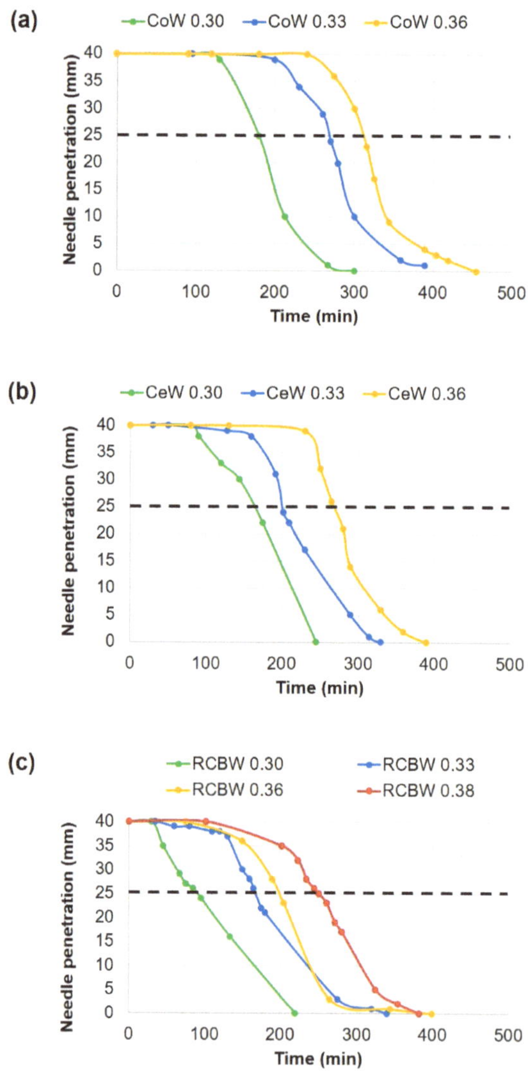

Figure 8. Curves of setting time of the mixes: (**a**) CoW, (**b**) CeW and (**c**) RCBW.

In general, it can be seen that the higher the L/S ratio, the longer the setting time of the mixes. It is also possible to identify that waste type influences initial and final setting times of the mixes. The shortest times are registered for RCBW mixes, followed by CeW mixes and CoW mixes, with the longest times. The RCBW 0.30, 0.33, 0.36 and 0.38 mixes recorded initial setting times of 80, 170, 200 and 250 min, respectively. The CeW 0.30, 0.33 and 0.36 mixes had initial setting times of 160, 200 and 270 min, respectively, while the CoW 0.30, 0.33 and 0.36 mixes reported respective initial setting times of 180, 270 and 310 min.

Ultrasonic pulse velocity monitoring, according to Uppalapati et al. [35], is sensitive to hydrate formation and microstructural changes associated with the setting—hardening—of cementitious materials. Figure 9 relates the ultrasonic pulse velocity of the CoW 0.30, CeW 0.33 and RCBW 0.36 mixes during their setting process. It should be recalled that these mixes were selected as optimal during the evaluation of their properties in the fresh state (Section 3.2.1) and featured initial setting times of 180 min (CoW 0.30), 200 min (CeW 0.33), and 250 min (RCBW 0.38). In Figure 9 a direct correlation is seen between the hardening process of the mixes and the ultrasonic pulse velocity reported, steadily increasing as the mix gradually sets. The ultrasonic pulse velocity for the RCBW 0.38, CeW 0.33 and CoW 0.30 mixes in the initial setting time (needle penetration = 25 mm) were ≈1365, ≈1340 and ≈1510 m/s, respectively. These values coincide with those reported elsewhere [35] for alkali-activated materials during the initial setting time (1450–1550 m/s). Values above 1650–1750 m/s meanwhile are normally associated with the final setting time of cementitious materials.

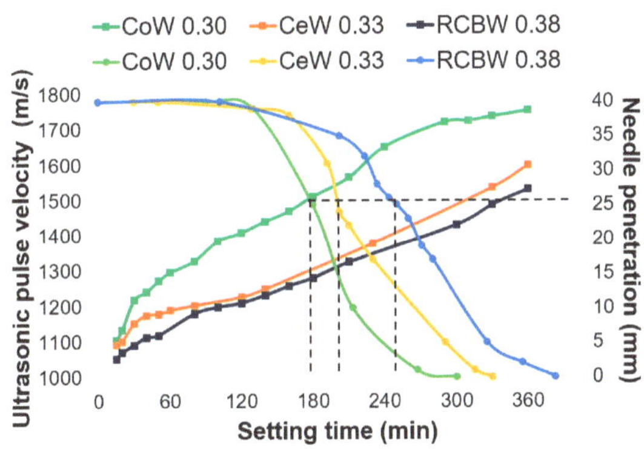

Figure 9. Evolution of ultrasonic pulse velocity as a function of setting time of the CoW 0.30, CeW 0.33 and RCBW 0.38 mixes.

Furthermore, it can be seen in Figure 9 that the ultrasonic pulse velocity curves have the greatest slope (acceleration) during the first minutes (10–40 min). This demonstrates that, before the initial setting, the mixes underwent important changes in their microstructure and properties in the fresh state (flowability), even though with the Vicat needle there were no changes in depth of needle penetration (≈40 mm) in that same time interval (10–40 min). This finding allows to conclude that the conventional setting time test (apart from Vicat) is not the most appropriate method to study the open time of mixes for 3D printing and that it is necessary to use other techniques, such as ultrasonic pulse, for more detailed monitoring of reaction kinetics and changes in the fresh state. In this regard, the changes in the velocity of the ultrasonic pulse during the first minutes can be associated with the stages of dissolution (Step I), flocculation (Step II), gelation (Step III) and polycondensation (Step IV) [36]. These stages occur during the hydration process of hybrid cements as consequence of the chemical activation [21].

As mentioned in [7], the open time is usually less than the initial setting time (needle penetration = 25 mm); its experimental evaluation, via tests of loss of flowability as a function of time, is important. Extending the mixing time during 3D printing tests extends the open time of the mixes, while maintaining the mixes static following the completion of the initial mixing process reduces their useful life. The effect of mixing time (up to 90 min) on the minislump and flowability index of the mixes is thus presented in Figures 10 and 11, respectively.

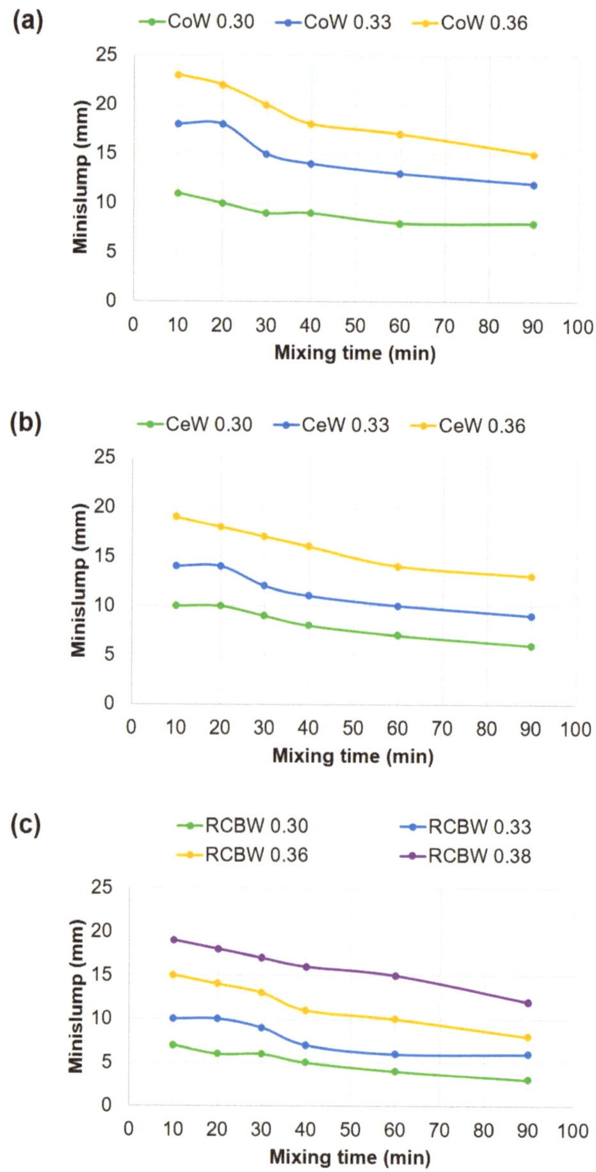

Figure 10. Effect of mixing time on the minislump of the mixes: (**a**) CoW, (**b**) CeW and (**c**) RCBW.

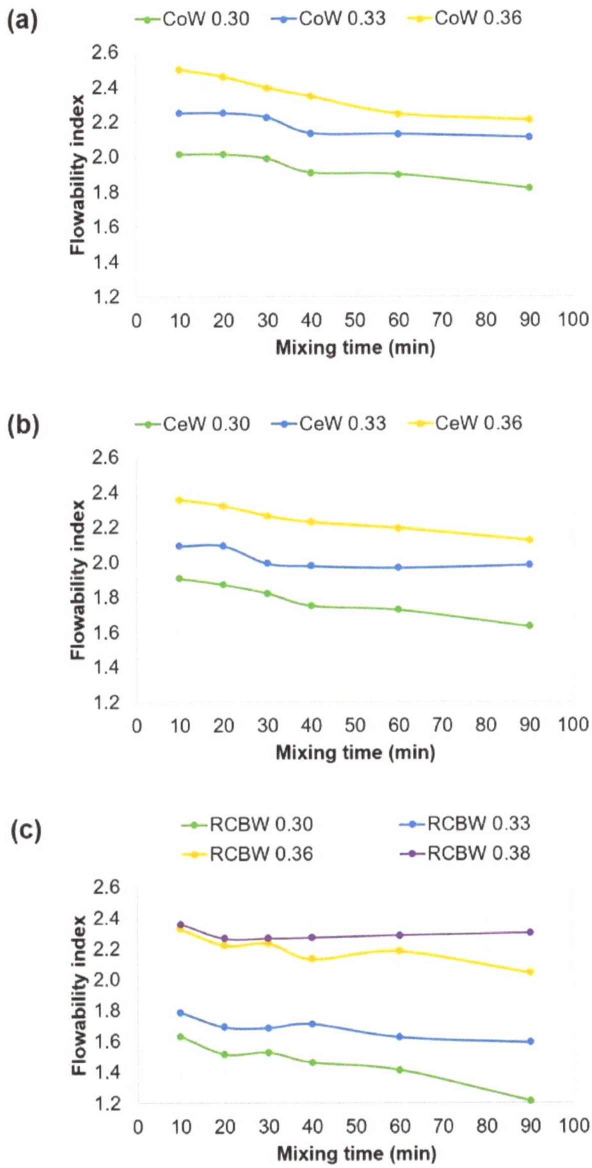

Figure 11. Effect of mixing time on the flowability index of the mixes: (**a**) CoW, (**b**) CeW and (**c**) RCBW.

It was generally observed that after 10–30 min of mixing, the mixes suffered a notable loss of workability (minislump and flowability) and therefore of their 3D printing capacity, results that agree with those reported by Ilcan et al. [25], Zhang et al. [37] and those of commercial products such as Sikacrete-751 3D, Sikacrete-752 3D and Sikacrete-7100 3D [38]. In conclusion, based on the results obtained in this study, it was established that the open times of the CoW 0.30, CeW 0.33 and RCBW 0.38 mixes were 30, 20 and 10 min, respectively.

3.3. Compressive Strength of the Mixes

The evolution of the compressive strength of the CoW, CeW and RCBW mixes is presented in Figure 12. In general, it is observed that the highest mechanical strengths are promoted at lower L/S ratios. Indeed, the maximum compressive strengths (90 days) of the CoW, CeW and RCBW mixes were obtained with an L/S ratio of 0.30 and achieved values of 30.7, 37.0 and 33.2 MPa, respectively.

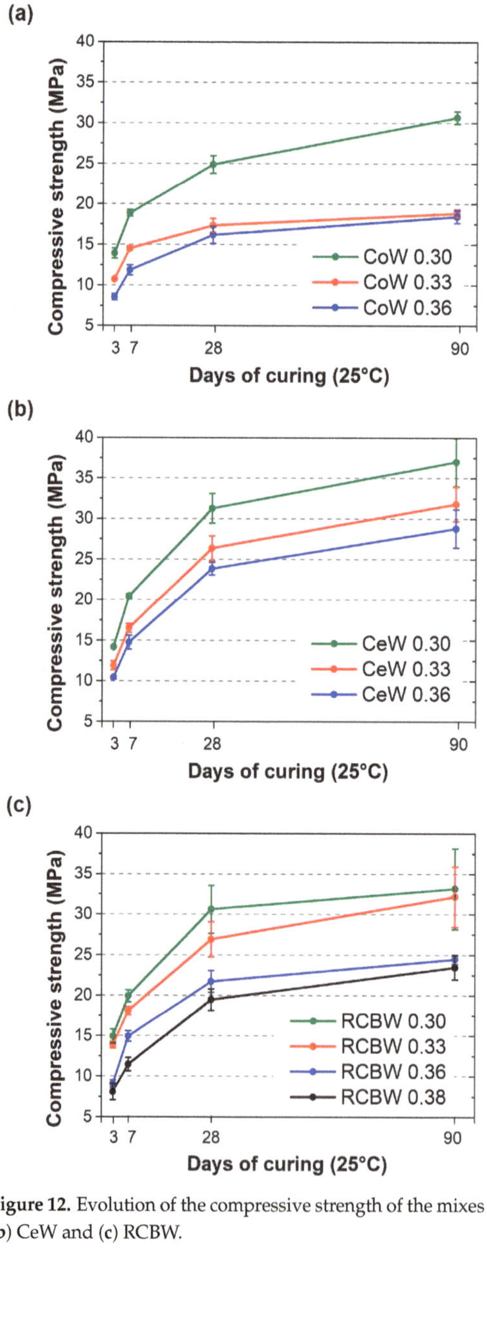

Figure 12. Evolution of the compressive strength of the mixes (3, 7, 28 and 90 days of curing): (**a**) CoW, (**b**) CeW and (**c**) RCBW.

It is evident meanwhile that waste type influences the mechanical performance of the mixes, and this may be related to the degree of chemical reactivity of each material. It should be noted that, in this study, the effect of particle size was controlled and a very similar average size and granulometric distribution was ensured for the three wastes (Figure 4). In this regard, controlling the particle size is considered fundamental for comparative purposes, since this property has a strong influence on the degree of reactivity of the precursor. Clarifying the above, the best mechanical performances for the same L/S ratio were produced by the CeW, followed by the RCBW and lastly the CoW. These mechanical results agree with those reported in other studies [22,23] using residues of the same nature (CDW) and alkaline activation processes.

Regarding the mechanical behaviour of the three mixes previously selected as optimal for the 3D printing process, their compressive strengths were 24.9 MPa (CoW 0.30), 26.4 MPa (CeW 0.33) and 19.5 MPa (RCBW 0.38) at 28 days of curing and 30.7 MPa (CoW 0.30), 31.8 MPa (CeW 0.33) and 23.5 MPa (RCBW 0.38) at 90 days of curing. In the case of the CeW and RCBW mixes, the optimal L/S ratios in the fresh state (0.33 and 0.38) do not coincide with the L/S ratio that promotes the best mechanical performance (0.30). This conflict has been described by other authors [2], recognizing that, in some cases, the need to obtain a fluid mix for 3D printing demands a high mixing water content (high L/S ratio) and this affects the compressive strength of the mixes. This was the case for the RCBW 0.38 mix, which reported the lowest mechanical performance among the mixes optimized for the 3D printing process.

3.4. *3D Printing Tests*

3.4.1. Printability and Buildability

In order to validate the application potential of the mixes previously defined as optimal (CoW 0.30; CeW 0.33; RCBW 0.38), the extrusion and 3D printing capacity at laboratory scale was evaluated according to the parameters described in the methodology (Section 2.3). Figure 13 shows the results of this 3D printing test, which correspond to the printing of hollow (without filling) cylinders of 50.8 mm in diameter × 101.6 mm in height (equivalent to 17 layers of 6 mm thickness). As can be seen, in general, all three mixes were found to have an adequate extrusion and 3D printing capacity, obtaining homogeneous portions and a good surface finish, without the presence of defects, discontinuities (breaks) and/or macro-pores that may compromise the aspect or appearance of the printed element. Some small defects can be seen on the surface of the specimens corresponding to CoW, however these do not compromise the final properties of the element.

Additionally, the buildability of the three mixes was verified by measuring the actual height of the 3D printed cylinders. The results show a high level of buildability (close to 97–99%), in accordance with the characterization in the fresh state previously reported (Figure 7). This result is considered important, since a high degree of fresh deformation (low buildability) can affect the final height of the element compared to the initial model. This difference would lead to an adjustment of the height of the element through the computerized design (CAD) of the part and the parameterization (G-code) necessary to execute the 3D printing process, altering the number of layers.

3.4.2. Physical-Mechanical Behaviour

Table 3 presents the density, absorption and porosity results for the 3D printed specimens corresponding to the CoW 0.30, CeW 0.33 and RCBW 0.38 mixes. As can be seen, the L/S ratio has a significant influence on the physical properties. The higher the L/S ratio, the higher the absorption and the porosity and therefore the lower the density of the 3D specimens obtained. The RCBW 0.38 mix reported the lowest apparent density value (1.92 g/cm^3), followed by CeW 0.33 (1.96 g/cm^3) and finally CoW 0.30 (1.97 g/cm^3) with the highest reported apparent density.

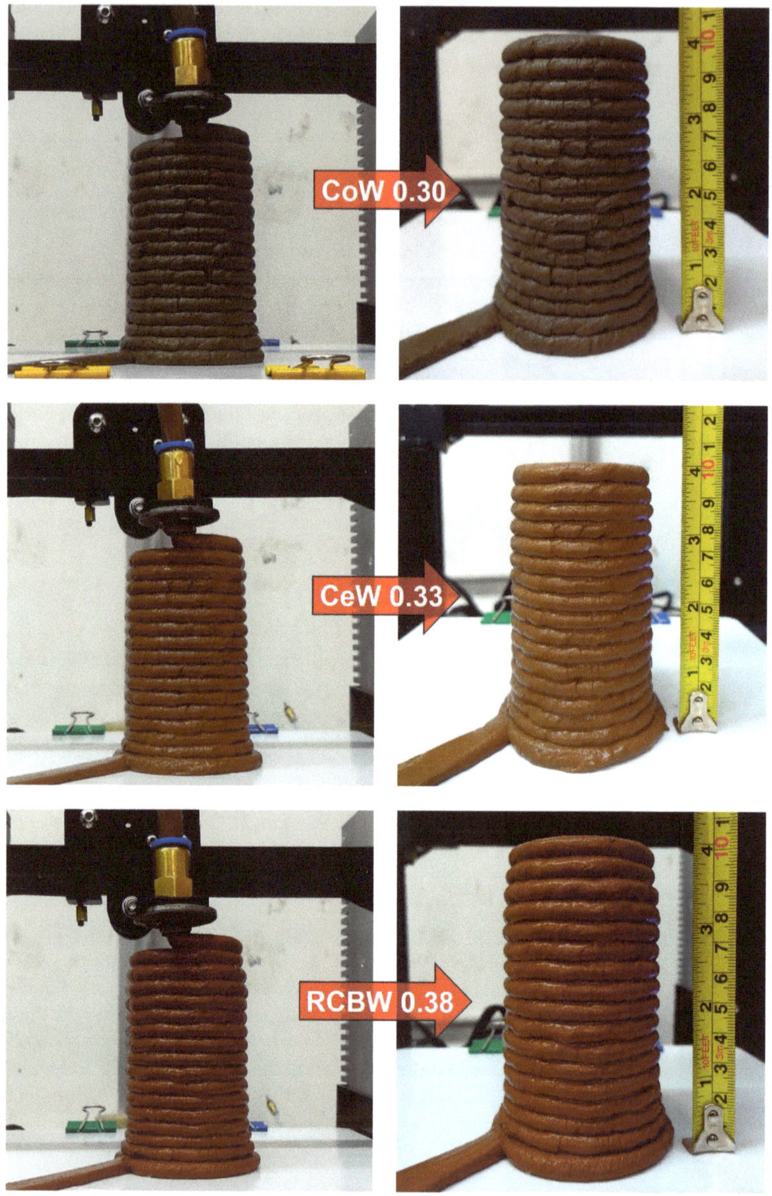

Figure 13. 3D printing and buildability tests of the CoW 0.30, CeW 0.33 and RCBW 0.38 mixes.

Table 3. Density, absorption and porosity of the 3D printed specimens (ASTM C642) [30].

Mix	Absorption (%)	Bulk Density, Dry (g/cm^3)	Apparent Density (g/cm^3)	Permeable Pore Volume (%)
CoW 0.30	26.9	1.55	1.97	42.4
CeW 0.33	28.5	1.52	1.96	43.9
RCBW 0.38	32.6	1.44	1.92	47.6

Regarding the mechanical characterization of the 3D printed specimens, Table 4 presents the results of flexural (7 days) and compressive strength (7 and 28 days). In this regard, the highest compressive strength was reported by the CeW 0.33 mix, achieving a value of 27.7 MPa at 28 days. The RCBW 0.38 and CoW 0.30 mixes meanwhile attained values of 21.7 MPa and 18.2 MPa at 28 days, respectively. It should be noted that these values are above the structural limit (17.5 MPa) established for concrete mixes according to the Colombian Regulations for Earthquake Resistant Construction (NSR-10).

Table 4. Flexural and compressive strength of the 3D printed specimens.

Mix	Flexural Strength (MPa)	Compressive Strength (MPa)	
	7 Days	7 Days	28 Days
CoW 0.30	4.9	11.5	18.2
CeW 0.33	4.6	12.7	27.7
RCBW 0.38	4.4	10.5	21.7

Meanwhile, the results of compressive strength of the 3D specimens coincide with the previously reported mechanical performance at the paste level (CeW > RCBW > CoW). It should be noted that, in the present study, the effect of the direction of the test on the compressive strength of the 3D printed specimens was not evaluated. However, some authors [26,39] show that in the direction of test used (perpendicular to the printing direction) the highest strength values are obtained.

Regarding the flexural strength (7 days) of the 3D printed specimens, it is evident that the reported values fluctuated between 4.4 and 4.9 MPa, these values being equivalent to 36.2–42.6% of the reported compressive strength of the 7 days of curing with these same mixes. In conventional concrete, this equivalence (flexural/compressive) is lower and normally ranges between 10–20% of the compressive strength. According to Kaliyavaradhan et al. [34], the perpendicular test direction, which corresponds to the one used in this study (Figure 4), promotes the best results for determining the flexural strength of 3D printed elements. In this regard, Demiral et al. [26] highlight that the flexural strength of 3D-printed specimens tested in the perpendicular direction may be even higher than that reported for conventional (mould-casted) specimens.

In relation to the above and to compare the quality of the 3D printed specimens versus conventional mould-casted specimens, an ultrasonic pulse test was performed 28 days after curing according to the procedure described in the methodology (Section 2.3).

The results are presented in Table 5, where it is highlighted that the 3D printed specimens achieve an ultrasonic pulse velocity (CoW = 98.9%; CeW = 98.7%; RCBW = 98.2%) very similar to that of conventional specimens. Indeed, the ultrasonic pulse velocity of the 3D printed specimens corresponding to the CoW 0.30, CeW 0.33 and RCBW 0.38 mixes was 3114, 3313 and 3238 m/s, respectively. In comparison, the conventional specimens reported values of 3149 m/s (CoW), 3356 m/s (CeW) and 3297 m/s (RCBW).

Table 5. Results of the ultrasonic pulse test of the 3D printed and conventional (mold-casted) specimens.

Mix	Cylinder	Ultrasonic Pulse Velocity (m/s)
CoW 0.30	3D printed	3114 ± 12
	Mould-casted	3149 ± 11
CeW 0.33	3D printed	3313 ± 10
	Mould-casted	3356 ± 23
RCBW 0.38	3D printed	3238 ± 12
	Mould-casted	3297 ± 13

According to the literature, the ultrasonic pulse velocity of concrete samples ranges from 3000 m/s (low quality concrete) to 5000 m/s (high quality concrete). In this regard, the speed of the ultrasonic pulse is directly related to the density of the material and to the mechanical properties. The quality shown by the 3D printed elements thus coincides with the physical-mechanical.

3.4.3. Microstructural Analysis

Additionally, the macroscopic (stereomicroscope) and microstructural (SEM) observation of the 3D printed specimens is presented in Figures 14–16 for the case of the CoW 0.30, CeW 0.33 and RCBW 0.38 mixes, respectively.

It is worth mentioning again that the specimens were printed with a fill percentage of 100%, obtaining a solid structure in which it is almost imperceptible to distinguish between one layer and another by visual inspection of the cross section of the specimens. Only at the edges of the specimen is it possible to distinguish the area of interface between layers and the superficial silhouette of each layer. The quality of the filling of the 3D specimens and their level of densification agrees with the results obtained using the ultrasonic pulse test (Table 5) and their similarity with the results reported by the conventional specimens (mould-casted).

A magnification (50×) of the area of interface between layers allowed to corroborate the perfect adhesion between the layers and the obtaining of a solid and homogeneous structure in the CoW and CeW mixes, consistent with the physicomechanical performance of these specimens. In the case of the RCBW 0.38 mix, some cracks were identified in the area of interface between layers (directed) and inside the layers (non-directed), which may be associated with shrinkage and drying shrinkage phenomena promoted by the high L/S ratio (0.38) that this mix required for its 3D printing, and which in turn could have affected the mechanical performance of the 3D specimens (interlayer adhesion). According to Nodehi et al. [40], shrinkage is one of the main causes of the concentration of tensile and shear stresses in the area of interface between layers, and therefore of the generation of fissures and cracks in this area.

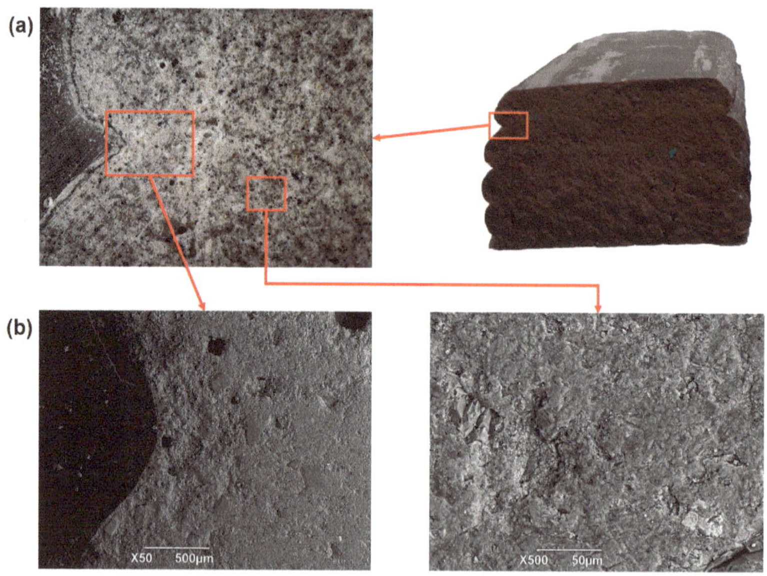

Figure 14. (**a**) Macroscopic (stereomicroscope), and (**b**) microstructural (SEM) observation of the interface zone between layers. Cross section of the CoW 0.30 3D printed specimen.

Figure 15. (**a**) Macroscopic (stereomicroscope), and (**b**) microstructural (SEM) observation of the interface zone between layers. Cross section of the CeW 0.33 3D printed specimen.

Figure 16. (**a**) Macroscopic (stereomicroscope), and (**b**) microstructural (SEM) observation of the interface zone between layers. Cross section of the RCBW 0.38 3D printed specimen.

4. Conclusions

The addition of 30% OPC and 4% Na_2SO_4 as chemical activator allowed the synthesis of hybrid cements based on 70% concrete waste (CoW), ceramic waste (CeW) and red clay brick waste (RCBW) with compressive strengths (25 °C, 90 days) up to 30.7, 37.0 and 33.2 MPa, respectively. Additionally, it was possible to demonstrate that the variation of the L/S ratio (0.30–0.38) affects the mechanical strength of hybrid cements, it being necessary to find a balance in relation to the properties in the fresh state necessary for its application in a 3D printing process.

The L/S ratio controls the properties in the fresh state (minislump, flowability index and buildability) of the mixes for 3D printing. A high L/S ratio promoted greater flowability and at the same time affected the buildability of the mixes. Very dry mixes meanwhile do not favour extrusion and 3D printing processes.

The type of waste influenced the rheological behavior of the mixes, being most fluid in the case of CoW, followed by CeW and finally RCBW. The RCBW demanded the greatest L/S ratio (0.38) among the waste studied to achieve the necessary flowability in 3D printing. The optimal L/S ratios for the CoW and CeW meanwhile were 0.30 and 0.33, respectively.

The results showed that the CoW, CeW and RCBW mixes that were found to have a minislump between 10–20 mm, flowability index between 2.0–2.4 and buildability greater than 80% were suitable for use in 3D printing processes.

The evaluation of the loss of workability (minislump and flowability) as a function of mixing time allowed us to determine that the open time of the CoW 0.30, CeW 0.33 and RCBW 0.38 mixes was 30, 20 and 10 min, respectively. These results were below the initial setting time reported for these same mixes (180–250 min). Moreover, the analysis of the ultrasonic pulse velocity during the fresh state made it possible to show that the mixes present microstructural changes before the initial setting time, which is consistent with the loss of workability (minislump and flowability) reported during the first minutes of reaction ($\leq$30 min).

The CoW 0.30, CeW 0.33 and RCBW 0.38 mixes presented an adequate extrusion and 3D printing capacity, allowing to obtain portions with a good surface finish, without the presence of defects, discontinuities and/or pores. Meanwhile, the 3D printed cylinders (17 layers) made it possible to establish the high level of buildability of the mixes (close to 97–99%), managing to corroborate the results obtained by the mixes in the fresh state tests.

The 3D printed specimens (100% infill) achieved adequate physical-mechanical performance, with compressive strengths of 18.2 MPa (CoW 0.30), 27.7 MPa (CeW 0.33) and 21.7 MPa (RCBW 0.38) after 28 days of curing (25 °C), values that exceed the structural limit ($\geq$17.5 MPa) established by the Colombian Regulations for Earthquake Resistant Construction (NSR-10) for concrete mixes.

The print quality of the mixes was confirmed using an ultrasonic pulse test (28 days). The CoW 0.30, CeW 0.33 and RCBW 0.38 mixes reported speed values of 3114, 3313 and 3238 m/s, respectively, results very similar to those obtained using conventional specimens (mold-casted). Microscopic observation (SEM) meanwhile revealed a dense interface and good quality interlayer adhesion for the case of the CoW 0.30 and CeW 0.33 mixes. In contrast, for the RCBW 0.38 mix, the presence of fissures and cracks in the interface between layers was identified to be a result of contraction and shrinkage phenomena due to drying, possibly promoted by the high L/S ratio of this mix (0.38). This finding suggests the possibility of studying (in future research) the control of this phenomenon through the incorporation of microfibers and particles.

Author Contributions: Conceptualization and methodology, R.R.-S. and R.M.d.G.; methodology and investigation, R.R.-S., F.M. and A.V.; writing—original draft preparation, R.R.-S.; supervision, project administration, funding acquisition, R.M.d.G.; writing—review and editing, R.R.-S. and R.M.d.G. All authors have read and agreed to the published version of the manuscript.

Funding: This research was funded by Ministry of Science Technology and Innovation (Minciencias) through Funding Call 6 of the 2021–2022 biennium of the General Royalties System (SGR) (BPIN 2020000100625).

Institutional Review Board Statement: Not applicable.

Informed Consent Statement: Not applicable.

Data Availability Statement: Not applicable.

Acknowledgments: The authors, members of the Composite Materials Group (CENM), thank the project "Development of a 3D printing system of sustainable non-conventional materials for the advancement of rural infrastructure in the department of Cauca" of the Universidad del Valle, financed by the Ministry of Science Technology and Innovation (Minciencias).

Conflicts of Interest: The authors declare no conflict of interest.

References

1. Grand View Research 3D Printing Construction Market Size, Share & Trends Analysis Report by Construction Method (Extrusion, Powder Bonding), by Material Type (Concrete, Metal), by End-User (Building, Infrastructure), and Segment Forecasts, 2022–2030. Available online: https://www.grandviewresearch.com/industry-analysis/3d-printing-constructions-market (accessed on 24 November 2022).
2. Ma, G.; Wang, L.; Ju, Y. State-of-the-Art of 3D Printing Technology of Cementitious Material—An Emerging Technique for Construction. *Sci. China Technol. Sci.* **2018**, *61*, 475–495. [CrossRef]
3. De Schutter, G.; Lesage, K.; Mechtcherine, V.; Nerella, V.N.; Habert, G.; Agusti-Juan, I. Vision of 3D Printing with Concrete—Technical, Economic and Environmental Potentials. *Cem. Concr. Res.* **2018**, *112*, 25–36. [CrossRef]
4. Alhumayani, H.; Gomaa, M.; Soebarto, V.; Jabi, W. Environmental Assessment of Large-Scale 3D Printing in Construction: A Comparative Study between Cob and Concrete. *J. Clean. Prod.* **2020**, *270*, 122463. [CrossRef]
5. Weng, Y.; Li, M.; Ruan, S.; Wong, T.N.; Tan, M.J.; Ow Yeong, K.L.; Qian, S. Comparative Economic, Environmental and Productivity Assessment of a Concrete Bathroom Unit Fabricated through 3D Printing and a Precast Approach. *J. Clean. Prod.* **2020**, *261*, 121245. [CrossRef]
6. Sanjayan, J.G.; Nematollahi, B. Chapter 1—3D Concrete Printing for Construction Applications. In *3D Concrete Printing Technology*; Sanjayan, J.G., Nazari, A., Nematollahi, B., Eds.; Butterworth-Heinemann: Oxford, UK, 2019; pp. 1–11, ISBN 978-0-12-815481-6.
7. Robayo-Salazar, R.; Mejía de Gutiérrez, R.; Villaquirán-Caicedo, M.A.; Delvasto Arjona, S. 3D Printing with Cementitious Materials: Challenges and Opportunities for the Construction Sector. *Autom. Constr.* **2023**, *146*, 104693. [CrossRef]
8. Bhattacherjee, S.; Basavaraj, A.S.; Rahul, A.V.; Santhanam, M.; Gettu, R.; Panda, B.; Schlangen, E.; Chen, Y.; Copuroglu, O.; Ma, G.; et al. Sustainable Materials for 3D Concrete Printing. *Cem. Concr. Compos.* **2021**, *122*, 104156. [CrossRef]
9. Gökçe, H.S.; Tuyan, M.; Nehdi, M.L. Alkali-Activated and Geopolymer Materials Developed Using Innovative Manufacturing Techniques: A Critical Review. *Constr. Build. Mater.* **2021**, *303*, 124483. [CrossRef]
10. Zhong, H.; Zhang, M. 3D Printing Geopolymers: A Review. *Cem. Concr. Compos.* **2022**, *128*, 104455. [CrossRef]
11. Raza, M.H.; Zhong, R.Y.; Khan, M. Recent Advances and Productivity Analysis of 3D Printed Geopolymers. *Addit. Manuf.* **2022**, *52*, 102685. [CrossRef]
12. Lazorenko, G.; Kasprzhitskii, A. Geopolymer Additive Manufacturing: A Review. *Addit. Manuf.* **2022**, *55*, 102782. [CrossRef]
13. Khan, M.A. Mix Suitable for Concrete 3D Printing: A Review. *Mater. Today Proc.* **2020**, *32*, 831–837. [CrossRef]
14. Xia, M.; Sanjayan, J. Method of Formulating Geopolymer for 3D Printing for Construction Applications. *Mater. Des.* **2016**, *110*, 382–390. [CrossRef]
15. Robayo-Salazar, R.A.; Mejía de Gutiérrez, R. Natural Volcanic Pozzolans as an Available Raw Material for Alkali-Activated Materials in the Foreseeable Future: A Review. *Constr. Build. Mater.* **2018**, *189*, 109–118. [CrossRef]
16. Zhao, J.; Tong, L.; Li, B.; Chen, T.; Wang, C.; Yang, G.; Zheng, Y. Eco-Friendly Geopolymer Materials: A Review of Performance Improvement, Potential Application and Sustainability Assessment. *J. Clean. Prod.* **2021**, *307*, 127085. [CrossRef]
17. Shehata, N.; Mohamed, O.A.; Sayed, E.T.; Abdelkareem, M.A.; Olabi, A.G. Geopolymer Concrete as Green Building Materials: Recent Applications, Sustainable Development and Circular Economy Potentials. *Sci. Total Environ.* **2022**, *836*, 155577. [CrossRef]
18. Ouellet-Plamondon, C.; Habert, G. Life Cycle Assessment (LCA) of Alkali-Activated Cements and Concretes. In *Handbook of Alkali-Activated Cements, Mortars and Concretes*; Woodhead Publishing: Sawston, UK, 2015; pp. 663–686, ISBN 9781782422761.
19. Valencia-Saavedra, W.; Robayo-Salazar, R.; Mejía de Gutiérrez, R. Alkali-Activated Hybrid Cements Based on Fly Ash and Construction and Demolition Wastes Using Sodium Sulfate and Sodium Carbonate. *Molecules* **2021**, *26*, 7572. [CrossRef]
20. Joseph, S.; Snellings, R.; Cizer, Ö. Activation of Portland Cement Blended with High Volume of Fly Ash Using Na_2SO_4. *Cem. Concr. Compos.* **2019**, *104*, 103417. [CrossRef]
21. Donatello, S.; Fernández-Jimenez, A.; Palomo, A. Very High Volume Fly Ash Cements. Early Age Hydration Study Using Na_2SO_4 as an Activator. *J. Am. Ceram. Soc.* **2013**, *96*, 900–906. [CrossRef]

22. Robayo-Salazar, R.; Valencia-Saavedra, W.; Gutiérrez, R.M. Reuse of Powders and Recycled Aggregates from Mixed Construction and Demolition Waste in Alkali-Activated Materials and Precast Concrete Units. *Sustainability* **2022**, *14*, 9685. [CrossRef]
23. Robayo-Salazar, R.; Valencia-Saavedra, W.; Mejía de Gutiérrez, R. Recycling of Concrete, Ceramic, and Masonry Waste via Alkaline Activation: Obtaining and Characterization of Hybrid Cements. *J. Build. Eng.* **2022**, *46*, 103698. [CrossRef]
24. Şahin, O.; İlcan, H.; Ateşli, A.T.; Kul, A.; Yıldırım, G.; Şahmaran, M. Construction and Demolition Waste-Based Geopolymers Suited for Use in 3-Dimensional Additive Manufacturing. *Cem. Concr. Compos.* **2021**, *121*, 104088. [CrossRef]
25. Ilcan, H.; Sahin, O.; Kul, A.; Yildirim, G.; Sahmaran, M. Rheological Properties and Compressive Strength of Construction and Demolition Waste-Based Geopolymer Mortars for 3D-Printing. *Constr. Build. Mater.* **2022**, *328*, 127114. [CrossRef]
26. Demiral, N.C.; Ozkan Ekinci, M.; Sahin, O.; Ilcan, H.; Kul, A.; Yildirim, G.; Sahmaran, M. Mechanical Anisotropy Evaluation and Bonding Properties of 3D-Printable Construction and Demolition Waste-Based Geopolymer Mortars. *Cem. Concr. Compos.* **2022**, *134*, 104814. [CrossRef]
27. *ASTM C230/230M-20*; Standard Specification for Flow Table for Use in Tests of Hydraulic Cement. American Society for Testing and Materials (ASTM): West Conshohocken, PA, USA, 2020.
28. *ASTM C191-21*; Standard Test Methods for Setting of Hydraulic Cement by Vicat Needle. American Society for Testing and Materials (ASTM): West Conshohocken, PA, USA, 2021.
29. *UNE EN 1015:2020*; Methods of Test for Mortar for Masonry-Part 11: Determination of Flexural and Compressive Strength of Hardened Mortar. AENOR: Madrid, Spain, 2020.
30. *ASTM C642-13*; Standard Test Method for Density, Absorption, and Voids in Hardened Concrete. American Society for Testing and Materials (ASTM): West Conshohocken, PA, USA, 2013.
31. *ASTM C597-16*; Standard Test Method for Pulse Velocity Through Concrete. American Society for Testing and Materials (ASTM): West Conshohocken, PA, USA, 2016.
32. Souza, M.T.; Ferreira, I.M.; de Moraes, E.G.; Senff, L.; Arcaro, S.; Pessôa, J.R.C.; Ribeiro, M.J.; de Oliveira, A.P.N. Role of Chemical Admixtures on 3D Printed Portland Cement: Assessing Rheology and Buildability. *Constr. Build. Mater.* **2022**, *314*, 125666. [CrossRef]
33. Tay, Y.W.D.; Qian, Y.; Tan, M.J. Printability Region for 3D Concrete Printing Using Slump and Slump Flow Test. *Compos. Part B* **2019**, *174*, 106968. [CrossRef]
34. Kaliyavaradhan, S.K.; Ambily, P.S.; Prem, P.R.; Ghodke, S.B. Test Methods for 3D Printable Concrete. *Autom. Constr.* **2022**, *142*, 104529. [CrossRef]
35. Uppalapati, S.; Vandewalle, L.; Cizer, Ö. Monitoring the Setting Process of Alkali-Activated Slag-Fly Ash Cements with Ultrasonic P-Wave Velocity. *Constr. Build. Mater.* **2021**, *271*, 121592. [CrossRef]
36. Ranjbar, N.; Mehrali, M.; Kuenzel, C.; Gundlach, C.; Pedersen, D.B.; Dolatshahi-Pirouz, A.; Spangenberg, J. Rheological Characterization of 3D Printable Geopolymers. *Cem. Concr. Res.* **2021**, *147*, 106498. [CrossRef]
37. Zhang, Y.; Zhang, Y.; She, W.; Yang, L.; Liu, G.; Yang, Y. Rheological and Harden Properties of the High-Thixotropy 3D Printing Concrete. *Constr. Build. Mater.* **2019**, *201*, 278–285. [CrossRef]
38. SIKA 3D CONCRETE PRINTING: Sikacrete® 3D Materials for Fast and Precisely Printed Concrete. Available online: https://www.sika.com/en/knowledge-hub/3d-concrete-printing.html#sika (accessed on 24 November 2022).
39. Che, Y.; Yang, H. Hydration Products, Pore Structure, and Compressive Strength of Extrusion-Based 3D Printed Cement Pastes Containing Nano Calcium Carbonate. *Case Stud. Constr. Mater.* **2022**, *17*, e01590. [CrossRef]
40. Nodehi, M.; Aguayo, F.; Nodehi, S.E.; Gholampour, A.; Ozbakkaloglu, T.; Gencel, O. Durability Properties of 3D Printed Concrete (3DPC). *Autom. Constr.* **2022**, *142*, 104479. [CrossRef]

sustainability

MDPI

Article

Sulfur Mortar Goes to Infinity: Mechanical Performance and Characterization of Sulfur Mortar Composed of Different Aggregates During Heating Cycles, Exploring Potential Sustainability, Recyclability, and Circularity

Qinjian Wang [1], Marie-Paule Delplancke [2] and Didier Snoeck [1,*]

[1] BATir Department, École Polytechnique de Bruxelles, Université Libre de Bruxelles (ULB), 50 F.D. Roosevelt Ave., CP 194/02, 1050 Brussels, Belgium; qinjian.wang@ulb.be

[2] 4MAT Department, École Polytechnique de Bruxelles, Université Libre de Bruxelles (ULB), 50 F.D. Roosevelt Ave., CP 165/63, 1050 Brussels, Belgium; marie-paule.delplancke@ulb.be

* Correspondence: didier.snoeck@ulb.be

Abstract: Sulfur mortar hardens quickly, shows a high chemical resistance, and can be recycled, making it ideal for construction and rehabilitation in extreme environments. Despite its potential for sustainability, current research lacks sufficient characterization of sulfur mortar's performance during recycling, particularly regarding the physical and chemical changes when iron oxide is introduced. This study investigates the replacement of conventional siliceous sand with high-iron-content sand in sulfur mortar, through a series of five break–recast cycles. The results demonstrate an 11% increase in compressive strength and a 26% increase in flexural strength after five recasting cycles. Optical microscopy and scanning electron microscopy (SEM) revealed that recasting improved the distribution of the sulfur binder, while the formation of iron sulfates filled the gaps between aggregates and the binder. X-ray diffraction (XRD) and Fourier transform infrared spectroscopy (FTIR) confirmed the presence of iron sulfates, and differential scanning calorimetry (DSC) showed that high-iron-content sulfur mortar narrowed the phase change temperature range, preventing uneven solidification within the samples. This study sheds light on the strengthening mechanisms that occur during the recycling process, enhancing the material's durability and recyclability. This aligns with circular economy principles, contributes to resource efficiency, and supports sustainable construction practices.

Keywords: sulfur mortar; recyclability; characterization; circularity; sustainability

Citation: Wang, Q.; Delplancke, M.-P.; Snoeck, D. Sulfur Mortar Goes to Infinity: Mechanical Performance and Characterization of Sulfur Mortar Composed of Different Aggregates During Heating Cycles, Exploring Potential Sustainability, Recyclability, and Circularity. *Sustainability* **2024**, *16*, 10803. https://doi.org/10.3390/su162410803

Academic Editor: Antonio Caggiano

Received: 29 October 2024
Revised: 3 December 2024
Accepted: 5 December 2024
Published: 10 December 2024

1. Introduction

Global warming has become a major focus of discussion and research in many countries. In March 2023, the Intergovernmental Panel on Climate Change (IPCC) released its Sixth Assessment Report (AR6), which emphasizes the scale and pace of the work being performed, as well as the need for the IPCC to provide a clearer understanding of the impacts of global warming as current programs are insufficient to address the daunting challenges of climate change [1]. According to studies, the human industry produces a quarter of all greenhouse gas emissions from materials synthesis [2]. Among these, the use of cement and concrete-related materials is challenged by environmental requirements: since the cement production process accounts for 8% [3] of the total greenhouse gas emissions, it is the most significant source of man-made greenhouse gas emissions [4]. In addition, urbanization is expected to continue to place significant demand on the cement industry for the next 100 years [5]. Indeed, since the COVID-19 pandemic came to an end, cement shipments surged again as a result of the economic recovery measures implemented in various countries and regions [6]. At the same time, various countries are exploring and practicing decarbonization models for the cement industry, as well as for sustainable economic development, such as Europe [7], USA [8], China [9], Switzerland [10], and Saudi Arabia [11].

Sulfur is one of the by-products of the petroleum refining process [12]. According to statistics, today, more than 60 million tons of sulfur are produced annually [13]. Sulfur has been studied as an alternative binder for concrete in North America since the 1970s as a way to deplete the large deposits of sulfur caused by mining development [14]. Sulfur converts from a solid to a liquid when heated to approximately 120 °C [15], and it can be mixed with aggregates and cooled to make sulfur mortar in the broadest sense of the word. In addition to consuming the redundancy of sulfur from the oil industry as an innovative building material, sulfur mortar is also seen as one of the potential building materials for extraterrestrial colonization and the construction of human habitats on exoplanets [16,17]; large amounts of sulfur-containing material were found on both the Moon [17] and Mars [18]. Even pure sulfur crystals were found on Mars [19]. Compared to traditional Portland concrete, sulfur mortar has comparable mechanical properties [12]; it is acid-resistant, corrosion-resistant, and water-resistant [20–23] and does not require any water during the casting process [24], which is extremely beneficial to preserve today's increasingly scarce water resources. Wang et Snoeck provided a detailed review of the history, current status, and potential future development of sulfur mortar [17].

Recycling of building materials is an important path to reach sustainability. As sulfur mortar materials can be reheated and remelted, it may show interesting circular features. In 2016, Wan et al. [25] suggested in their study that the recasting of sulfur mortars, which were made with Martian regolith acting as aggregates, increased the compressive strength of the recasts by 20–30% compared to the virgin samples; this discovery generated interest in the recyclability of sulfur mortars. Gulzar et al. [26] took it one step further by investigating sulfur mortars using normal sand as aggregate and found that the compressive strength of normal-sand sulfur mortar increases by about 30% after the first recasting but decreases after the second recasting and that recasting has no effect on acid and corrosion resistance [26].

Currently, a few studies have been conducted on the recyclability of sulfur mortar, and the properties are generally not the primary purpose of the studies; they do not detail properties such as changes in mechanical performances exhibited by sulfur mortar during recycling or characterize the structural features of the material before and after recycling. Undoubtedly, research dedicated to the recyclability of sulfur mortar is essential to further advance the study of sulfur as an emerging sustainable building material and expand the scope of its applications. A profound knowledge of the possible crystal formation during recycling is of paramount importance.

In this paper, normal sand and high-iron-content sand were used as aggregates for sulfur mortar, because many studies have illustrated that metallic ions, such as iron, aluminum, etc., react with sulfur [25,27,28]. This may have a potential impact on the strength of sulfur mortar upon recycling. However, current research stopped at limited times of recycling sulfur mortar, and the microscopical characterization is lacking. In this study, the sulfur mortar with two different aggregate types was subjected to five destruction–recasting cycles, and samples were selected when significant mechanical property changes during the recycling process were observed. On these obtained samples, microscopic and chemical characterization was performed to fundamentally investigate the reasons for the changes in the mechanical properties of sulfur mortar in the process of multiple recycling and use, and to study the influence of the different aggregate types on the recyclability of sulfur mortar. At the same time, experiments were conducted to determine whether new chemical structures are generated during multiple (10 in this study) heating–cooling cycles, which may affect the properties of the sulfur mortar samples. Figure 1 shows the methodology of this study.

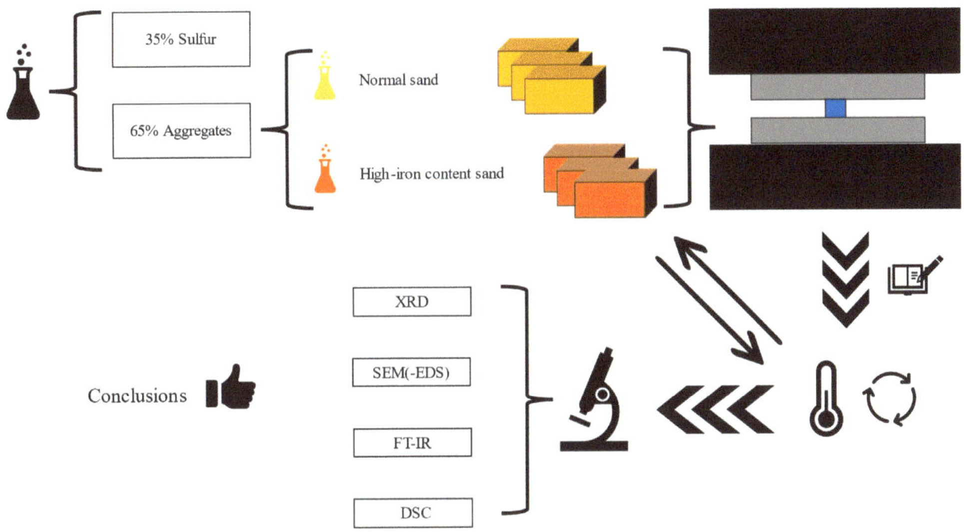

Figure 1. Schematic diagram of the study.

2. Materials and Methods

2.1. Raw Materials

In this study, the sulfur was obtained from Norbert König e. Kfm, Wittenberg, Germany, in the form of semicircular granules/beads/pellets. The elemental composition of this material is offered by the company and is shown in Table 1(a). It is mostly pure elemental sulfur and is stable at room temperature. High-iron-content sand from Rodesco, Belgium, was used. Because the main components are iron (III) oxides, it has a reddish-brown color. It is divided into lumps and powder and has a wide particle size distribution. The elemental composition is shown in Table 1(b). At the same time, to investigate the influence of different aggregates on the performance, normal sand was also used. Its chemical composition is also shown in Table 1(c).

Table 1. Elemental composition of raw materials: (a) Sulfur, (b) high-iron-content sand, (c) normal sand [29].

Components	Sulfur	Organic	Ash	Moisture	Acid(H_2SO_4)			
Wt. % (a)	99.95	0.02	0.015	0.5	0.007			
Components	Fe	O	Si	Al	Mn	K	Others	Total
Wt. % (b)	63.81	31.8	1.69	1.23	0.18	0.17	Ca, Mn	99.45
Components	SiO_2	Al_2O_3	CaO					
Wt. % (c)	97.12	1.84	0.96					

2.2. Sample Preparation and Mixing

Previous studies revealed the effect of changes in aggregate particle size distribution on the mechanical properties of sulfur mortars. This study focuses on the changes in compressive strength of sulfur mortars with different aggregates during recasting cycles, and analyzes the possible relationship with the chemical composition. Nevertheless, according to the research literature [25], too large aggregates can lead to insufficiently smooth surfaces of the samples, which can harm the mechanical tests. Therefore, in this research, the two aggregates were sieved and the fraction below 2.37 mm was selected. All the raw materials are shown in Figure 2a.

(**a**)

(**b**)

Figure 2. (**a**) Raw materials in containers with inner diameters as follows: left—9 cm; middle—6 cm; right—6 cm. (**b**) Heating procedure and samples containing high-iron-content sand.

The aggregates were mixed with sulfur and stirred for three minutes before being placed in the metal mixing vessel to ensure homogenous dispersion. Since the process of recycling sulfur mortar inevitably results in the loss of raw materials—for example, materials solidified in molds or heating ovens are difficult to recycle—10% more material was used than necessary. A preliminary study by the authors investigated variations of the sulfur content of 15, 20, 25, 30, and 35 wt. % of total volume, linked to the mechanical properties and overall workability and stability. In this study, a sulfur ratio of 35 wt. % was selected, which is slightly higher than the recommended value of 30 wt. % found in previous experiments [25–27], to ensure that the loss of raw materials during the recycling process does not induce a loss in the workability of the sulfur mortar at the end of the experiment. At the same time, metallic prismatic casting molds of size 40 × 40 × 160 mm^3 were placed in the heating furnace along with casting and plastering tools, and compaction tools. The oven was heated from an ambient temperature of about 20 °C to 135 °C at an average heating rate of about 12 °C/min, and kept constant at this temperature for about one hour, during which the raw materials were stirred using the heated stirring tool every half an hour, to ensure that the dissipated heat was uniform, and that the melted sulfur was well mixed with the aggregates. The total time from heating the sulfur mixture to casting was 1 h. Heating the molds and mixing tools was necessary so that when the sulfur mortar was poured, it did not solidify immediately on the mold walls and tools and did not lead to an uneven internal structure of the samples, with delamination and cracks. The arrangement of the heating oven and samples is shown in Figure 2b. The sulfur beads provided by the supplier contained organic material. Researchers have been trying to reduce temperature contraction by using organic additives to avoid the formation of larger sulfur crystals during solidification [22]. In preliminary tests conducted by the authors, in this research, no significant dimensional changes or shrinkage was observed during the cooling and solidification of the samples. However, it is believed that the inclusion of an additional modifier would enhance the properties of larger elements to be cast.

2.3. Recast Strength Test Procedure

The poured specimens were allowed to cool at an ambient temperature of 20 ± 1 °C in standard laboratory conditions. After 24 h, the specimens were de-molded, visually verified, and tested for mechanical strength using a mortar prism test apparatus: the specimens were first subjected to a three-point bending test and then to a compressive test, with a loading pattern following the EN 196-1 standard [30], to avoid any influence on the results of the experiments from an excessively fast loading rate. After the strength test, the pieces were recovered, placed again in a stirring vessel, put into the heating furnace together with the casting mold and the stirring tool, and cast again with the same experimental conditions to obtain the recast samples. Subsequently, the samples were resubmitted to the mechanical test, a total of five times. Three samples of each group of both aggregates were prepared, and all values were averaged over the three samples to ensure the reliability of the experimental data.

2.4. Characterization Tests

To investigate whether the internal structure and/or chemical composition of sulfur mortar changes during the recycling process and whether such changes can explain the changes in mechanical properties, sampling was performed for the initial cast and the first, second, and fifth recasts. The following physical/chemical characterization tools were used.

2.4.1. Optical Microscopic Analysis

To study the formed porosity and overall features of the sulfur mortar samples, microscopic analysis was performed using a Leica Emspira 3 microscope (Leica Microsystems GmbH, Wetzlar, Germany). The analysis of the micrographs was performed afterward using the ImageJ software, version 1.54m.

2.4.2. Scanning Electron Microscopy (SEM) and Energy Dispersive X-Ray Spectroscopy (EDS)

Samples for SEM-EDS testing were obtained directly from the samples that were mechanically loaded till failure. The obtained pieces were sealed in sample vials for storage to avoid humidity effects, using an ion sputter coater to evenly deposit a thin layer of carbon film on the sample's surface. The fifth recast sample was embedded in resin and polished up to 1 μm precision. The fracture surfaces were observed in all other samples. A Hitachi SU 70 instrument (Hitachi High-Tech Corporation, Tokyo, Japan) was used, using the secondary electron mode.

2.4.3. X-Ray Diffraction (XRD)

In addition to the SEM-EDS tests, XRD tests were performed as a complement to see if the sulfur mortar produced new crystalline compounds after multiple recycling. Samples were taken directly from the broken samples and pulverized using a ball mill at a speed of 500 rpm for 3 min, using the top filling method to prepare the samples. A D8Eco from Bruker (Bruker AXS GmbH, Karlsruhe, Germany) was used in the Bragg–Brentano configuration. A 2θ range spreading from 5° to 70° was chosen, and the data were interpreted with the software Diffrac Eva, version 7.

2.4.4. Fourier Transform Infrared (FTIR) Spectrum

It is worth exploring whether the various components of sulfur mortar interact during the recycling process. This research therefore also used FTIR to characterize the bonds between elements. Samples were made by mechanically grinding specimens, and 3 mg of sample was mixed with 100 mg of potassium bromide and compacted to form a pellet. A Vertex 70 using a Hyperion 2000 MIR source (Bruker Optik GmbH, Ettlingen, Germany) was used to investigate the absorption between 4200 and 400 cm^{-1}, and the resolution was 2 cm^{-1}.

2.4.5. Differential Scanning Calorimetry (DSC)

The thermal properties of sulfur mortar were investigated by DSC equipment (Mettler Toledo, model DSC821e, Greifensee, Switzerland). A sample of about 50 mg was placed in a crucible, and the temperature range was settled between 20 °C and 140 °C at a heating rate of 10 °C/min. Ten heating–cooling cycles were performed under a nitrogen atmosphere. The raw materials were assessed in addition to the raw mix of the studied materials.

3. Results and Discussion

This section presents the changes in the compressive strength of sulfur mortar with two aggregates during recycling. The results of optical microscopy, SEM, XRD, and FTIR characterization of the samples during the cyclic recycling process are also shown, as well as the experimental results of the ten thermal cycles of the DSC test. The aim is to link the strength variation with the physico-chemical characteristics obtained with microstructural, chemical composition, and thermodynamic characterization techniques.

3.1. Mechanical Properties

The mechanical performance of the two series after five damage–recast cycles is shown in Figure 3. The compressive strength (Figure 3a) of sulfur mortar with both aggregates exceeds 30 MPa, which is comparable to conventional cement concrete. For normal-sand sulfur mortar, the compressive strength increased from 32 MPa to 34 MPa on average after the first recasting. However, in subsequent recasts, its strength gradually declined. By the fifth recasting, the compressive strength was still comparable to the first casting and within the standard deviation. Therefore, the differences are not statistically significant. The sulfur mortar samples with high iron content consistently showed higher compressive strength than normal-sand sulfur mortar, approximately 15% higher. Notably, the compressive strength of high-iron-content sulfur mortar increased during the recasting process. It reached an average of 41 MPa after the fifth recasting, an 11% increase compared to 37 MPa after the first casting.

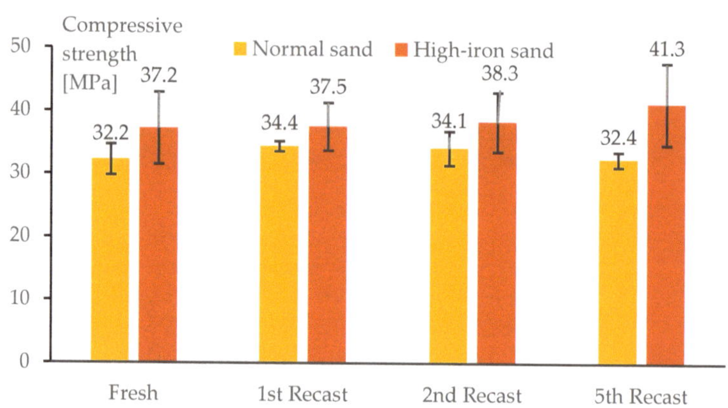

(**a**)

Figure 3. *Cont.*

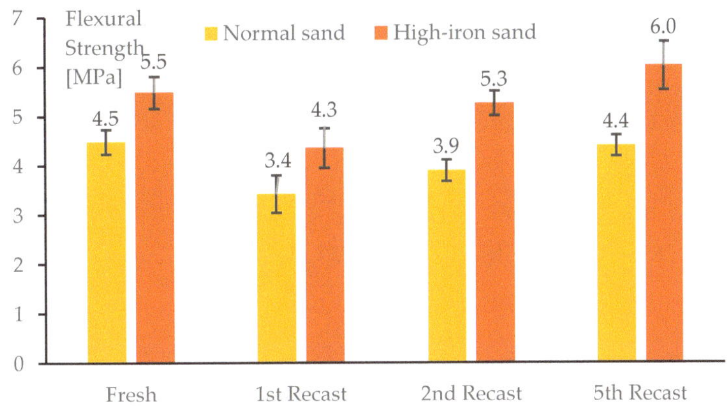

(**b**)

Figure 3. (**a**) Compressive strength [MPa], (**b**) Flexural strength [MPa]; variation during the recycling of the sulfur mortar containing normal and high-iron sand types. The results show the average value with the standard deviation (n = 3).

The three-point bending test results (Figure 3b) followed a similar trend as the compressive strength data. The flexural strength also peaked after the fifth recasting.

3.2. Optical Microscopy Characterization

Figure 4 shows optical microscope images of freshly cast and recast sulfur mortar with normal sand and high-iron sand aggregates. In the freshly cast normal-sand sulfur mortar (Figure 4a), some sulfur remained in crystalline form, covering the surface of the aggregates. Larger aggregates could be rubbed off by hand, leaving pores about 1 mm in diameter (marked by red circles in the figure). This indicates a weak bond between the sulfur and the aggregates. In contrast, after the first recast, normal-sand sulfur mortar, the sulfur was more uniformly distributed over the sand particles, and the larger-grain-size sand particles were tightly bonded to the sulfur. The tighter bond formed by the sulfur to the aggregate is believed to be one of the reasons for its increase [26]. The same analogy can be applied to high-iron-sand sulfur mortar. In the freshly cast samples, the aggregate coverage by sulfur was incomplete, and large portions of the aggregate remained exposed. Sulfur crystals, 0.4 to 0.5 mm in diameter, were observed and marked by a red circle in Figure 4b. After the fifth recast, most of the aggregate surface was covered by sulfur crystals. From a microstructural perspective, recasting improved the connection between the aggregates and the sulfur binder in both samples. It also reduced segregation and porosity.

(**a**)

Figure 4. *Cont.*

(**b**)

Figure 4. Optical microscopy images of sulfur mortar samples containing (**a**) normal sand, (**b**) high-iron-content sand, from the fresh cast, and the 5th recast sample series. The scale bar amounts to 1 mm.

3.3. SEM-EDS Characterization

SEM-EDS images of sulfur mortar samples with both aggregates are shown in Figure 5. The normal-sand sulfur mortar forms a dense structure by encapsulating aggregate particles in molten sulfur. The EDS analysis shows that sulfur and silica are consistently present as separate components. From the fresh cast to the fifth recast, the results are in line with Yahia and Muhammad's research [31]: sulfur mortar cured under dry conditions forms a uniform distribution and texture between the sulfur binder and aggregates. In high-iron-sand sulfur mortar, freshly poured samples show that sulfur is not tightly attached to the aggregate. A distinct black gap is visible between the sulfur and the aggregate. EDS analysis reveals that oxygen is bonded to both iron and silicon. To accurately present the results, a sample of high-iron-sand sulfur mortar was set in resin and polished flat to observe their cross sections. As shown in Figure 5b (bottom), the sulfur was tightly bound to the aggregate without significant cracking after the fifth recasting. There were only small black parts in the EDS results, in Figure 5b (bottom right). This represents the discontinuities after the 5th recast, as these were filled by the resin and carbon film. According to the literature [27,32,33], the strength is normally inversely proportional to porosity. From the morphology, the porosity of normal sand mortar exhibited limited variation, while the one with high-iron-content sand exhibited a huge difference after the 5th recast; thus, the strength increased more obviously than the normal sand mortar. This corresponds logically to the correlation between porosity and strength.

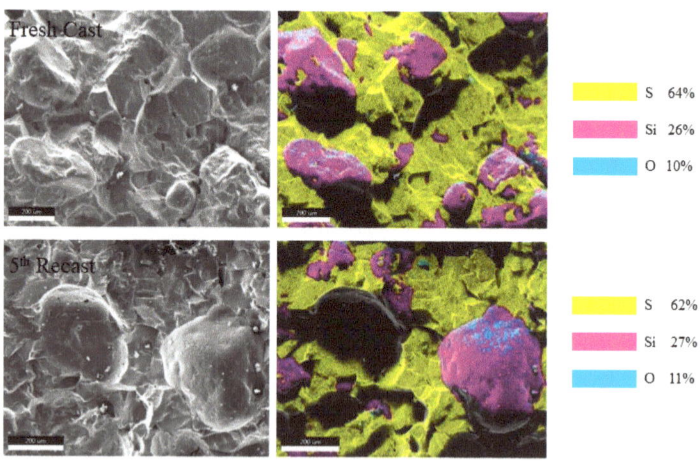

(a)

Figure 5. *Cont.*

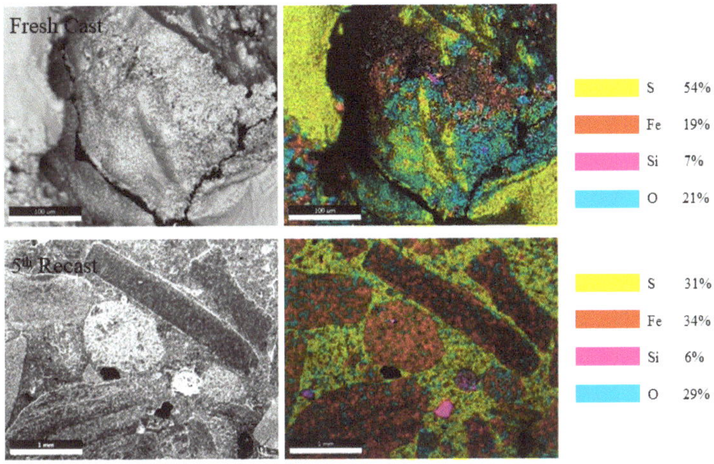

(**b**)

Figure 5. SEM-EDS characterization of sulfur mortar using (**a**) normal sand, (**b**) high-iron-content sand. The scale bars amount to 200 μm (**a**) and 1 mm (**b**).

Additionally, the EDS results of the high-iron-sand sample show an important observation. Iron, sulfur, and oxygen are simultaneously present in the russet areas. These areas indicate the high-iron-content aggregates (Figure 5b, bottom right). The EDS results also exhibited a co-existence of sulfur, iron, and oxygen atoms. This could indicate that during the recasting process, the iron oxides reacted with sulfur to form iron sulfate crystals. The oxygen in the air reacted with liquid sulfur, forming SO_2 [27], which would be absorbed chemically and physically at the basic points and acid points of Al_2O_3 [34]. Meanwhile, the iron oxide would significantly absorb SO_2 in an oxygen-rich environment, forming iron oxide [35]. Those formations of aluminum and iron sulfates are also verified in XRD results in the following text.

3.4. X-Ray Diffraction Characterization

Both normal-sand sulfur mortars and high-iron-content-sand sulfur mortars were characterized using XRD (Figure 6). According to the literature [12], the strength development of normal-sand sulfur mortar mainly relies on the solidification of liquid sulfur on the aggregate. The microscopic image in Figure 4 reveals that the crystalline sulfur is merely distributed around the aggregates. The EDS analysis in Figure 5 shows that the chemical composition is mainly elemental sulfur and silica. There is no clear evidence of other phases forming connections with the gravel. This was confirmed by XRD (Figure 6a) analysis of the normal-sand sulfur mortar, and all diffraction peaks were allocated to sulfur and silica, and no other new crystalline chemical compounds were formed, even after five heating recasts. The chemical composition of the material remained stable. In the high-iron-sand sulfur mortar, XRD analysis revealed additional features. Along with the major sulfur and iron oxide peaks, a more intense peak appeared around $2\theta = 15°$. This peak partly corresponds to elemental sulfur. It also included a contribution from aluminum sulfate, as confirmed by additional peaks at 25°, 35°, and 40° [36]. Additionally, the characteristic diffraction peak of iron sulfate was observed near 25°. Even if less than 5% of the SEM-EDS images show a combination of metallic elements, sulfur, and oxygen (Figure 5b), X-ray patterns confirm the formation of the metal sulfates. It is obvious that there is no significant peak pattern for iron or aluminum sulfide, and it is due to the reaction temperature between iron and sulfur starting at least around 300 °C, while the reaction between sulfur and aluminum starts at 565 °C [37]. Both are much higher than the temperature in this study.

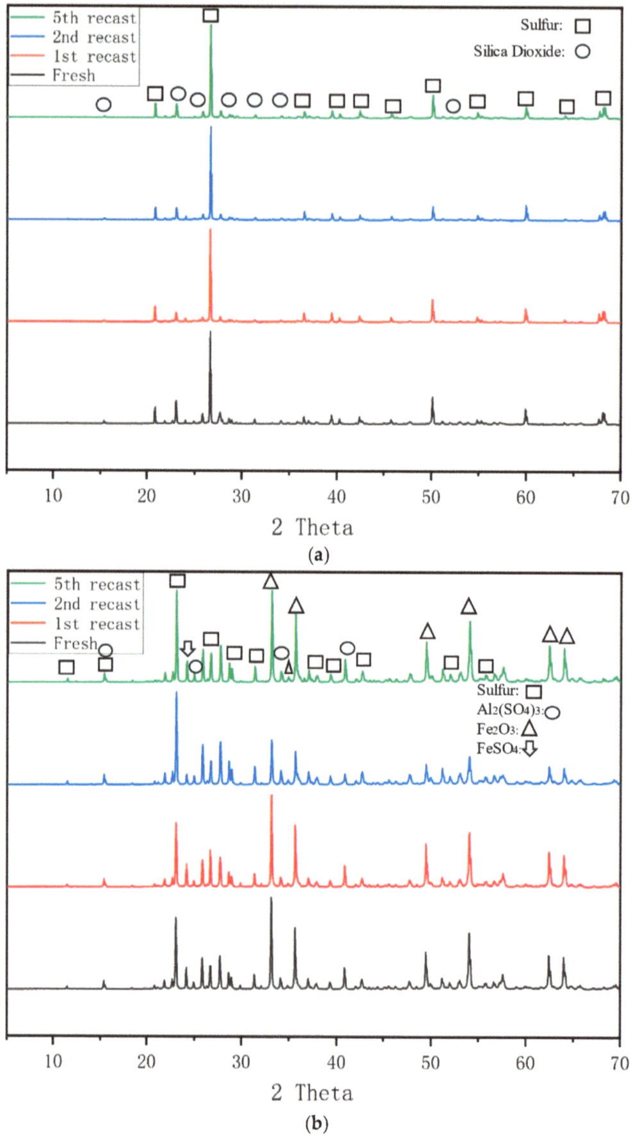

Figure 6. XRD characterization of different recycled samples: (**a**) Normal sand, (**b**) high-iron-content-sand sulfur mortars.

3.5. FTIR Characterization

To further investigate the inertness of normal sand toward sulfur and the reaction of sulfur with the high-iron-content aggregates, samples were characterized by FTIR. The experimental results are shown in Figure 7.

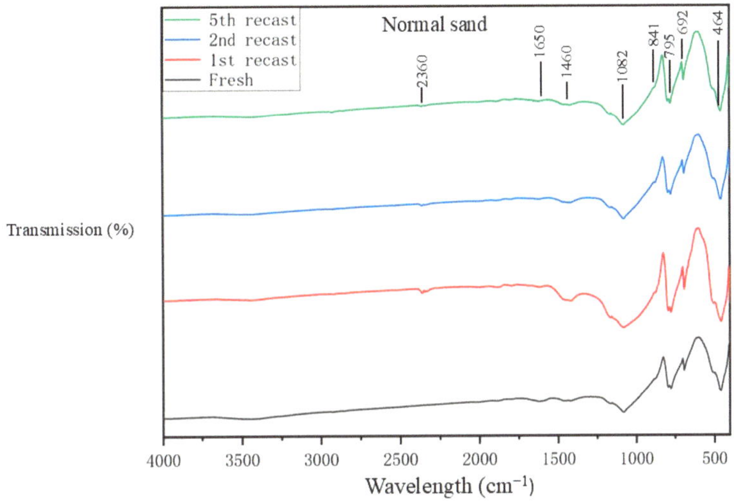

(**a**)

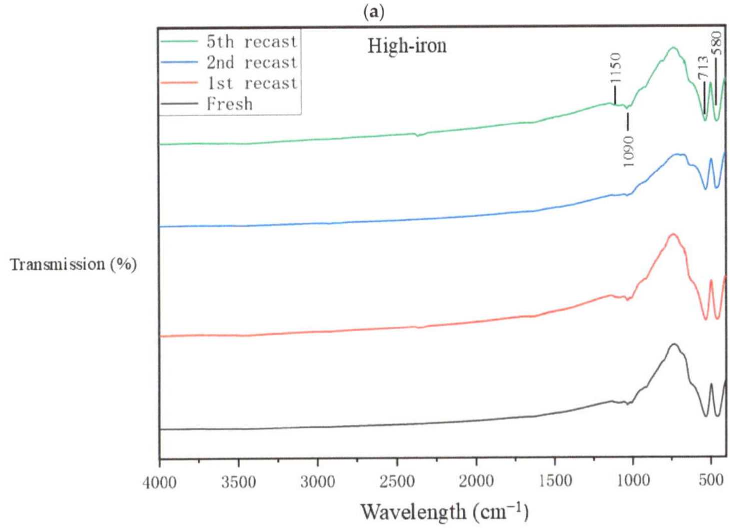

(**b**)

Figure 7. FTIR characterization of (**a**) normal sand, (**b**) high-iron-content sulfur mortars.

Figure 7a depicts the FTIR analysis of normal-sand sulfur mortar after recasting. The absorption peaks in the bands near 841, 1082, 1460, 1650, and 2360 cm^{-1} indicate the characteristic profile of sulfur monomers [38]. In contrast, the absorption peaks near the bands 464, 692, and 795 cm^{-1} indicate the Si-O bonds and the absorption peaks near 1082 cm^{-1} characterize the presence of Si-O-Si bonds [39]. This aligns with the XRD and SEM-EDS results. These analyses show that the main components of the normal-sand sulfur mortar remained unchanged during recasting. The primary components are sulfur monomers and silicon oxide. The absorption peaks near the 580 cm^{-1} band are attributed to the Fe-O-Fe bond [40,41], and the vibrations near 713 and 1150 cm^{-1} are attributed to the Al-O-Al bond [42,43], which confirms the presence of the metal oxides. Furthermore, a small stretching band of the S-O bond near the 1090 cm^{-1} is also noted, corresponding to inorganic sulfate, and a shoulder near 980 cm^{-1}, indicating a symmetric vibration of $(SO4)^{2-}$ [44]. This evidence indicates that during the recasting of high-iron-content samples, a small portion of the sulfur reacted with oxygen, producing gas sulfur dioxide, which is partly soluble in liquid sulfur [45]. This soluble sulfur dioxide reacted with metal oxides; sulfites were produced first, finally leading to sulfates, because of the reaction with oxygen from the air [46,47]. The sulfate crystals closely adhered to the surface and the aggregate's interior. Initially, the strength came from the physical connection between sulfur crystals and aggregates. Later, the formation of sulfates added a chemical connection, further increasing the strength of the high-iron-content sulfur mortar.

3.6. *Differential Scanning Calorimetry Characterization*

To better understand the thermodynamic behavior of the materials during recasting, the normal-sand and high-iron-content sand aggregates were mixed with sulfur, respectively, in the same proportions (35 wt. % sulfur/65 wt. % aggregates) in a DSC crucible. Fifty mg samples were subjected to ten cycles of heating and cooling. The heat flow was followed by differential scanning calorimetry. The experimental results are presented in Figure 8. The red arrows and blue arrows present the heating and cooling path of the DSC tests. The temperature range is from 20 to 140 °C, and the heating/cooling speed is 10 °C/min.

According to the reference, the crystal properties of sulfur in the range of 20–140 °C are summarized [14,48,49] as follows: Sulfur is stable in the form of orthorhombic crystals (Sα) at room temperature. At 110 °C, Sα transforms into monoclinic crystals (Sβ). At higher temperatures (120 °C), the transformation of Sβ into liquid sulfur appears. Figure 8a shows the results of ten heating–cooling cycles on normal-sand sulfur mortar. When the mixture was heated for the first time (1st light blue line), heat flow peaks at around 105 °C and 118 °C, indicating the transition of Sα to Sβ and Sβ to liquid sulfur, respectively. During subsequent cycles, all curves showed a single peak around 120 °C. During the cooling process from 140 °C to 20 °C, exothermic peaks were investigated between 105 °C and 75 °C. The nucleation and growth of sulfur crystals were initiated at different undercooling levels depending on the cycle. Figure 8b shows four main peak groups. During cooling, the exothermic peaks exhibited by the high-iron-content-sand samples around 118 °C correspond to similar enthalpy change. This suggests that the crystallization process exhibited more stable and homogeneous thermodynamic characteristics than the normal-sand sulfur mortar samples. Some interesting exothermic peaks were observed in the remaining experiments, occurring near 51 °C; however, no exothermic peak was observed in the first heating, nor the normal sand heating. Wang et al. [50] investigated the loss of water of crystallization from iron sulfate hydrate, using the DSC method, and this loss of bound water is an exothermic reaction that also occurred at 53 °C. Therefore, this is believed to be the characteristic curve of loss of bound water after the metal sulfate crystals have combined with moisture in the air to form a hydrate, thereby confirming the existence of sulfate crystals.

Compared those two DSC curves, the main differences are in the temperature where the phase change of sulfur happened: the normal sulfur mortar presented two phase changes at around 110 °C and 120 °C during the 1st heating; however, during the following recycling process, the first phase change absorption peak shifted to around 80 °C. More than that, it is obvious that during the 10-cycle cooling process, the phase change started from 110 °C to 80 °C. It could be due to the different products after every cycle, leading to an unstable phase change temperature. In contrast, the sample of high iron content presents a stable, repeatable, and consistent phase change curve, except for the 1st recycle. These concentrated phase change temperatures illustrate that the sulfur mortar composites with high iron content always produced analogous products. This is an advantage for sulfur mortar, whose quality is highly dependent on the homogeneity of temperature variation during solidification, thus preventing shrinkage cracks due to the uneven change in temperature.

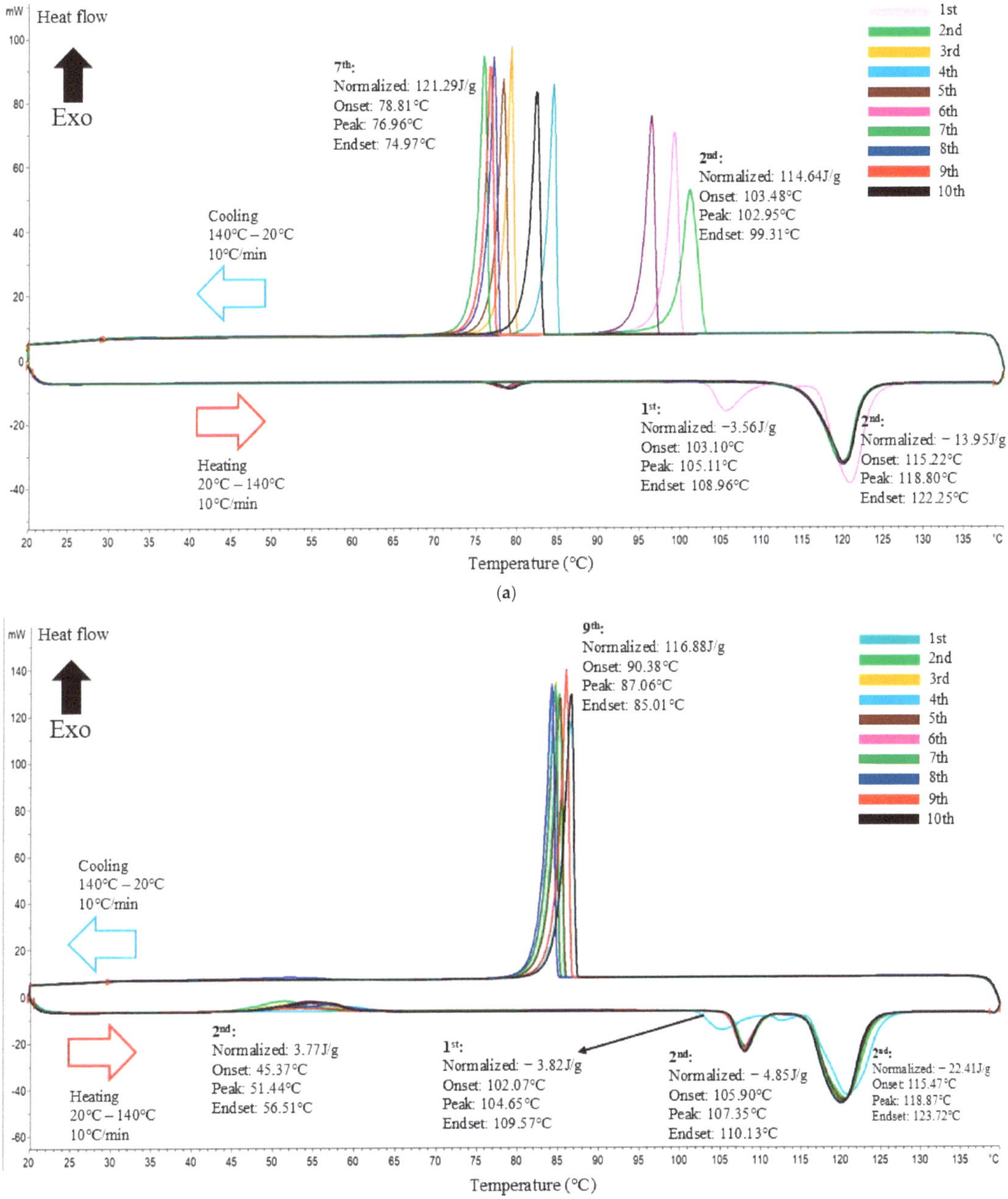

Figure 8. DSC characterization of (**a**) normal sand, (**b**) high-iron-content-sand sulfur mortar.

4. Conclusions

In this study, the mechanical performance variation of sulfur mortar during the recast process was investigated. High-iron-content sand was used as a replacement aggregate for normal sand. Different chemical characterization techniques were used to systematically investigate the microstructure, chemical composition, and thermal characterizations. The main conclusions are summarized as follows:

- Compared to normal-sand sulfur mortar, high-iron-sand sulfur mortar showed a greater increase in both compressive and bending strength during five recasts. The compressive strength of high-iron sulfur mortar increased by 10.8%, while its bending strength increased by 9.8%. In contrast, normal-sand sulfur mortar showed insignificant changes in mechanical performance.
- After the 5th recast, optical microscopy showed a more uniform distribution of sulfur between the aggregates in both types of sulfur mortar. This is identified as the reason for the strength increase in normal-sand sulfur mortar. However, SEM-EDS analysis revealed additional features in high-iron-content samples. A combination of sulfur, iron, and oxygen formed crystals around the aggregates. These crystals filled voids and discontinuities inside the sulfur mortar. This improved the bond between the sulfur binder and the aggregates, resulting in a greater increase in strength.
- Using XRD and FTIR techniques, the existence of iron sulfate was confirmed. Its presence was identified through its crystal structure and the chemical bond of S=O. It is concluded that during the recasting process, high-iron-content aggregates reacted with liquid sulfur. This reaction generated new crystal structures, which significantly contributed to the strength improvement.
- Using the DSC technique, the high-iron-content sulfur mortar showed a more regular crystalizing temperature compared to the normal-sand sulfur mortar. DSC curves also proved that high-iron-content sulfur mortar gives better thermal dynamics.
- This study highlights the potential of sulfur mortar as a sustainable construction material. The ability to recycle sulfur mortar through multiple break–recast cycles without significant loss of mechanical performance supports resource efficiency and waste reduction, key principles of sustainability. Additionally, the formation of iron sulfates during recycling reduces material segregation and porosity, contributing to durability and prolonged service life. These findings align with the goals of achieving a circular economy by promoting the reuse of materials and minimizing environmental impact. This research provides a foundation for the development of sustainable construction practices using sulfur-based composites in extreme and resource-constrained environments.
- Further investigation is required regarding the manner and degree of involvement of metal oxides in the actual production process of sulfur mortar, such as adding catalysts to stimulate more metal oxides to react with sulfur, observing whether the resulting products will positively or negatively affect the performance of sulfur mortar. This can be related to the inclusion of an additional modifier.

Author Contributions: Conceptualization, Q.W. and D.S.; Methodology, Q.W. and D.S.; Validation, Q.W.; Formal analysis, Q.W.; Investigation, Q.W.; Resources, D.S.; Data curation, Q.W.; Writing—original draft preparation, Q.W.; Writing—review and editing, M.-P.D. and D.S.; Visualization, Q.W.; Supervision, D.S.; Project administration, D.S.; Funding acquisition, D.S. All authors have read and agreed to the published version of the manuscript.

Funding: This research was funded by Université Libre de Bruxelles, the ARC Project "Actions de Recherche Concertées" Consolidation 2022–2025, grant number: HeSCon: Sulphur Concrete, to enhance the structural integrity and durability of the material.

Institutional Review Board Statement: Not applicable for studies not involving humans or animals.

Informed Consent Statement: Not applicable.

Data Availability Statement: All the data are shared by reasonable request.

Acknowledgments: The authors would like to thank Université Libre de Bruxelles (ULB) for funding the ARC Project HeSCon. The authors would like to thank Tiriana Segato and Patrizio Madau, both from the 4MAT department at ULB, for their assistance and timely help in the preparation of the samples and chemical characterization. The authors also thank Brice Delsaute, from the CRIC-OCCN, for the high-iron-content sand used in this study.

Conflicts of Interest: All the authors identify and declare that the funders had no role in the design of the study; in the collection, analyses, or interpretation of data; in the writing of the manuscript; or in the decision to publish the results.

References

1. AR6 Synthesis Report. Available online: https://www.ipcc.ch/report/ar6/syr/resources/press (accessed on 10 March 2024).
2. Pauliuk, S.; Heeren, N.; Berrill, P.; Fishman, T.; Nistad, A.; Tu, Q.; Wolfram, P.; Hertwich, E.G. Global scenarios of resource and emission savings from material efficiency in residential buildings and cars. *Nat. Commun.* **2021**, *12*, 5097. [CrossRef] [PubMed]
3. Pearce, F. *Can the World's Most Polluting Heavy Industries Decarbonize?* Yale School of Environment: New Haven, CT, USA, 2021.
4. Dahanni, H.; Ventura, A.; Le Guen, L.; Dauvergne, M.; Orcesi, A.; Cremona, C. Life cycle assessment of cement: Are existing data and models relevant to assess the cement industry's climate change mitigation strategies? A literature review. *Constr. Build. Mater.* **2024**, *411*, 134415. [CrossRef]
5. Habert, G.; Miller, S.A.; John, V.M.; Provis, J.L.; Favier, A.; Horvath, A.; Scrivener, K.L. Environmental impacts and decarbonization strategies in the cement and concrete industries. *Nat. Rev. Earth Environ.* **2020**, *1*, 559–573. [CrossRef]
6. Belaïd, F. How does concrete and cement industry transformation contribute to mitigating climate change challenges? *Resour. Conserv. Recycl. Adv.* **2022**, *15*, 200084. [CrossRef]
7. The European Green Deal—European Commission. Available online: https://commission.europa.eu/strategy-and-policy/priorities-2019-2024/european-green-deal_en (accessed on 27 March 2024).
8. Galik, C.S.; DeCarolis, J.F.; Fell, H. Evaluating the US mid-century strategy for deep decarbonization amidst early century uncertainty. *Clim. Policy* **2017**, *17*, 1046–1056. [CrossRef]
9. Wang, Y.; Yi, H.; Tang, X.; Wang, Y.; An, H.; Liu, J. Historical trend and decarbonization pathway of China's cement industry: A literature review. *Sci. Total Environ.* **2023**, *891*, 164580. [CrossRef]
10. Obrist, M.D.; Kannan, R.; Schmidt, T.J.; Kober, T. Decarbonization pathways of the Swiss cement industry towards net zero emissions. *J. Clean. Prod.* **2021**, *288*, 125413. [CrossRef]
11. Belaïd, F.; Al Sarihi, A. Energy transition in Saudi Arabia: Key initiatives and challenges. In Proceedings of the International Association for Energy Economics Energy Forum, Athens, Greece, 21–24 September 2022; pp. 8–13.
12. Mohamed, A.-M.O.; El-Gamal, M. *Sulfur Concrete for the Construction Industry: A Sustainable Development Approach*; J. Ross Publishing: Fort Lauderdale, FL, USA, 2010; ISBN 1-60427-005-5.
13. Boyd, D.A. Sulfur and its role in modern materials science. *Angew. Chem. Int. Ed.* **2016**, *55*, 15486–15502. [CrossRef]
14. Fediuk, R.; Mugahed Amran, Y.H.; Mosaberpanah, M.A.; Danish, A.; El-Zeadani, M.; Klyuev, S.V.; Vatin, N. A critical review on the properties and applications of sulfur-based concrete. *Materials* **2020**, *13*, 4712. [CrossRef]
15. Lewandowski, M.; Kotynia, R. Assessment of sulfur concrete properties for use in civil engineering. *MATEC Web Conf.* **2018**, *219*, 03006. [CrossRef]
16. Naser, M.Z. Extraterrestrial construction materials. *Prog. Mater. Sci.* **2019**, *105*, 100577. [CrossRef]
17. Wang, Q.; Snoeck, D. To boldly go where no one has gone before: Sulfur concrete, a promising construction material fulfilling the demands for a sustainable future on celestial objects: A review. *Mater. Today* **2023**, *72*, 301–317. [CrossRef]
18. Franz, H.B.; King, P.L.; Gaillard, F. Sulfur on Mars from the Atmosphere to the Core. In *Volatiles in the Martian Crust*; Elsevier: Amsterdam, The Netherlands, 2019; pp. 119–183, ISBN 978-0-12-804191-8. [CrossRef]
19. NASA's Curiosity Rover Discovers a Surprise in a Martian Rock—NASA 2024. Available online: https://www.nasa.gov/missions/mars-science-laboratory/curiosity-rover/nasas-curiosity-rover-discovers-a-surprise-in-a-martian-rock/ (accessed on 25 October 2024).
20. Ghasemi, S.; Nikudel, M.; Zalooli, A.; Khamehchiyan, M.; Yousefvand, F.; Ghasemi, A. Durability Assessment of Sulfur Concrete and Portland Concrete in Laboratory Conditions and Marine Environments. *J. Mater. Civ. Eng.* **2022**, *34*, 04022167. [CrossRef]
21. Gumeniuk, A.; Hela, R.; Polyanskikh, I.; Gordina, A.; Yakovlev, G. Durability of Concrete with Man-made Thermoplastic Sulfur Additive. *IOP Conf. Ser. Mater. Sci. Eng.* **2020**, *869*, 032012. [CrossRef]
22. Vlahovic, M.M.; Martinovic, S.P.; Boljanac, T.D.; Jovanic, P.B.; Volkov-Husovic, T.D. Durability of sulfur concrete in various aggressive environments. *Constr. Build. Mater.* **2011**, *25*, 3926–3934. [CrossRef]
23. Shin, M.; Kim, K.; Gwon, S.-W.; Cha, S. Durability of sustainable sulfur concrete with fly ash and recycled aggregate against chemical and weathering environments. *Constr. Build. Mater.* **2014**, *69*, 167–176. [CrossRef]
24. Khademi, A.G.; Sar, H.I.K. Comparison of sulfur concrete, cement concrete and cement-sulfur concrete and their properties and application. *Curr. World Environ. J.* **2015**, *10*, 63–68. [CrossRef]
25. Wan, L.; Wendner, R.; Cusatis, G. A novel material for in situ construction on Mars: Experiments and numerical simulations. *Constr. Build. Mater.* **2016**, *120*, 222–231. [CrossRef]
26. Gulzar, M.A.; Rahim, A.; Ali, B.; Khan, A.H. An investigation on recycling potential of sulfur concrete. *J. Build. Eng.* **2021**, *38*, 102175. [CrossRef]
27. Shahsavari, M.H.; Karbala, M.M.; Iranfar, S.; Vandeginste, V. Martian and lunar sulfur concrete mechanical and chemical properties considering regolith ingredients and sublimation. *Constr. Build. Mater.* **2022**, *350*, 128914. [CrossRef]
28. Troemner, M.; Cusatis, G. Martian Material Sourcing Challenges Propel Earth Construction Opportunities. *Matter* **2019**, *1*, 547–549. [CrossRef]

29. Soundarya, N. A review on the physical & chemical properties of sea sand to be used a replacement to fine aggregate in concrete. *Mater. Today Proc.* **2022**, *51*, 1527–1531. [CrossRef]
30. *EN 196-1:2016*; Methods of Testing Cement—Part 1: Determination of Strength. European Committee for Standardization (CEN): Brussels, Belgium, 2016.
31. Abdel-Jawad, Y.; Al-Qudah, M. The combined effect of water and temperature on the strength of sulfur concrete. *Cem. Concr. Res.* **1994**, *24*, 165–175. [CrossRef]
32. Lian, C.; Zhuge, Y.; Beecham, S. The relationship between porosity and strength for porous concrete. *Constr. Build. Mater.* **2011**, *25*, 4294–4298. [CrossRef]
33. Kearsley, E.P.; Wainwright, P.J. The effect of porosity on the strength of foamed concrete. *Cem. Concr. Res.* **2002**, *32*, 233–239. [CrossRef]
34. Mitchell, M.B.; Sheinker, V.N.; White, M.G. Adsorption and Reaction of Sulfur Dioxide on Alumina and Sodium-Impregnated Alumina. *J. Phys. Chem.* **1996**, *100*, 7550–7557. [CrossRef]
35. Fu, H.; Wang, X.; Wu, H.; Yin, Y.; Chen, J. Heterogeneous Uptake and Oxidation of SO_2 on Iron Oxides. *J. Phys. Chem. C* **2007**, *111*, 6077–6085. [CrossRef]
36. Kloprogge, J.T.; Geus, J.W.; Jansen, J.B.H.; Seykens, D. Thermal stability of basic aluminum sulfate. *Thermochim. Acta* **1992**, *209*, 265–276. [CrossRef]
37. Wada, H.; Takada, K.; Sasaki, T. DSC studies on reactions of the elements with sulfur. *Solid State Ion.* **2004**, *172*, 421–424. [CrossRef]
38. He, H.; Zhang, C.-G.; Xia, J.; Peng, A.-A.; Yang, Y.; Jiang, H.; Zheng, L.; Ma, C.-Y.; Zhao, Y.-D.; Nie, Z.-Y.; et al. Investigation of Elemental Sulfur Speciation Transformation Mediated by Acidithiobacillus ferrooxidans. *Curr. Microbiol.* **2009**, *58*, 300–307. [CrossRef]
39. Herth, E.; Zeggari, R.; Rauch, J.-Y.; Remy-Martin, F.; Boireau, W. Investigation of amorphous SiOx layer on gold surface for Surface Plasmon Resonance measurements. *Microelectron. Eng.* **2016**, *163*, 43–48. [CrossRef]
40. Najafpour, M.; Jameei Moghaddam, N. Iron oxide deposited on metallic nickel for water oxidation. *Sustain. Energy Fuels* **2017**, *1*, 658–663. [CrossRef]
41. Qazi, U.; Iftikhar, R.; Ikhlaq, A.; Riaz, I.; Jaleel, R.; Nusrat, R.; Javaid, R. Application of Fe-RGO for the removal of dyes by catalytic ozonation process. *Environ. Sci. Pollut. Res.* **2022**, *29*, 89485–89497. [CrossRef]
42. Aluminium Oxide—Database of ATR-FT-IR Spectra of Various Materials. Available online: https://spectra.chem.ut.ee/paint/fillers/aluminium-oxide/ (accessed on 24 April 2024).
43. Yinghao, W.; Zhao, W.; Wang, W.; Sui, W. Fabricating binary anti-corrosion structures containing superhydrophobic surface and sturdy barrier layer for Al alloys. *RSC Adv.* **2016**, *6*, 5100–5110. [CrossRef]
44. Ramaswamy, V.; Vimalathithan, R.M.; Ponnusamy, V. Synthesis and Characterization of $BaSO_4$ Nano-particles Using Micro-emulsion Technique. *Adv. Appl. Sci. Res.* **2010**, *1*, 197–204.
45. Touro, F.J.; Wiewiorowski, T.K. Molten Sulfur Chemistry. II. The Solubility of Sulfur Dioxide in Molten Sulfur. *J. Phys. Chem.* **1966**, *70*, 3531–3534. [CrossRef]
46. Baltrusaitis, J.; Cwiertny, D.M.; Grassian, V.H. Adsorption of sulfur dioxide on hematite and goethite particle surfaces. *Phys. Chem. Chem. Phys.* **2007**, *9*, 5542–5554. [CrossRef]
47. Heterogeneous Reactions of Sulfur Dioxide on Typical Mineral Particles. *J. Phys. Chem. B* **2006**, *110*, 12588–12596. [CrossRef]
48. Moon, J.; Kalb, P.D.; Milian, L.; Northrup, P.A. Characterization of a sustainable sulfur polymer concrete using activated fillers. *Cem. Concr. Compos.* **2016**, *67*, 20–29. [CrossRef]
49. Currell, B.R.; Williams, A.J. Thermal analysis of elemental sulphur. *Thermochim. Acta* **1974**, *9*, 255–259. [CrossRef]
50. Wang, T.; Debelak, K.A.; Roth, J.A. Dehydration of iron(II) sulfate heptahydrate. *Thermochim. Acta* **2007**, *462*, 89–93. [CrossRef]

sustainability

Article

Durable Structural Concrete Produced with Coarse and Fine Recycled Aggregates Using Different Cement Types

Carla Vintimilla, Miren Etxeberria * and Zongyang Li

Department of Civil and Environmental Engineering, Universitat Politècnica de Catalunya·Barcelona TECH, Campus Nord, 08034 Barcelona, Spain; carla.vintimilla@upc.edu (C.V.); zongyang.li@estudiantat.upc.edu (Z.L.)
* Correspondence: miren.etxeberria@upc.edu; Tel.: +34-934-011-788

Abstract: The durability properties of structural recycled aggregate concrete (RAC) produced with 50% coarse recycled concrete aggregates and up to 20% fine recycled concrete aggregates were analysed and compared to those of conventional concrete (NAC). Both the RAC and NAC mixtures achieved the same compressive strength when using an effective water–cement ratio of 0.47 and 0.51, respectively. All the concretes were produced using three types of cement: CEM II A/L 42.5 R, CEM II A/S 42.5 N/SRC and CEM III/B 42.5 N-LH/SR. The properties of drying shrinkage, chloride permeability, and accelerated carbonation coefficient of the concretes were determined experimentally, and the obtained results were compared with the values estimated by specific standards of exposure to XC1–XC4 (corrosion induced by carbonation can happen due to the presence of humidity) and XS1 (corrosion caused by chlorides from seawater) environments. The results showed that all the concretes achieved maximum drying shrinkage for use in structural concrete. Any concretes produced with CEM IIIB, including the RAC-C50-F20 concrete, achieved very low chloride ion penetrability, ranging between 500 to 740 Coulombs. In addition, all concretes manufactured with CEM IIAL and CEM IIAS, including RAC-C50-F20, were suitable for use in XC3 and XC4 exposure environments, both with 50- and 100-year lifespans.

Keywords: coarse and fine recycled aggregates; supplementary cementitious materials; CEM IIAL; CEM IIAS; CEM IIIB; concrete durability; compressive strength; drying shrinkage; carbonation; chloride penetration

Citation: Vintimilla, C.; Etxeberria, M.; Li, Z. Durable Structural Concrete Produced with Coarse and Fine Recycled Aggregates Using Different Cement Types. *Sustainability* **2023**, *15*, 14272. https://doi.org/10.3390/su151914272

Academic Editor: Syed Minhaj Saleem Kazmi

Received: 6 August 2023
Revised: 15 September 2023
Accepted: 25 September 2023
Published: 27 September 2023

1. Introduction

Concrete is widely employed in construction due to its remarkable strength and durability. However, the durability of concrete as a permeable material depends significantly on the quality of its constituent materials [1]. Limited investigations have been carried out studying the durability characteristics of concrete made with fine recycled concrete aggregate (FRCA) and coarse recycled concrete aggregate (CRCA) [2–6]. However, in general, the durability of recycled aggregate concrete (RAC) is lower than that of natural aggregate concrete (NAC) due to the influence of factors such as the connectivity of the porous network and water content and the type of cementitious materials (SCM) used having an important influence [2,6–9]. In addition, SCM such as blast furnace slag (BFS) further enhances the sustainability of RAC by reducing carbon dioxide emissions and increasing the circular economy [10,11].

The use of structural RAC in chloride-containing environments has sparked debate among researchers [12]. Some authors indicate that the increase in RCA content could lead to a higher diffusion of chlorides in RAC due to its high porosity [13,14]. However, some studies have also demonstrated that by reducing the amount of adhered cement mortars, the resistance of RAC to chloride ion penetration can be improved, particularly when RCA originates from higher-strength concrete [15]. The penetration of chloride ions is a major contributor to the corrosion of steel reinforcements. The results of numerous studies [7,16–20] conducted on this topic have revealed the following: the diffusion coefficient of chloride ions exhibits a linear increase with the proportional increase in recycled aggregate use; FRCA influences more than the CRCA in concrete diffusion coefficients; and, similar to NAC, chloride ion migration can be reduced by decreasing the water-to-binder ratio or incorporating SCM such as fly ash, silica fume or blast furnace slag (BFS) [7]. According to Li et al. [13], following the ASTM C1202 classification, while the concrete produced with up to 50% of RCA achieved low chloride ion penetrability, the RAC produced with a higher percentage than 50% was classified with medium penetrability, consequently needing the use of SCM to improve the resistance to chloride ion penetration. As confirmed, BFS cement enhances chloride penetration resistance in concrete due to its ability to immobilize chloride ions [21,22]. This enhancement is achieved through physical and chemical mechanisms, through chloride ion adsorption on the C-S-H surface [21] and the formation of Friedel's salt due to the higher aluminate content in the BFS cement [21,22].

Carbonation in concrete is a physicochemical process in which CO_2 penetrates the cement paste and reacts with Portlandite, forming calcite and reducing the concrete pH from 13 to 8–9. The carbonation rate is influenced by the permeability and moisture content of the concrete [2,7]. As a result of carbonation, steel reinforcement loses its protection, and corrosion begins when adequate oxygen and water levels are present [7].

Extensive research has been conducted on the carbonation resistance analysis of RAC [16,18,23,24]. Different factors can influence the carbonation depth of the RAC. These factors include the replacement ratio of recycled aggregates, the origin and quality of the RCA, the crushing technique employed for RCA production, the cement type and quantity used in concrete production, the curing process, and the use of superplasticisers to reduce the water–cement ratio [7,9,25]. Conclusions regarding the impact of RCA on the carbonation resistance of concrete can be ambiguous and conflicting. According to Pedro et al. [26], the carbonation depth increases as the concrete's compressive strength decreases. Certain researchers [8,25] have argued that RAC mixtures produced with coarse CRCA exhibit similar or even higher carbonation resistance than NAC due to aged adhered mortar. Zeng et al. [27] suggested that the optimal replacement percentage of natural aggregates (NAs) with RCA is 50%, which prevents a decrease in carbonation resistance. Etxeberria et al. [24] also reached a similar conclusion when employing a 50% replacement of uncarbonated CRCA. Loti et al. [18] found that RAC produced with up to 50% coarse RCA met the current European standards, thus supporting its use in structural application when up to 50% of CRCA is employed in concrete production. However, more investigation is needed to evaluate the influence of FRCA recycled aggregates on carbonation resistance and, in general, on concrete durability.

Furthermore, the service life of concrete structures against carbonation strongly depends on the type of cement used in concrete production [28,29]. The carbonation depth of concrete mixtures produced using SCM was higher due to the reduction of Portlandite during cement hydration, reducing Ca availability [29,30] and, consequently, causing less resistance to carbonation. Although SCMs reduce alkali reserve usually leads to a reduction in pore size, they can decrease the permeability of cementitious matrices [31]. However, carbonation not only lowers the overall pH but may also result in the coarsening of the pore structure, potentially diminishing its durability and susceptibility to various forms of degradation, including chemical and physical attacks [30]. Consequently, the carbonation concrete's service life decreases as more SCM is used to replace clinker [28,32,33]. However, using limited mineral admixtures in RAC production can improve carbonation resistance. The RCA produced using CEM IIAS achieved a higher carbonation resistance than that produced with CEM IIAL due to the addition of available CaO in the slag cement. In addition, the use of up to 50% of CRCA had little influence on the carbonation depth [34].

According to the shrinkage values obtained, RAC concrete produced with higher percentages of RCA achieved higher shrinkage values [35,36]. Gonzalez and Etxeberria [37] studied the drying shrinkage of RACs produced using only CRCA obtained from different sources. They concluded that the CRCA produced from a lower-strength parent concrete achieved the highest drying shrinkage value. The increase in the RAC shrinkage value is due to the high water absorption of RCAs, which are porous and contain old cement paste [7]. Vintimilla and Etxeberria [36] determined that all concretes with up to 60% CRCA achieved shrinkage values similar to NAC. In addition, they concluded that the use of FRCA increased the shrinkage value when compared to concrete made only with CRCA. Nevertheless, the concretes produced with up to 60% CRCA and 20% FRCA also obtained adequate values ranging from between −200 and −800, following American Concrete Institute (ACI) standards [38]. Simsek et al. [39] conducted a study to evaluate the influence of using 20%, 40%, 60%, 80%, and 100% FRCA or CRCA in the substitution of natural aggregates. They concluded that after 90 days, the RAC with up to 20% FRCA achieved adequate properties. Moreover, several recent studies [36,40] have confirmed the existence of a slight influence of FRCA on structural concrete performance, however not being detrimental and consequently technically viable for their use.

The type of cement used in concrete production also influences the drying shrinkage value. It has been determined that concretes produced using Portland clinker-based cement have very high strength due to the fact that it increases the hydration heat and, as a consequence, leads to higher drying shrinkage [41]. In contrast, the early stage shrinkage caused in SCM cement significantly contributes to final shrinkage, raising the risk of concrete cracking in later stages [42,43].

The main objective of this research work was to conduct a comprehensive analysis of the durability properties of concrete produced using 50% CRCA and up to 20% FRCA. To achieve this objective, all conventional and recycled concretes were designed to obtain the same compressive strength. Thus, all the RAC concretes were manufactured with an effective water–cement value of 0.47 and the NAC with 0.51. Three types of cement, CEM II A/L 42.5 R, CEM II A/S 42.5 N/SRC and CEM III/B 42.5 N-LH/SR, were used for concrete production. The properties of drying shrinkage, chloride permeability, and accelerated carbonation coefficient of the concrete were determined experimentally. The obtained results were compared with the values estimated by specific standards for exposure to XC1–XC4 and XS1 environments.

2. Materials and Methods

2.1. Materials

2.1.1. Cement and Chemical Admixtures

Three different cement types, CEM II A/L 42.5 R, CEM II A/S 42.5 N/SRC and CEM III/B 42.5 N-LH/SR, defined by the European standard EN 197-1 [44], were employed. The three types of cement are sustainable (produced using SCM) and available in Barcelona: (1) CEM II A/L 42.5 R (88% clinker, 12% limestone, excluding the set regulator, added in 5%), with high initial strength, ideal for applications requiring rapid setting; (2) CEM IIAS 42.5 N/SRC (83% clinker, 12% blast furnace slag (BFS) and 5% minority component), providing moderate sulphate resistance and enhanced durability and (3) CEM IIIB 42.5 N-LH/SR (27% clinker, 70% BFS and 3% minority component) with low heat development and sulphate resistance. The composition details of the three cement types are illustrated in Table 1.

Table 1. Composition of cement as a percentage of the total weight.

Cement	CaO	SiO_2	Al_2O_3	Fe_2O_3	SO_3	MgO	K_2O	TIO_2	Na_2O
CEM II A-L 42.5 R	61.47	17.87	3.61	2.64	3.69	1.45	0.736	0.183	0.228
CEM II A-S 42.5 N/SRC	59.97	21.66	4.35	3.75	3.47	2.1	0.395	0.327	0.314
CEM III/B 42.5 N-LH/SR	49.4	27.8	8.41	1.96	3.96	4.65	0.48	0.457	0.365

Two chemical admixtures were used in the concrete manufacturing process: (1) a superplasticiser (S) based on polycarboxylate ether (PAE) polymer technology and (2) a multifunctional admixture (P) based on Modified Lignin Sulfonate. The manufacturer's recommendations for these admixtures ranged from 0.3% to 2% S and 0.5% to 1.5% P, based on the cement weight.

2.1.2. Natural Aggregates

Natural limestone aggregates were used for concrete production. One fine fraction (0/4 mm, FNA) and two coarse fractions (4/10 mm CNA1 and 8/20 mm CNA2) were employed. Figure 1 shows the geometrical characteristics of the aggregate fractions. Figure 2 describes the grading distribution of each fraction (determined and classified following EN 933-1 [45] and EN12620 [46] specifications). In addition, the recommended upper and lower limits for fine aggregate, as stipulated in the Structural Concrete Code (SC-BOE) [47], are described. The dry density and water absorption were determined under the EN 1097-6 [48] standard. The obtained property values are shown in Table 2.

Figure 1. Image of all the fractions: (**a**) raw aggregates; (**b**) recycled aggregates.

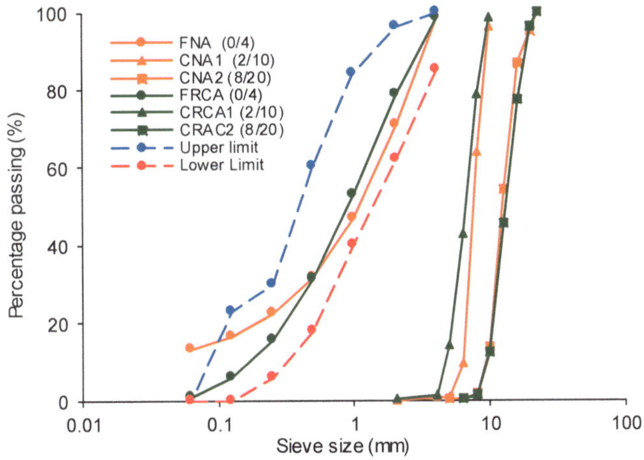

Figure 2. Particle size distribution of natural and recycled aggregates.

Table 2. Properties of natural and type A RCA aggregates studied.

Property (Standard)	Specification	FNA (0/4)	CNA 1 (4/10)	CNA 2 (8/20)	FRCA (0/4)	CRCA-1 (2/10)	CRCA-2 (8/20)	SC-BOE
Density (Kg/m^3)	EN 1097-6 [48]	2.67	2.65	2.68	2.32	2.33	2.36	2.1 [1]
Water Absorption (%)	EN 1097-6 [48]	0.95	0.77	0.73	5.73	5.62	5.16	<7
Humidity (%)		0.37	0.16	0.1	2.73	4.50	4.55	
Sand equivalent (%)	EN 933-8 [49]				100			>70
Los Angeles coefficient (wt%)	EN 1097-2 [49]						35.77	<40
Flakiness index (wt%)	EN 933-3 [50]						12.81	<35
Alkali–aggregate reaction (%)	UNE 146508 [51]				0.042			<0.10

[1] Property defined in EN 206.

2.1.3. Recycled Aggregates

The production of RCA involved crushing, cleaning with water, and sieving construction and demolition waste (CDW) at a recycling plant in Barcelona, Spain [52]. Concrete mixtures were manufactured using one fine fraction (0/4 mm, FRCA) and two coarse fractions (2/10 mm, CRCA-1 and 8/20 mm, CRCA-2). Figures 1 and 2 illustrate the shapes and size distribution of the three RCA fractions, respectively.

The components of the CRCA-2 (8/20 mm) fraction were characterized in accordance with the EN 933-11 [53] specification, and the obtained values are described in Table 3. According to the EN 206 [54] specification, the RA employed in this work were classified as type A (RC90, RCU95, Rb10-, Ra1, FL2- and XRg1-), with concrete (RC) and natural stone (Ru) components representing over 90% of the total content, while the ceramic content constituted less than 10%. Specifically, the aggregates were classified as RCU95 [36].

Table 3. Constituents of type A CRCA-2 (8/20 mm) aggregates.

Type	Concrete, Concrete Products, Mortar (Rc)	Unbound Mortar, Nature Stone (Ru)	Mansory (Rb)	Asphalt (Ra)	Glass (Rg)	Other (x)
CRCA-2	68.3%	28.0%	1.9%	0.98%	0.0%	0.39%
EN 12620	Rc + Ru > 95		≤10%	≤1%	≤1%	-

Due to the presence of adhered mortar in RCAs, the porosity of RCA was higher than that of natural aggregates (NAs), reducing their density and increasing their absorption capacity. As previously documented in scientific studies [2,26,36,55].

Table 2 shows that the obtained dry density value of the different fractions of RCA was higher than the minimum requirement value of 2.1 kg/dm^3 established by the European standard EN 206 [54] for their use in concrete production. Moreover, although the water absorption value obtained by the RCA aggregates was higher than that of the NA aggregates, it was lower than the maximum value of 7% specified by the structural code SC-BOE. Various studies have reported that the absorption capacities of coarse and fine fraction type A RCA could reach 3.9–9.6% [3,56] and 2.4–19.3% [57–59], respectively.

The RCA aggregates also achieved adequate property values of the Los Angeles coefficient, sand equivalent, and flakiness index (see Table 2) for concrete production. Alkali–aggregate reactivity analysis was conducted on the FRCA 0/4 mm fraction. It was established that the specimens exhibited an expansion of less than 0.1% after 14 days, indicating that they could be classified as non-reactive materials.

2.2. Concrete Production and Test Procedures

2.2.1. Concrete Production

All the concrete mixtures were designed to be exposed to XC1–XC4 (corrosion induced by carbonation can happen due to humidity presence) and XS1 (corrosion caused by chlorides from seawater) environments [47]. Those concretes require a minimum characteristic design strength (fck) of 30 MPa (C30/37), using a total water–cement ratio of 0.50 and 300 kg of the three types of cement described in Table 1.

The effective water–cement of 0.47 was maintained constant in all the produced concrete (Table 4). The effective water–cement ratio of 0.47 was determined in conventional concrete (NAC-0.47 concrete) after the effectively absorbed water by aggregates was removed from the total water–cement ratio of 0.50.

Table 4. Mix proportions of concretes were produced with CEM IIAL, CEM IIAS and CEM IIIB.

Materials (kg)	Concrete Types				
	NAC-0.51	NAC-0.47	RAC-C50	RAC-C50-F10	RAC-C50-F20
Cement	300	300	300	300	300
Total water	165	150	175.8	179.5	182.3
CNA 1	354.5	360.1	180.5	180.5	180.5
CNA 2	723.68	737.2	369.3	369.3	369.3
FNA	954.1	971.9	1014.2	875.7	778.4
CRCA 1	-	-	165.8	165.2	165.6
CRCA 2	-	-	338.8	339.1	337.2
FRCA	-	-	-	87.1	174.1
P (%)	1/0.7 [1]	1	1/0.6 [1]	1/0.5 [1]	1/0.3 [1]
S (%)	1	1	1/1.5 [1]	1/1.5 [1]	1/1.5 [1]
effective w/c	0.51	0.47	0.47	0.47	0.47
Slump-IIAS (mm)	175	145	150	155	150
Slump-IIIB (mm)	175	160	135	150	150
Slump-IIAL (mm)	175	150	190	200	195

[1] Plasticizer content utilised in CEM IIAL.

Specifically, the effective absorption capacities of fine and coarse aggregates natural aggregates were 70% and 20% of the total absorption capacities, respectively. In comparison, the effective absorption capacities of FRCA and CRCA were 100% and 70%, respectively, of their absorption capacities in 24 h. Table 2 also presents the average humidity (%) of the RCA when they were employed for concrete production. All coarse fractions of RCA were used with high humidity (between 80 and 90% of their absorption capacity) during manufacturing. The total water content in the concrete was calculated by combining the effective water and the water within the aggregates (humidity plus the amount of effectively absorbed water). Table 4 illustrates how the total water–cement ratio increased with higher volumes of RCA incorporated in the concrete mixes. To achieve the same compressive strength of RAC as NAC, a new control mixture of NAC was formulated with a total water-to-cement ratio of 0.55 and an effective water-to-cement ratio of 0.51 (NAC-0.51).

Table 4 shows all the concrete mixtures produced. In addition to the two conventional concretes, NAC-0.47 and NAC-0.51, the RAC was produced using 50% CRCA and 0% FRCA (RAC-C50), 50% CRCA and 10% FRCA (RAC-C50-F10), and 50% CRCA and 20% FRCA (RAC-C50-F20). The five types of concrete were produced using the three types of cement described previously.

The slump values of the concrete samples were determined following the EN 12350-2 [60] specification. It was found that the concretes achieved a slump range between 135–200 mm, which, according to the Structural Concrete Code (SC-BOE) [47], is considered fluid (100–150 mm) and liquid consistency (160–200 mm), and defined as adequate property for building construction. As Table 4 shows, to achieve adequate workability, 1% of S and 1% of P (CEM IIAS and CEM IIIB) were employed. The RAC produced with CEM IIAL showed a slightly higher slump, which can be attributed to the higher dosage of superplasticiser used.

The concrete mixtures were produced employing a vertical axis mixer, adding the materials following a fixed sequence. Initially, aggregates were added, starting with coarser aggregates and terminating with finer ones. They were mixed for 1 min. After the cement was added, water was gradually added while the mixing process continued. Then, the chemical admixtures were added. The complete mixture was then mixed for one additional minute.

The concrete samples underwent manual compaction using a steel rod. Subsequently, a plastic sheet was employed to cover the concrete specimens, which were then subjected to a 24 h air-curing period. After 24 h of casting, the samples were demoulded and stored under controlled conditions at a temperature of 21 °C with a humidity of 95% until the testing ages.

2.2.2. Test Procedure

The concrete's compressive strength was determined using a 3000-kN capacity loading machine. The compressive strength was determined at 7, 28, and 56 days following the UNE-EN 12390-3 [61] specifications. For each testing age, three cubic specimens measuring 100 × 100 × 100 mm were utilised.

The drying shrinkage of all the produced concretes was determined following the EN 12390-16 [62] specification. Each concrete mixture used two specimens of 75 × 75 × 280 mm. After a 24 h casting, they were demoulded. Their initial lengths and weights were measured, and the two specimens of each concrete were placed in a controlled climatic chamber (temperature of 20 ± 2 °C and relative humidity of 50 ± 5%). Length and weight measurements were recorded at intervals of 1, 7, 14, 28, 56 and 91 days.

The chloride permeability in the concrete was assessed following the ASTM C1202 [63] "Standard Test Method for Electrical Indication of Concretes Ability to Resist Chloride Ion Penetration". Two cylindrical concrete samples of 200 mm in length were employed for each mixture, and from those, two disc specimens of 100 mm in diameter and 50 mm in thickness were obtained. Two disc specimens, one of each sample, were used to determine the concrete's chloride ion penetration after 28 and 56 days of curing. The chloride penetrability of the produced concretes was quantified by measuring the total charge (in Coulombs) passed during a 6 h testing period. A potential difference of 60 V was applied across each side of the specimen, which was immersed, one side in solutions containing sodium hydroxide (NaOH) and the other side in sodium chloride (NaCl).

The accelerated carbonation method following the UNE-EN 12390-12 [64] specification was employed in order to assess the carbonation resistance of the produced concrete mixtures. Each concrete mixture used two prismatic 100 × 100 × 300 mm samples. All concrete specimens underwent a curing process in a humidity chamber for 28 days, followed by a 14-day pre-conditioning period under laboratory conditions (CO_2 concentration of 425 ppm, 20 ± 2 °C, and 50–55% relative humidity, RH). Subsequently, the samples were stored in a chamber with an environment consisting of 3% CO_2, 57% HR and at 20 °C. The carbonation depth of each specimen was measured at specific intervals of 0, 14, 28, 56, 70, and 91 days of exposure to the chamber. In order to determine the carbonation depth, a solution containing phenolphthalein indicator was applied to the freshly fractured surface of the concrete. The solution contained 1 g of phenolphthalein dissolved in 70 g of ethanol and 30 g of water following the UNE-EN 14630 [65] specification.

3. Results

3.1. Compressive Strenght

Table 5 shows the obtained compressive strength values ($fcm_{,cub100}$) of cubic concrete specimens (100 × 100 × 100 mm) and their standard deviation at 7, 28 and 56 days. The concrete specimens were designed for exposure in XC1–XC4 and XS1 environments, with strength class C30/37. Consequently, according to the Structural Concrete Code (SC-BOE) [47], the minimum characteristic and average strength values of 30 MPa and 38 MPa, respectively, were established for cylindrical specimens of 150 mm and 300 mm in length. According to the calculations made by Vintimilla and Etxeberria et al. [36] following the specifications given in the Structural Concrete Code (SC-BOE) [47], the minimum average compressive strength of 100 mm cube ($fcm_{,cub100}$) should achieve 46 MPa. In addition, it was found that the standard deviations in the compressive strength results of all samples were acceptable. The dispersion was more noticeable at 7 days but decreased at 28 and 56 days. This pattern indicated that the obtained results closely aligned with the average value, ensuring measurement reliability.

Table 5. Compressive strength and its standard deviation (between brackets values) in all produced concretes.

Concrete Reference	IIAL 7d	IIAL 28d	IIAL 56d	IIAS 7d	IIAS 28d	IIAS 56d	IIIB 7d	IIIB 28 d	IIIB 56d
NAC_{47}	52.5 (1.3)	62.9 (1.3)	65.5 (1.0)	54.5 (1.3)	69.8 (1.0)	71.3 (1.9)	53.1 (0.8)	67.2 (0.4)	69.9 (0.9)
NAC_{51}	45.2 (2.0)	56.2 (1.6)	58.8 (1.0)	54.1 (2.0)	59.2 (0.5)	64.7 (0.2)	51.5 (0.1)	57.2 (1.2)	59.9 (0.9)
RAC-C50	48.6 (2.5)	57.3 (1.0)	59.9 (1.7)	53.9 (2.5)	59.2 (2.3)	62.8 (1.2)	53.7 (0.3)	61.4 (2.0)	62.0 (1.0)
RAC-C50-F10	46.9 (1.8)	56.3 (1.5)	57.5 (0.3)	52.4 (1.8)	59.7 (1.3)	59.9 (0.2)	53.4 (2.8)	60.6 (1.7)	61.6 (0.7)
RAC-C50-F20	44.9 (2.4)	52.7 (0.3)	53.8 (0.4)	50.2 (2.4)	60.7 (0)	63.7 (1.2)	53.6 (1.5)	62.8 (1.4)	62.9 (1.3)

Compressive strength values for 100 mm cube specimens. ( ) Standard deviation.

Table 5 shows that the RAC achieved 18% lower compressive strength than that of NAC-0.47 (all concretes were made using the same effective water–cement ratio of 0.47). However, the RAC achieved similar strength to NAC-0.51 concrete, which was made with an effective water–cement ratio of 0.51 (see Figure 3). These findings highlight the influence of recycled aggregates on the mechanical properties of concrete and emphasise the importance of adjusting the water-to-cement ratio to achieve comparable compressive strength to that of NAC.

Although all the RACs had the same effective water–cement ratio, RACs manufactured with type CEM IIAL cement exhibited slightly lower compressive strength than those obtained with type CEM IIAS and type CEM IIIB cement. This difference could be attributed to environmental temperature; the RACs made with CEM IIAL were produced in spring/summer, while the others were produced in autumn/winter [66]. These factors could influence the setting and curing process of the concrete, directly affecting its final strength.

Figure 3a–c describes the ratio of compressive strength value obtained by each concrete with respect to that of NAC-0.51 at 7, 28 and 56 days. In general, NAC-0.47 achieved between 3% and 17% higher strength than NAC-0.51 concrete at different ages.

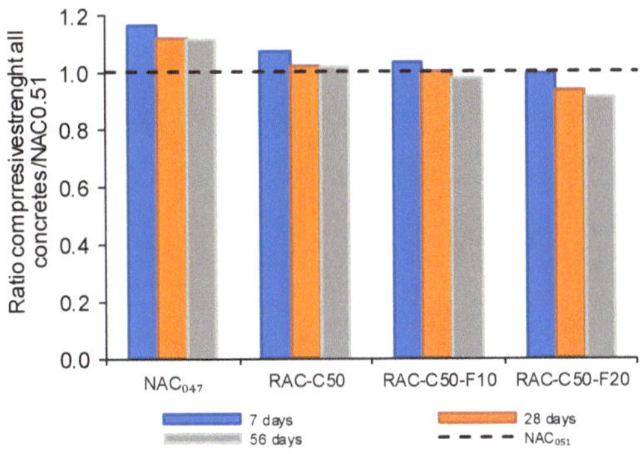

(**a**) CEM-IIAL

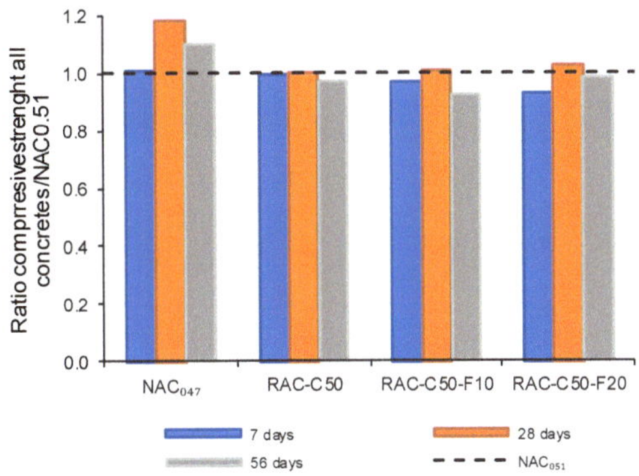

(**b**) CEM-IIAS

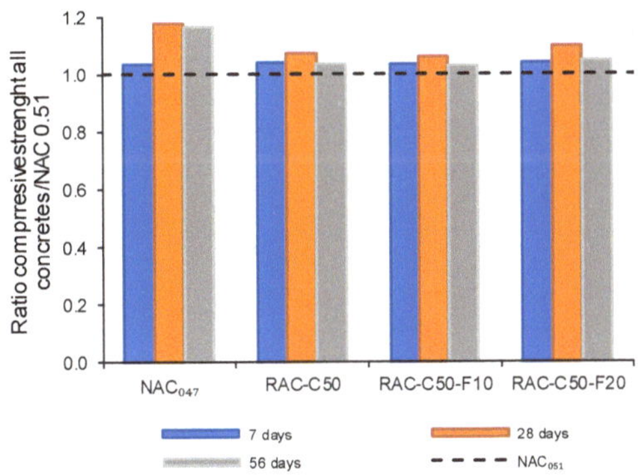

(**c**) CEM-IIIB

Figure 3. Relative compressive strength at all concrete ages produced with cements: (**a**) type CEM IIAL; (**b**) type CEM IIAS; (**c**) type CEM III-B.

Figure 3a describes the ratio when the concretes were produced using the cement CEM IIAL. While the RAC-C50 and RAC-C50-F10 achieved similar values to NAC-0.51 at any age, the RAC-C50-F20 obtained a decrease of up to 8.5% at 56 days. Figure 3b shows the results obtained for concrete produced with CEM IIAS. All the RAC, including the RAC-C50-F20, achieved a similar strength to the NAC-0.51 concrete. Similarly, Figure 3c shows that RAC made with CEM IIIB cement achieved a slightly higher strength than NAC-0.51. The results (Figure 3) show that the different substitution levels of RCA (50% CRCA and up to 20% FRCA) did not have a significant detrimental impact on compressive strength. In addition, as discovered in previous work [36], it was verified that the concrete produced with 50% CRCA and 20% FRCA (RAC-C50-F20) achieved similar compressive strength to concrete containing only CRCA (RAC-C50) and NAC when the RAC and NAC were made with an effective water–cement ratio of 0.47 and 0.52, respectively.

Gao and Wang [67] reported that concrete produced with a higher percentage of FRCA caused a reduction of the compressive strength value. However, several researchers suggested that incorporating 30% FRCA as a replacement for natural sand could still achieve satisfactory properties [5,68]. Evangelista and Brito [68] and Pedro and Brito [4] also confirmed the use of up to 30% FRCA for structural concrete production.

According to the results, the NAC and RAC concrete (produced with 50% CRCA and up to 20% FRCA) achieved similar compressive strength when an effective water–cement ratio of 0.51 and 0.47 was used in concrete production, respectively. They all achieved a suitably designed compressive strength of C30/37 for structural applications.

3.2. Drying Shrinkage

Figures 4a–c and 5a–c illustrate the drying shrinkage (με) and mass loss (%) values, respectively, over 91 days for the NAC-0.51 and RAC concretes produced using cement CEM IIAL (a), CEM IIAS (b) and CEM IIIB (c).

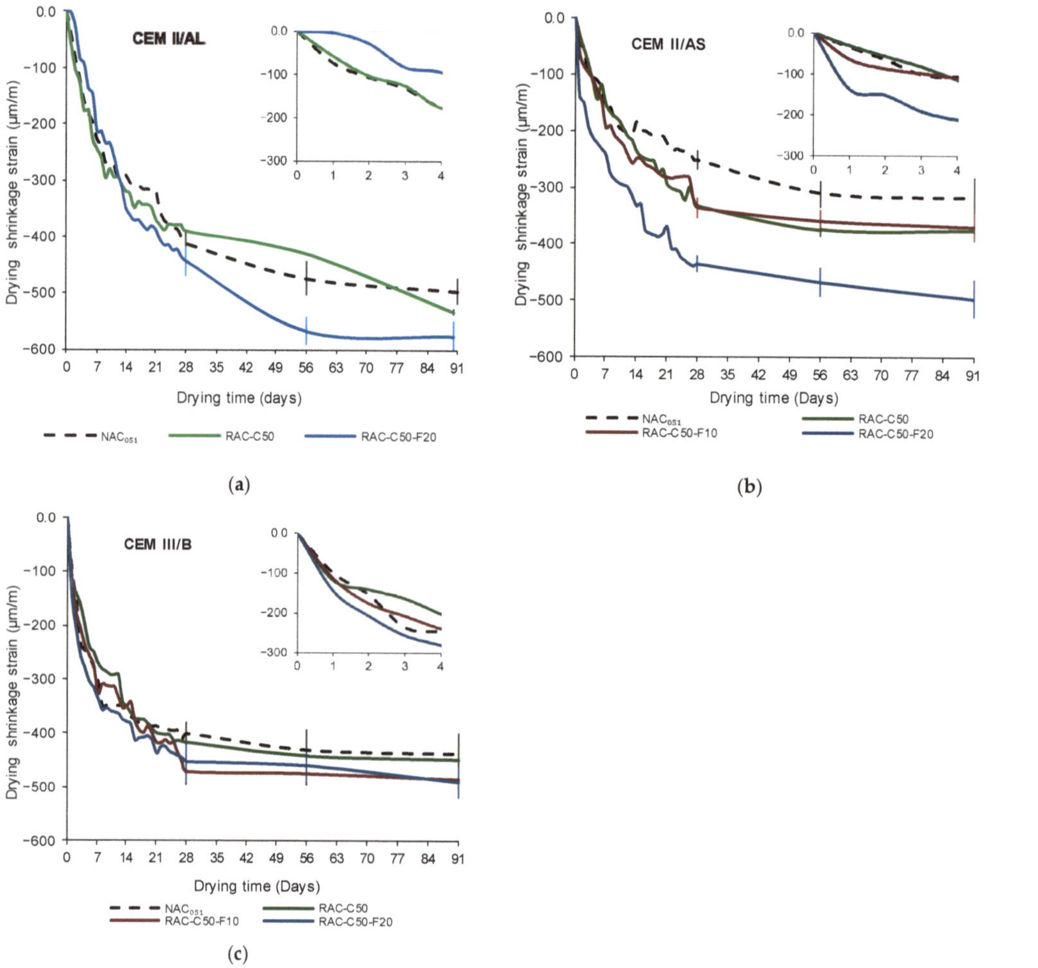

Figure 4. Drying shrinkage development at 91 days: (**a**) CEM II/AL, (**b**) CEM II/AS, (**c**) CEM III/B.

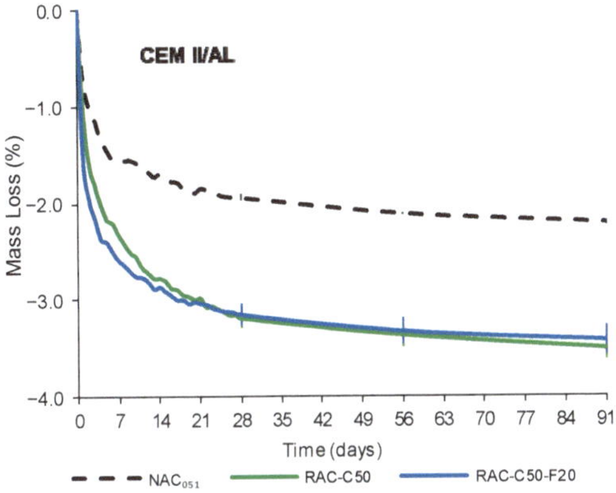

(a)

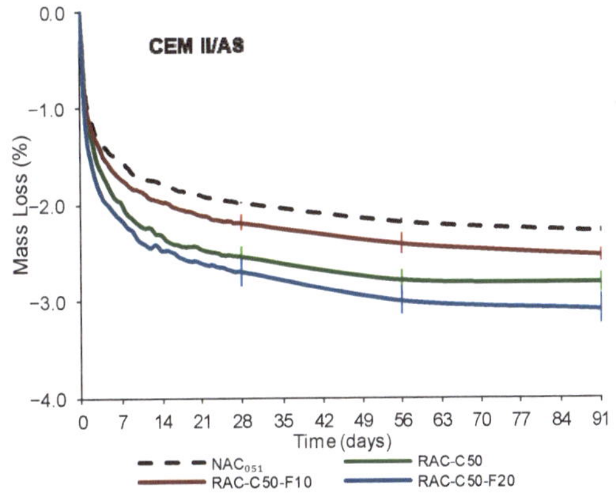

(b)

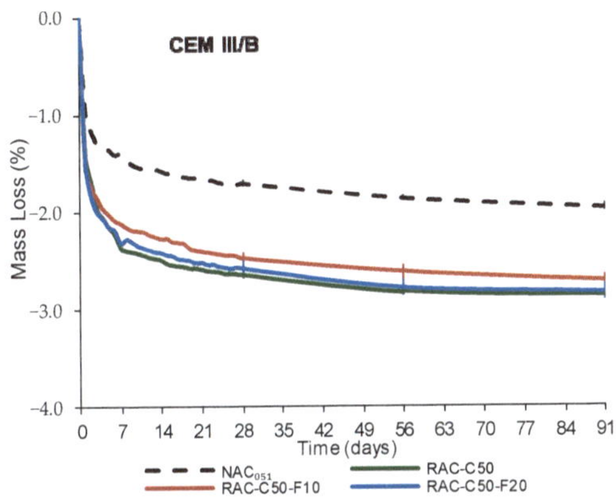

(c)

Figure 5. Mass loss development at 91 days: (**a**) CEM II/AL, (**b**) CEM II/AS, (**c**) CEM III/B.

According to shrinkage values, the concretes produced with CEM IIAL achieved a drying shrinkage value between −496.8 and −562.7 μm/m at 91 days (see Figure 4a). The RAC-C50 concrete achieved a 7.3% higher drying shrinkage value than that of the NAC. Furthermore, the RAC-C50-F20 concrete achieved a 13.3% higher drying shrinkage than NAC concrete (the RAC-C50-F10 concrete was not tested).

According to the results obtained from the concretes produced using CEM IIAS (see Figure 4b), the obtained shrinkage values were between −318.82 and −496.52 μm/m. Figure 4b shows that RAC-C50 and RAC-C50-F10 had similar drying shrinkage values, which were 18% and 16%, respectively, higher than NAC-0.51. Moreover, the RAC-C50-F20 concrete had 56% higher drying shrinkage than NAC-0.51.

In accordance with the results obtained from the concretes produced with CEM IIIB (see Figure 4c), all the concretes achieved similar drying shrinkage values, between −437.3 and −489.6 μm/m. Figure 4c shows that RAC-50 only exceeds 3% of the value obtained by NAC-0.51 concrete, while RAC-C50-10 and RAC-C50-20 were 11% and 12% higher than NAC-0.51, respectively. The use of RAC reduced stiffness caused by the amount of adhered mortar in the recycled aggregate [36,69,70]. This property is closely associated with the modulus of elasticity, which is the principal mechanical indicator of material stiffness [69]. Vintimilla and Etxeberria [36] determined that RAC concrete produced using

CEM IIAL with 50% CRCA and 20% FRCA achieved a 19% lower modulus elasticity and a 14% higher drying shrinkage value than those of NAC. Bendimerad et al. [71] confirmed that the increase in drying shrinkage was associated with a decrease in modulus. According to the achieved results, FRCA strongly influenced and increased the shrinkage value; consequently, the concrete produced with 20% FRCA reached the highest shrinkage value regardless of the obtained compressive strength, as all the concretes exhibited similar compressive strength. Despite this increase in shrinkage, all the values were considered acceptable according to ACI [38], which states that the typical drying shrinkage values of NAC range from −200 to −800 when a high water–cement ratio is employed. In addition, in general, RAC presents deviations similar to those NAC, but between them showed a moderate disperse in most cases.

During the first four days of curing within a drying chamber, the concretes produced with lower clinker (CEM IIIB cement, see Figure 4c) achieved the highest shrinkage value. Drying shrinkage mainly occurs during the early ages and tends to stabilise over time [4]. However, this behaviour is more apparent when SCM is employed as a concrete binder [42,43]. The NAC-0.51 produced with CEM IIIB, CEM IIAL and CEM IIAS cement reached −200 μm/m, −170 μm/m and −110 μm/m, respectively. In addition, the shrinkage value increased as the percentage of RCA used increased. The RAC-C50-F20 produced with CEM IIIB cement achieved a drying shrinkage value of −280 μm/m in the first four days of drying. However, as mentioned above, the shrinkage values stabilised over 28 days.

Figure 5a–c show the mass loss (in %) of each concrete produced with CEM IIAL, CEM IIAS and CEM IIIB, respectively. The three NAC-0.51 mixes achieved a similar mass loss of 2.2%, 2.3%, and 1.9%, respectively. As expected, the concrete mass loss increased as the percentage of recycled aggregates rose. Similarly, for the drying shrinkage value, the RAC-C50-F20 produced with CEM IIAL achieved the highest mass loss with 3.5%, followed by the RAC-C50-F20 produced with CEM IIAS and lastly, CEM IIIB with a mass loss of 3.1% and 2.86%, respectively (see Figure 5). These values are consistent with the results found by other researchers [36,37,52,72].

In all cases, higher drying shrinkage was closely associated with a higher mass loss when comparing concretes that employed the same type of cement. All control concretes exhibited an average mass loss of approximately 2%, while the incorporation of fine and coarse recycled aggregates resulted in an increment of approximately 3% to 3.5%.

The formulations provided by the Structural Concrete Code (SC-BOE) [47] and the Eurocode 2: EN 1992-1-1 (EC-02) [73] were used to predict drying shrinkage in RAC concretes. The calculation method used was described in a previous paper [36].

In order to determine the shrinkage value following the Structural Concrete Code (SC-BOE) [47], the following factors should be considered: the compressive strength at 28 days, concrete specimen size, ambient RH, and the type of cement (the CEM IIAL was considered high early strength (Class CR); the CEM II/AS; and CEM IIIB cements were considered ordinary early strength (Class CN)). However, it must be noted that the SC-BOE does not consider the use of RCA.

However, the use of RCA to estimate the shrinkage value is considered in Eurocode 2: EN 1992-1-1 (EC-02) [73]. The influence of CRCA and FRCA was calculated by applying a specific factor (ηshRA) in the formula to determine the drying shrinkage of RAC. The ηshRA is described as 1+0.8 αRA, where αRA represents the ratio between the recycled aggregates quantity (CRCA and FRCA) and the total quantity of aggregates (coarse and fine aggregates) employed. This factor (ηshRA) is applied when the RCA is employed in replacement of 20–40% of NAs ($0.20 < \alpha RA \leq 0.40$) [36,73]. In this research work, the αRA factors were defined by 0.27, 0.31, and 0.36 for the RAC-C50, RAC-C50-F10, and RAC-C50-F20 concretes, respectively.

To demonstrate the effectiveness of the codes in predicting drying shrinkage in concrete, Figure 6 illustrates the ratio between the experimentally obtained drying shrinkage value of each concrete and the value determined using the following standards: (a) Structural Concrete Code (SC-BOE) and (b) Eurocode 2: EN 1992-1-1 (EC-02).

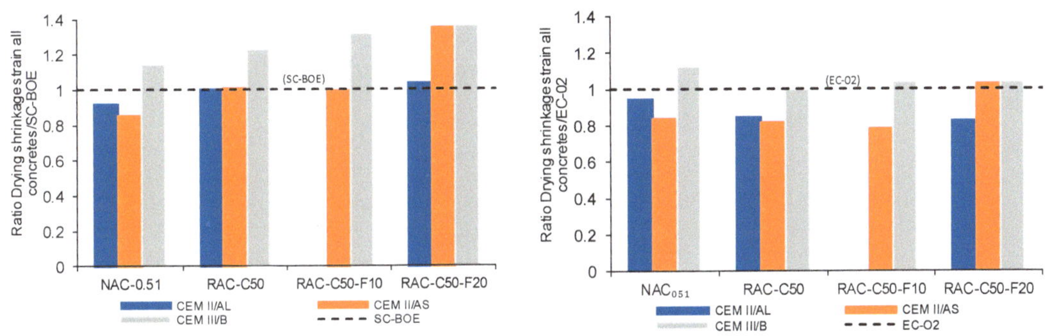

(a) (b)

Figure 6. Shrinkage estimation analysis depicted through ratios of (**a**) experimental results/numerical estimation SC-BOE and (**b**) experimental results/numerical estimation EC-02.

According to Figure 6a, the Spanish Structural Concrete Code (SC-BOE) is not exact in the estimation of NAC-0.51 concrete's shrinkage value. It has been observed that the concrete produced with CEM IIAL and CEM IIAS cements achieved a 10–15% lower shrinkage value than the value estimated by SC-BOE. However, the concrete produced with CEM IIIB cement achieved a 15% higher shrinkage value than that of the value estimated for SC-BOE cement. Although the type of cement was considered in the drying shrinkage calculation for the SC-BOE, it was not considered in a higher early shrinkage caused by high BFS content cement (CEM IIIB), which can influence total drying shrinkage [42,43]. Moreover, the compressive strength at 28 days is the primary parameter considered in SC-BOE estimation; this proved to be similar in all NAC-0.51 concretes. However, as Figure 6a, indicates more parameters besides the compressive strength should be considered. Revilla-Cuesta [41] suggested that a partial correction coefficient should be used for every change in concrete composition, including aspects such as the type of concrete (vibrated, high-performance, or self-compacting), the content of RA, the maturity of the RA, and the addition of an alternative binder.

The SC-BOE adequately estimates the shrinkage values for RAC-C50 and RAC-C50-F20 concretes made with CEM IIAL as well as the RAC-C50 and RAC-C50-F10 concrete made with CEM IIAS as the use of RCA slightly increased the shrinkage value of concretes. However, the RCA-C50-F20 made with CEM IIAS achieved a 38% higher shrinkage value than the value estimated by SC-BOE. The SC-BOE adequately estimated the shrinkage value of RCA-C50-F20 made with CEM IIAL as it achieved a lower strength than any concrete produced with this cement. Consequently, it can be stated that the SC-BOE can adequately estimate the drying shrinkage of concrete produced with 50% CRCA and up to 10% FRCA. However, it estimates a lower shrinkage rate than the value obtained experimentally when 50% CRCA and 20% FRCA are employed in concrete production.

In addition, all the concretes made using CEM IIIB reached a higher shrinkage rate than estimated by SC-BOE. Several researchers have reported that the code estimations could create a ±30% dispersion in the results [41,74]. Moreover, this difference increased when recycled aggregates were used. As mentioned previously, concrete with a high BFS content exhibits higher early shrinkage, which can influence total drying shrinkage [42,43]. Furthermore, the specimens were placed in a climatic chamber after a short period of curing (after 1 day of casting) [42], which also influenced the increase in experimentally obtained shrinkage values.

Figure 6b describes the ratio between the experimental results and the values determined by EC-02. The values estimated by EC-02 for concrete produced with CEM IIAL and CEM IIAS were higher than those obtained experimentally. However, similar to SC-BOE, the NAC-0.51 concrete produced using CEM IIIB, EC-02 estimated a lower shrinkage value than it achieved experimentally. Moreover, as mentioned above, EC-02 considers shrinkage increase as a factor due to the use of recycled aggregates. Consequently, the EC-02 prediction of RAC drying shrinkage is more accurate for the experimental results than the values obtained by the SC-BOE.

3.3. Chloride Ion Penetration

Table 6 describes the chloride ion penetrability values and their standard deviation (values given between brackets) of produced concrete mixtures measured at 28 and 56 days of curing. The ASTMC1202 test classified the chloride ion penetrability as low (1000–2000 Coulomb), moderate (2000–4000 Coulomb), and high (>4000 Coulombs of total passed charge) [75]. This research found that chloride ion penetrability varied significantly according to the type of cement used, as several researchers have stated [32,76–78]. In addition, a direct correlation was observed between the percentage of recycled aggregate replacement ratio and chloride ion penetrability. This is a fact also defined in previous research works [2,4,17].

Table 6. Chloride ion penetrability and the standard deviation (described in brackets) determined in Charge pass in coulombs.

Concrete Types	IIAL (Coulombs)		Δ (%)	IIAS (Coulombs)		Δ (%)	IIIB (Coulombs)		Δ (%)
	28d	56d		28d	56d		28d	56d	
NAC-0.51	5314 (2)	4096 (271)	23	2897 (111)	1976 (129)	32	674 (15)	501 (12)	26
RAC-C50	4479 (441)	4065 (71)	9	2535 (136)	1962 (80)	23	610 (9.0)	503 (8)	18
RAC-C50-F10	6038 (596)	4448 (97)	26	3130 (58)	2293 (5)	27	626 (16)	531 (15)	15
RAC-C50-F20	6401 (569)	4944 (178)	23	4515 (91)	2866 (66)	37	740 (40)	532 (18)	28

( ) Standard deviation. Δ (increase in resistance).

Figure 7 shows the ratio between the charge passed from each concrete produced with respect to 4000 coulombs (the maximum value considered a moderate corrosion risk concrete). Figure 7a,b describe the data at 28 and 56 days, respectively. Figure 7a shows that all concretes manufactured with CEM IIAL exhibit high values of chloride ion penetration. The addition of BFS to cement reduced the ion penetrability of the concrete, as Kopecký and Balázs et al. [78] stated.

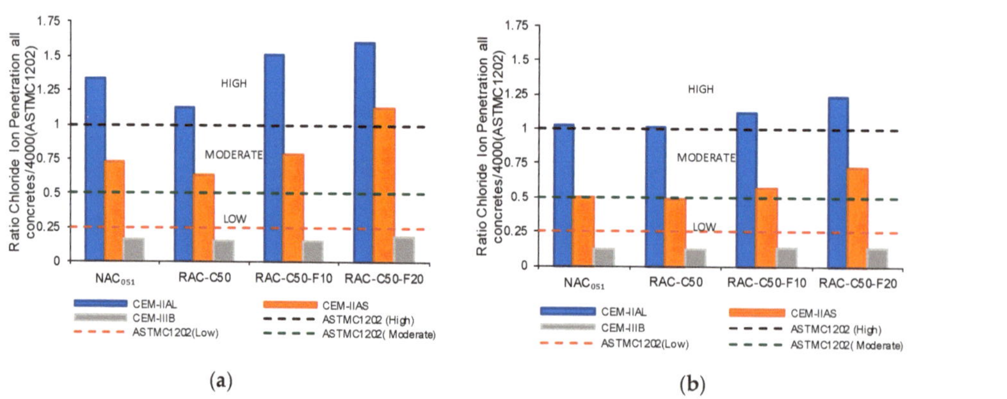

Figure 7. The ratio of chloride ion penetrability (determined in charge pass) for all concretes with respect to maximum value of 4000 Coulombs: (**a**) 28 days and (**b**) 56 days.

Based on the influence of RCA use, the RAC-C50 achieved lower chloride ion penetrability than that of NAC, independent of cement type. However, it must be mentioned that the RAC-C50 and NAC-0.51 were produced with effective water–cement ratios of 0.47 and 0.51, respectively. In agreement with the study conducted by Kopeckó and Balázs et al. [78], it was shown that an increase in the w/c ratio leads to an increase in the depth of chloride penetration while keeping the same cement content constant. Moreover, when FRCA was employed for concrete production, and more evidently with the use of 20% FRCA in the replacement of natural sand, the chloride ion penetrability increased. In addition, this was more evident when cement without BSF (CEMII AL) or low BSF (CEM IIAS) was used for concrete production. The high porosity and microcracks of the old mortar are present on the RCA surface, resulting in the increased permeability of chloride ions [75,79].

Researchers have demonstrated that RAC exhibits more capillary channels than NAC; these are primarily attributed to the introduction of interfacial transition zones (ITZs) between natural aggregates and old cement mortars, as well as the presence of microcracks in the RCA [2,15]. However, Etxeberria et al. [80] have demonstrated that the total charge passed value for all concretes mixed using CEM IIIB cement with different percentages of recycled mixed aggregates (volumes of 0%, 25%, 50% and 100%) ranged from 800 to 1400 coulombs. The authors have also demonstrated that an adequate cement type was necessary to increase chloride ion penetration resistance in concrete production. Sim and Park [20] concluded that the incorporation of FRCA had a minimal impact on chloride ion penetration. They observed that the type of cement used had a more significant influence on concrete performance than the quantity of recycled aggregates. In addition, Table 6 shows that the standard deviation of concretes produced with CEM IIAL was higher than that produced with CEM IIAS. In addition, the concretes produced with CEM IIIB achieved the lowest deviation standard. These findings highlight variability in concrete properties due to different cement types. Notably, CEM IIAL and CEM IIAS exhibited relatively high standard deviations; however, to ensure accurate values, more than two should be used.

After a curing period of 56 days (see Figure 7b), the chloride penetration resistance increased in all the concretes. However, all the concretes produced using CEM IIAL, including NAC, still had very high chloride ion penetrability values. This fact can be attributed to the limestone base of CEM IIAL concrete [81], which had higher chloride ion permeability than those mixes with a higher replacement of SCM [81,82]. As a consequence, it was concluded that CEM IIAL cement was unsuitable for defined application due to its limited ability to resist chloride ion penetration. The obtained results of chloride penetrability in this work were slightly lower than those determined by Etxeberria and Castillo [83], in which the concrete produced with 50% coarse RCA and the same type of cement with effective water–cement ratio of 0.50 and a cement content 350 kg/m^3 obtained 8799 C at 28 days and 6377 C at 56 days. In concrete produced using CEM IIAS, an improvement in chloride ion penetration resistance was observed from 28 to 56 days, with a range between 23% and 37% in all samples. This fact demonstrates that all concrete mixtures achieved a moderate level of resistance in terms of chloride ion penetration. In addition, all the concretes produced using type CEM IIIB cement had low chloride permeability at 28 and 56 days, independently of the percentage of RCA employed. As mentioned above, BFS cement enhances chloride penetration resistance in concrete due to its ability to immobilize chloride ions [21,22].

3.4. Carbonation Resistance

Table 7 summarises the carbonation depth (in mm) and its standard deviation (between brackets), which was determined by testing each produced concrete after 91 days of exposure to 3% CO_2, 57% RH and 20 °C. Although the NAC-0.51 concretes achieved the lowest carbonation depths (in each type of cement concretes), the RAC-C50 and RAC-C50-F10 concretes reached similar values to that of NAC-0.51 concrete, with the exception of the RAC-C50-F10 concrete produced with CEM IIAL, which had a 12.9% higher carbonation depth than the corresponding NAC-0.51. According to Guo et al. [2], the RAC and NAC achieved similar resistance and carbonation depth when the RAC was produced with a lower w/c ratio.

Table 7. Carbonation depth, their standard deviation, and the accelerated and theoretical natural carbonation coefficient of all concretes.

Concrete Types	Carbonation Depth (mm) at 90 Days			Carbonation Coefficient					
				kacc (mm/day 0.5)			knatTHEO (mm/year 0.5)		
	II AL	II AS	III B	II AL	II AS	III B	II AL	II AS	III B
NAC-0.51	7.7 (0.1)	6.0 (0.1)	12.0 (0.4)	0.81	0.65	1.22	1.84	1.48	2.79
RAC-C50	8.0 (0.4)	6.1 (0.2)	12.1 (0.2)	0.84	0.68	1.27	1.9	1.55	2.89
RAC-C50-F10	8.7 (0.2)	6.3 (0)	12.1 (0.1)	0.97	0.68	1.28	2.19	1.55	2.92
RAC-C50-F20	9.8 (0.0	7.2 (0.2)	12.9 (0.1)	1.04	0.78	1.39	2.37	1.77	3.16

Moreover, the use of 20% FRCA in natural sand replacement proved to reduce the carbonation resistance of concrete. The RAC-C50-F20 concrete achieved the highest carbonation depth in each cement type concrete.

The accelerated carbonation coefficient (*Kacc*) of each concrete was calculated under a steady state condition based on Fick's first law of diffusion, represented by Equation (1).

$$Xc(t) = Kacc{\cdot}(t)^{0.5} \tag{1}$$

where *Xc* is the determined carbonation depth (mm), *Kacc* is the carbonation coefficient (mm/day$^{0.5}$), and *t* is time (days). The carbonation depth was determined at 0, 14, 28, 56, 70, and 91 days.

As shown in Table 7, the concretes produced with CEM IIAS cement achieved the lowest Kacc values, followed by those made with the CEM IIAL and CEM IIIB. The RAC-C50-F20 concrete produced with CEM IIAS also achieved a lower Kacc than that of the NAC-0.51 produced with CEM IIAL and CEM IIIB. These findings are consistent with Etxeberria et al. [34], who demonstrated that concretes with CEM IIAS display the lowest values of carbonation depth, regardless of the aggregates used. In addition, the test proved that any concrete produced with CEM IIAL achieved a lower carbonation rate than NAC-0.51 produced with CEM IIIB. Several researchers [33,84] have noted that this increase in the carbonation coefficient in concrete made with CEM IIIB is directly related to a reduced clinker content when compared to CEM IIAL and CEM IIAS cement types, as well as the reduced CO_2 buffering capacity. Consequently, the carbonation resistance of recycled concrete decreases when the employed cement was composed of a high volume of mineral admixtures, reducing the CaO content [20,33,85], resulting in the coarsening of the pore structure and potentially diminishing its durability [30].

In addition, the accelerated carbonation coefficient of RAC-C50 concretes increased by less than 4% compared to that of NAC, regardless of the cement type employed. These findings are in line with several research studies [24,27,33,83]. Moreover, the Kacc of concrete produced with 10% FRCA in the replacement of natural sand, using CEM IIAS and CEM IIIB, were 5.2% and 4.8%, higher, respectively, than that of NAC-0.51. However, the concrete produced with CEM IIAL reached a 19.4% higher value than that of NAC. Furthermore, the use of 20% FRCA in replacement of natural sand increased the Kacc value. The RAC-C50-F20 concretes produced with CEM IIAL, CEM IIAS, and CEM IIIB cements achieved 29.1%, 19.7%, and 13.3% higher *Kacc* values, respectively, than the corresponding NAC-0.51 concrete. The findings demonstrate that incorporating FRCA replacements can lead to notable increases in the carbonation depth.

It is important to note that even when concretes achieved the same compressive strength at 28 days, there were variations in the carbonation depth values depending on the type of cement used.

The theoretical natural carbonation coefficient (*knatTHEO*) is related to the *kacc*, and it can be determined using Equation (2) [86,87]. The obtained values of *knatTHEO* are described for each in Table 7. According to previous wok [34], it was determined that the *knatTHEO* of NAC and RAC was 1.6 and 1.8 times higher, respectively, than knat (natural carbonation rate obtained experimentally), guaranteeing similar behaviour in both types of concretes. Leemann et al. [33] supports the use of accelerated carbonation tests as a method for evaluating resistance under natural conditions. Nevertheless, further research is needed to enhance the predictive capability of carbonation depth in RAC and its accuracy with different levels of RCA [88].

$$\frac{Kacc}{KnatTHEO} = \frac{(\varnothing acc)^{0.5}}{(\varnothing natTHEO)^{0.5}}, \tag{2}$$

where $\varnothing acc$ and $\varnothing natTHEO$ are the CO_2 concentrations in the accelerated carbonation (3%) and natural carbonation processes (425 ppm, in Barcelona), respectively.

Table 8 describes the carbonation depth values obtained by each produced concrete, calculated based on the *knatTHEO* rate, over a lifespan of 50 and 100 years. According to The Spanish Structural Concrete Code (SC-BOE) [47], concrete produced for use under XC3 exposure conditions must have a minimum cover of 20 mm for 50 years and 30 mm for 100 years lifespans. Additionally, for concrete XC4 environment conditions, a minimum cover depth of 25 mm for 50 year and 35 mm for 100 years lifespans are obligatory.

Table 8. Carbonation depth after lifespan of 50 and 100 years.

Concrete Types		Carbonation Depth (50 Years)			Carbonation Depth (100 Years)		
		II AL	II AS	III B	II AL	II AS	III B
NAC-051		13.0	10.4	19.7	18.4	14.8	27.9
RAC-C50		13.5	11.0	20.5	19.0	15.5	28.9
RAC-C50-F10		15.5	11.0	20.6	21.9	15.5	29.2
RAC-C50-F20		16.8	12.5	22.3	23.7	17.7	31.6
Min. Cover (mm) [47]	XC3		20			25	
	XC4		30			35	

According to the obtained results, it was determined that all the concretes manufactured with CEM IIAL and CEM IIAS, including RAC-C50-F20, were suitable to be used in XC3 and XC4 exposure environments, both over 50 and 100-year lifespans.

Based on concretes produced using CEM IIIB, while all the concretes were acceptable to be exposed to the XC4 environment for 50 and 100 years, none of the concrete could be considered adequate for exposure to an XC3 environment, even for 50 years. In addition, the theoretical carbonation depth value of NAC-0.51 was 19.7mm in 50 years, and according to the Structural Concrete Code (SC-BOE), the minimum cover is 20 mm.

Moreover, according to Silva et al. [23], in order to prevent corrosion of concretes exposed to environmental conditions classified as XC3 and XC4 (as specified in the EN 206-1), the maximum accelerated carbonation coefficient should be 35 mm/year$^{0.5}$ for XC3 and 50 mm/year$^{0.5}$ for XC4 when 50 years of service life is considered. Consequently, according to those limits, all the concretes produced complied with the minimum requirements established for XC3 and XC4 environments during a 50-year service life.

4. Conclusions

The results of this study lead to the following conclusions:

- The compressive strength of RAC using 50% CRCA and up to 20% FRCA was lower than that of NA when being produced with the same w/c ratio. Consequently, the RAC must have a 0.04 lower effective water–cement ratio than that of the NAC to achieve the same compressive strength.

- All the RAC using 50% CRCA and up to 20% FRCA achieved a suitably designed compressive strength of C30/37 for structural applications.

The durability properties of RAC concretes, (RAC and NAC proved to have similar compressive strength):

- RAC produced with 50% CRCA and up to 10% FRCA achieved a similar shrinkage value to that of NAC, independent of the cement type employed. Although the use of 20% FRCA increased the drying shrinkage values of the concretes, the total drying shrinkage values obtained by RAC-C50-F20 concretes were acceptable for a structural concrete application.
- A comparative study of both the EC-02 and SC-BOE standards determined that the first provided higher accuracy in predicting drying shrinkage of RAC than the latter, independent of the type of cement employed. However, there are no standards which precisely estimate the shrinkage value of NAC produced with CEM IIIB. This is probably due to the fact that the standards are mainly based on considering the compressive strength at 28 days as the prime factor instead of the initial shrinkage value.
- The use of BFS cement reduced ion chloride penetrability independently of the type of aggregates used. Concretes made with CEM IIIB, including RAC-C50-F20, reached a very low ion penetrability, suitable for structural applications. In addition, the CEM IIAS concretes achieved moderate ion penetrability, except for the RAC-C50-F20, which achieved a high chloride ion penetrability due to the use of 20% FRCA. Moreover, all of the concretes, including NAC with CEM IIAL cement, achieved high penetrability and were unsuitable for structural applications.
- All concrete produced with CEM IIAS, including the RAC-C50-F20, achieved a lower carbonation coefficient than NAC with CEM IIAL cement. However, concretes manufactured with CEM IIAL and CEM IIAS were suitable for use in XC3 and XC4 exposure environments at 50 and 100-year lifespans.
- The carbonation resistance of RAC decreased when the cement employed had a high BFS. In addition, incorporating 20% FRCA can lead to notable increases in carbonation depth.

This study also underscores the critical importance of cement type selection in the durability and strength of structural recycled aggregate concrete, particularly under specific conditions. The concrete using CEM IIAS cement has been shown to meet durability requirements in accordance with shrinkage value, chloride penetrability and carbonation resistance, even with replacement rates of up to 50% CRCA-10% FRCA. However, while the concrete using CEM IIAL achieved low chloride penetration resistance, the concrete produced with CEM IIIB achieved low carbonation resistance, independently of the type of aggregates used for concrete production.

As futures research lines, it is recommended to conduct long-term research to assess the durability and resistance to factors such as corrosion and carbonation in structures built with recycled aggregates.

Author Contributions: Conceptualization, M.E. and C.V.; methodology, M.E.; validation, M.E., C.V. and Z.L.; investigation M.E. and C.V.; resources, M.E.; writing—original draft preparation, C.V.; writing, review and editing, M.E.; project administration, M.E.; funding acquisition, M.E. All authors have read and agreed to the published version of the manuscript.

Funding: First author is granted by Generalitat de Catalunya (GENCAT) and L'Agència de Gestió d'Ajuts Universitaris i de Recerca (AGAUR) for the scholarship "Ajust de suport a departaments i unitats de recerca universitaris per a la contractació de personal investigador predoctoral en formació FI SDUR 2020 (Ref: 2020 FISDU 00576).

Institutional Review Board Statement: Not applicable.

Informed Consent Statement: Not applicable.

Data Availability Statement: No data were used for the research described in the article.

Acknowledgments: The authors wish to thank Hercal Diggers company for their interest and support in the project, LafargeHolcin for the types of cement supply and especially the staff of the Laboratory of Technology of Structures and Materials "Lluis Agulló" of the UPC for their support.

Conflicts of Interest: The authors declare no conflict of interest.

References

1. Alabi, S.A.; Mahachi, J. Chloride ion penetration performance of recycled concrete with different geopolymers. *Mater. Today Proc.* **2020**, *38*, 762–766. [CrossRef]
2. Guo, H.; Shi, C.; Guan, X.; Zhu, J.; Ding, Y.; Ling, T.-C.; Zhang, H.; Wang, Y. Durability of recycled aggregate concrete—A review. *Cem. Concr. Compos.* **2018**, *89*, 251–259. [CrossRef]
3. Plaza, P.; del Bosque, I.S.; Frías, M.; de Rojas, M.S.; Medina, C. Use of recycled coarse and fine aggregates in structural eco-concretes. Physical and mechanical properties and CO_2 emissions. *Constr. Build. Mater.* **2021**, *285*, 122926. [CrossRef]
4. Pedro, D.; de Brito, J.; Evangelista, L. Structural concrete with simultaneous incorporation of fine and coarse recycled concrete aggregates: Mechanical, durability and long-term properties. *Constr. Build. Mater.* **2017**, *154*, 294–309. [CrossRef]
5. Guo, Z.; Chen, C.; Lehman, D.E.; Xiao, W.; Zheng, S.; Fan, B. Mechanical and durability behaviours of concrete made with recycled coarse and fine aggregates. *Eur. J. Environ. Civ. Eng.* **2020**, *24*, 171–189. [CrossRef]
6. Berredjem, L.; Arabi, N.; Molez, L. Mechanical and durability properties of concrete based on recycled coarse and fine aggregates produced from demolished concrete. *Constr. Build. Mater.* **2020**, *246*, 118421. [CrossRef]
7. Le, H.-B.; Bui, Q.-B. Recycled aggregate concretes—A state-of-the-art from the microstructure to the structural performance. *Constr. Build. Mater.* **2020**, *257*, 119522. [CrossRef]
8. Thomas, C.; Setién, J.; Polanco, J.; Alaejos, P.; de Juan, M.S. Durability of recycled aggregate concrete. *Constr. Build. Mater.* **2013**, *40*, 1054–1065. [CrossRef]
9. Kwan, W.H.; Ramli, M.; Kam, K.J.; Sulieman, M.Z. Influence of the amount of recycled coarse aggregate in concrete design and durability properties. *Constr. Build. Mater.* **2011**, *26*, 565–573. [CrossRef]
10. Sereewatthanawut, I.; Prasittisopin, L. Environmental evaluation of pavement system incorporating recycled concrete aggregate. *Int. J. Pavement Res. Technol.* **2020**, *13*, 455–465. [CrossRef]
11. Xing, W.; Tam, V.W.; Le, K.N.; Hao, J.L.; Wang, J. Life cycle assessment of recycled aggregate concrete on its environmental impacts: A critical review. *Constr. Build. Mater.* **2022**, *317*, 125950. [CrossRef]
12. Velardo, P.; del Bosque, I.S.; de Rojas, M.S.; De Belie, N.; Medina, C. Durability of concrete bearing polymer-treated mixed recycled aggregate. *Constr. Build. Mater.* **2022**, *315*, 125781. [CrossRef]
13. Li, X. Recycling and reuse of waste concrete in China. *Resour. Conserv. Recycl.* **2008**, *53*, 36–44. [CrossRef]
14. Adessina, A.; Ben Fraj, A.; Barthélémy, J.-F. Improvement of the compressive strength of recycled aggregate concretes and relative effects on durability properties. *Constr. Build. Mater.* **2023**, *384*, 131447. [CrossRef]
15. Zhan, B.J.; Xuan, D.X.; Zeng, W.; Poon, C.S. Carbonation treatment of recycled concrete aggregate: Effect on transport properties and steel corrosion of recycled aggregate concrete. *Cem. Concr. Compos.* **2019**, *104*, 103360. [CrossRef]
16. Bravo, M.; De Brito, J.; Pontes, J.; Evangelista, L. Durability performance of concrete with recycled aggregates from construction and demolition waste plants. *Constr. Build. Mater.* **2015**, *77*, 357–369. [CrossRef]
17. Evangelista, L.; de Brito, J. Durability performance of concrete made with fine recycled concrete aggregates. *Cem. Concr. Compos.* **2010**, *32*, 9–14. [CrossRef]
18. Lotfi, S.; Eggimann, M.; Wagner, E.; Mróz, R.; Deja, J. Performance of recycled aggregate concrete based on a new concrete recycling technology. *Constr. Build. Mater.* **2015**, *95*, 243–256. [CrossRef]
19. Sucic, A.; Lotfy, A. Effect of new paste volume on performance of structural concrete using coarse and granular recycled concrete aggregate of controlled quality. *Constr. Build. Mater.* **2016**, *108*, 119–128. [CrossRef]
20. Sim, J.; Park, C. Compressive strength and resistance to chloride ion penetration and carbonation of recycled aggregate concrete with varying amount of fly ash and fine recycled aggregate. *Waste Manag.* **2011**, *31*, 2352–2360. [CrossRef]
21. Pinto, S.R.; da Luz, C.A.; Munhoz, G.S.; Medeiros-Junior, R.A. Resistance of phosphogypsum-based supersulfated cement to carbonation and chloride ingress. *Constr. Build. Mater.* **2020**, *263*, 120640. [CrossRef]
22. Nishida, T.; Otsuki, N.; Ohara, H.; Garba-Say, Z.M.; Nagata, T. Some Considerations for Applicability of Seawater as Mixing Water in Concrete. *Sustain. Constr. Mater. Technol.* **2013**, 27. [CrossRef]
23. Silva, R.; Neves, R.; de Brito, J.; Dhir, R. Carbonation behaviour of recycled aggregate concrete. *Cem. Concr. Compos.* **2015**, *62*, 22–32. [CrossRef]
24. Etxeberria, M. Evaluation of Eco-Efficient Concretes Produced with Fly Ash and Uncarbonated Recycled Aggregates. *Materials* **2021**, *14*, 7499. [CrossRef]
25. Xiao, J.; Lei, B.; Zhang, C. On carbonation behavior of recycled aggregate concrete. *Sci. China Technol. Sci.* **2012**, *55*, 2609–2616. [CrossRef]
26. Pedro, D.; de Brito, J.; Evangelista, L. Influence of the use of recycled concrete aggregates from different sources on structural concrete. *Constr. Build. Mater.* **2014**, *71*, 141–151. [CrossRef]
27. Zeng, X. Progress in the research of carbonation resistance of RAC. *Constr. Build. Mater.* **2020**, *230*, 116976. [CrossRef]

28. Sideris, K.K.; Manita, P. Influence of Blended Cements on the Service Life of Reinforced Concrete Structures against Carbonation Induced Corrosion. In *International RILEM Conference on Synergising Expertise towards Sustainability and Robustness of CBMs and Concrete Structures*; Springer Nature Switzerland: Cham, Switzerland, 2023; pp. 1048–1054. [CrossRef]
29. Kirthika, S.; Singh, S. Durability studies on recycled fine aggregate concrete. *Constr. Build. Mater.* **2020**, *250*, 118850. [CrossRef]
30. von Greve-Dierfeld, S.; Lothenbach, B.; Vollpracht, A.; Wu, B.; Huet, B.; Andrade, C.; Medina, C.; Thiel, C.; Gruyaert, E.; Vanoutrive, H.; et al. Understanding the carbonation of concrete with supplementary cementitious materials: A critical review by RILEM TC 281-CCC. *Mater. Struct.* **2020**, *53*, 136. [CrossRef]
31. Borges, P.H.; Costa, J.O.; Milestone, N.B.; Lynsdale, C.J.; Streatfield, R.E. Carbonation of CH and C–S–H in composite cement pastes containing high amounts of BFS. *Cem. Concr. Res.* **2010**, *40*, 284–292. [CrossRef]
32. Flegar, M.; Bašić, A.D.; Bukvić, O.; Serdar, M. Carbonation of Concretes with Different Binder Chemistry–A Comparative Analysis. In *International RILEM Conference on Synergising Expertise towards Sustainability and Robustness of Cement-Based Materials and Concrete Structures. SynerCrete 2023*; Jędrzejewska, A., Kanavaris, F., Azenha, M., Benboudjema, F., Schlicke, D., Eds.; RILEM Bookseries; Springer: Cham, Switzerland, 2023; Volume 44. [CrossRef]
33. Leemann, A.; Loser, R. Carbonation resistance of recycled aggregate concrete. *Constr. Build. Mater.* **2019**, *204*, 335–341. [CrossRef]
34. Etxeberria, M. Carbonation resistance of recycled aggregate concrete using different cement types. In *RILEM Bookseries*; Springer: Cham, Switzerland, 2023; Volume 44, pp. 1065–1076. [CrossRef]
35. Silva, R.; de Brito, J.; Dhir, R. Prediction of the shrinkage behavior of recycled aggregate concrete: A review. *Constr. Build. Mater.* **2015**, *77*, 327–339. [CrossRef]
36. Vintimilla, C.; Etxeberria, M. Limiting the maximum fine and coarse recycled aggregates-Type A used in structural concrete. *Constr. Build. Mater.* **2023**, *380*, 131273. [CrossRef]
37. Gonzalez-Corominas, A.; Etxeberria, M. Effects of using recycled concrete aggregates on the shrinkage of high performance concrete. *Constr. Build. Mater.* **2016**, *115*, 32–41. [CrossRef]
38. Mcdonald, D.B.; Brooks, J.J.; Burg, R.G.; Daye, M.A.; Gardner, N.J.; Novak, L.C. Report on Factors Affecting Shrinkage and Creep of Hardened Concrete Reported by ACI Committee 209, 2014. Available online: https://scholar.google.es/scholar?hl=es&as_sdt=0%2C5&q=Mcdonald%2C+D.B.%3B+Brooks%2C+J.J.%3B+Burg%2C+R.G.%3B+Daye%2C+M.A.%3B+Gardner%2C+N.J.%3B+Novak%2C+L.C.+Report+on+Factors+Affecting+Shrinkage+and+Creep+of+Hardened+Concrete+Reported+by+ACI+Committee+20 (accessed on 14 September 2023).
39. Şimşek, O.; Sefidehkhan, H.P.; Gökçe, H. Performance of fly ash-blended Portland cement concrete developed by using fine or coarse recycled concrete aggregate. *Constr. Build. Mater.* **2022**, *357*, 129431. [CrossRef]
40. Sosa, M.E.; Zega, C.J. Experimental and Estimated Evaluation of Drying Shrinkage of Concrete Made with Fine Recycled Aggregates. *Sustainability* **2023**, *15*, 7666. [CrossRef]
41. Revilla-Cuesta, V.; Evangelista, L.; de Brito, J.; Skaf, M.; Manso, J.M. Shrinkage prediction of recycled aggregate structural concrete with alternative binders through partial correction coefficients. *Cem. Concr. Compos.* **2022**, *129*, 104506. [CrossRef]
42. Yang, J.; Wang, Q.; Zhou, Y. Influence of Curing Time on the Drying Shrinkage of Concretes with Different Binders and Water-to-Binder Ratios. *Adv. Mater. Sci. Eng.* **2017**, *2017*, 2695435. [CrossRef]
43. Hooton, R.; Stanish, K.; Angel, J.; Prusinski, J. The Effect of Ground Granulated Blast Furnace Slag (Slag Cement) on the Drying Shrinkage of Concrete—A Critical Review of the Literature. In *SP-263 Slag Cement Concrete*; American Concrete Institute: Indianapolis, IN, USA, 2009; pp. 79–94. [CrossRef]
44. *EN 197-1:2001*; Cement Part 1: Composition, Specifications and Conformity Criteria for Common Cements. European Committee for Standardization: Brussels, Belgium, 2011.
45. UNE-EN 933-1; Tests for Geometrical Properties of Aggregates Part 1: Determination of Particle Size Distribution—Sieving Method. 2012. Available online: www.aenor.es (accessed on 14 September 2023).
46. UNE-EN 12620: 2003+A1; Aggregates for Concrete, 2009. Available online: www.aenor.es (accessed on 14 September 2023).
47. Boletín Oficial del Estado (BOE), Structural Concrete Code. 2021. Available online: https://www.boe.es/eli/es/rd/2021/06/29/470 (accessed on 14 September 2023).
48. UNE-EN 1097-6; Determination of Particle Density and Water Absorption, 2014. Available online: www.aenor.es (accessed on 14 September 2023).
49. UNE-EN 933-8: 2012+A1; Assessment of Fines—Sand Equivalent Test, 2015. Available online: www.aenor.es (accessed on 14 September 2023).
50. UNE-EN 933-3; Determination of Particle Shape—Flakiness Index, 2012. Available online: www.aenor.es (accessed on 14 September 2023).
51. *UNE 146508*; Determination of the Alkali-Silica and Alkali-Silicate Potential Reactivity of Aggregates. Accelerated Mortar bar Test. Spanish Standard: Madrid, Spain, 2018.
52. Etxeberria, M.; Konoiko, M.; Garcia, C.; Perez, M. Water-Washed Fine and Coarse Recycled Aggregates for Real Scale Concretes Production in Barcelona. *Sustainability* **2022**, *14*, 708. [CrossRef]
53. UNE-EN 933-11; Classification Test for the Constituents of Coarse Recycled Aggregate, 2009. Available online: www.aenor.es (accessed on 14 September 2023).
54. *UNE-EN 206: 2013+A1*; Specification, Performance, Production and Conformity. Spanish Standard: Madrid, Spain, 2016.

55. Nedeljković, M.; Visser, J.; Šavija, B.; Valcke, S.; Schlangen, E. Use of fine recycled concrete aggregates in concrete: A critical review. *J. Build. Eng.* **2021**, *38*, 102196. [CrossRef]
56. Silva, S.; Evangelista, L.; de Brito, J. Durability and shrinkage performance of concrete made with coarse multi-recycled concrete aggregates. *Constr. Build. Mater.* **2021**, *272*, 121645. [CrossRef]
57. Wagih, A.M.; El-Karmoty, H.Z.; Ebid, M.; Okba, S.H. Recycled construction and demolition concrete waste as aggregate for structural concrete. *HBRC J.* **2013**, *9*, 193–200. [CrossRef]
58. Zega, C.J.; Di Maio, Á.A. Use of recycled fine aggregate in concretes with durable requirements. *Waste Manag.* **2011**, *31*, 2336–2340. [CrossRef]
59. Sosa, M.E.; Zaccardi, Y.A.V.; Zega, C.J. A critical review of the resulting effective water-to-cement ratio of fine recycled aggregate concrete. *Constr. Build. Mater.* **2021**, *313*, 125536. [CrossRef]
60. *UNE EN 12350-2*; Testing Fresh Concrete—Part 2: Slump Test. Spanish Standard: Madrid, Spain, 2019.
61. *UNE-EN-12390-3*; Testing Hardened Concrete—Part 3: Compressive Strength of Test Specimens. Spanish Standard: Madrid, Spain, 2019.
62. *UNE EN 12390-16*; Determination of the Shrinkage of Concrete. Spanish Standard: Madrid, Spain, 2019.
63. *ASTM C1202*; Standard Test Method for Electrical Indication of Concrete's Ability to Resist Chloride Ion Penetration. American Society for Testing and Materials: West Conshohocken, PA, USA, 2012; pp. 1–8.
64. *UNE-EN12390-12*; Determination of the Carbonation Resistance of Concrete—Accelerated Carbonation Method. Spanish Standard: Madrid, Spain, 2020.
65. *UNE EN 14630:2007*; Determination of Carbonation Depth in Hardened Concrete by the Phenolphthalein Method. Spanish Standard: Madrid, Spain, 2007.
66. Ortiz, J.; Aguado, A.; Agulló, L.; García, T. Influence of environmental temperatures on the concrete compressive strength: Simulation of hot and cold weather conditions. *Cem. Concr. Res.* **2005**, *35*, 1970–1979. [CrossRef]
67. Gao, D.; Wang, F. Effects of recycled fine aggregate and steel fiber on compressive and splitting tensile properties of concrete. *J. Build. Eng.* **2021**, *44*, 102631. [CrossRef]
68. Evangelista, L.; de Brito, J. Mechanical behaviour of concrete made with fine recycled concrete aggregates. *Cem. Concr. Compos.* **2007**, *29*, 397–401. [CrossRef]
69. Kim, J. Influence of quality of recycled aggregates on the mechanical properties of recycled aggregate concretes: An overview. *Constr. Build. Mater.* **2022**, *328*, 127071. [CrossRef]
70. Lye, C.-Q.; Dhir, R.K.; Ghataora, G.S. Shrinkage of recycled aggregate concrete. *Mater. Struct.* **2018**, *51*, 129. [CrossRef]
71. Bendimerad, A.; Delsaute, B.; Rozière, E.; Staquet, S.; Loukili, A. Advanced techniques for the study of shrinkage-induced cracking of concrete with recycled aggregates at early age. *Constr. Build. Mater.* **2020**, *233*, 117340. [CrossRef]
72. Etxeberria, M.; Gonzalez-Corominas, A. The assessment of ceramic and mixed recycled aggregates for high strength and low shrinkage concretes. *Mater. Struct.* **2018**, *51*, 129. [CrossRef]
73. *EN 1992-1-1:2021*; Eurocode 2: Design of Concrete Structures—Part 1-1: General Rules—Rules for Buildings, Bridges and Civil Engineering Structures. European Union: Brussels, Belgium, 2021.
74. Wang, Q.; Geng, Y.; Wang, Y.; Zhang, H. Drying shrinkage model for recycled aggregate concrete accounting for the influence of parent concrete. *Eng. Struct.* **2019**, *202*, 109888. [CrossRef]
75. Zhu, P.; Hao, Y.; Liu, H.; Wang, X.; Gu, L. Durability evaluation of recycled aggregate concrete in a complex environment. *J. Clean. Prod.* **2020**, *273*, 122569. [CrossRef]
76. Vázquez, E.; Barra, M.; Aponte, D.; Jiménez, C.; Valls, S. Improvement of the durability of concrete with recycled aggregates in chloride exposed environment. *Constr. Build. Mater.* **2014**, *67*, 61–67. [CrossRef]
77. Mohammed, T.U.; Rahman, M. Effects of cement types on chloride ingress in concrete. In Proceedings of the 3rd ACF Symposium on Assessment and Intervention of Existing Structures, Sapporo, Japan, 10–11 September 2019.
78. Kopecskó, K.; Balázs, G.L. Concrete with Improved Chloride Binding and Chloride Resistivity by Blended Cements. *Adv. Mater. Sci. Eng.* **2017**, *2017*, 7940247. [CrossRef]
79. Wu, J.; Zhang, Y.; Zhu, P.; Feng, J.; Hu, K. Mechanical Properties and ITZ Microstructure of Recycled Aggregate Concrete Using Carbonated Recycled Coarse Aggregate. *J. Wuhan Univ. Technol. Mater. Sci. Ed.* **2018**, *33*, 648–653. [CrossRef]
80. Etxeberria, M.; Gonzalez-Corominas, A. Properties of Plain Concrete Produced Employing Recycled Aggregates and Sea Water. *Int. J. Civ. Eng.* **2018**, *16*, 993–1003. [CrossRef]
81. Githachuri, K.; Alexander, M.G. Durability performance potential and strength of blended Portland limestone cement concrete. *Cem. Concr. Compos.* **2013**, *39*, 115–121. [CrossRef]
82. Otieno, M.; Beushausen, H.; Alexander, M. Effect of chemical composition of slag on chloride penetration resistance of concrete. *Cem. Concr. Compos.* **2013**, *46*, 56–64. [CrossRef]
83. Etxeberria, M.; Castillo, S. How the Carbonation Treatment of Different Types of Recycled Aggregates Affects the Properties of Concrete. *Sustainability* **2023**, *15*, 3169. [CrossRef]
84. Leemann, A.; Nygaard, P.; Kaufmann, J.; Loser, R. Relation between carbonation resistance, mix design and exposure of mortar and concrete. *Cem. Concr. Compos.* **2015**, *62*, 33–43. [CrossRef]
85. Pedro, D.; de Brito, J.; Evangelista, L. Performance of concrete made with aggregates recycled from precasting industry waste: Influence of the crushing process. *Mater. Struct.* **2015**, *48*, 3965–3978. [CrossRef]

86. Parrott, L.J. Cement and Concrete Association; Building Research Establishment. A Review of Carbonation in Reinforced Concrete; Cement and Concrete Association. 1987. Available online: http://worldcat.org/isbn/0721013651 (accessed on 15 June 2023).
87. Heede, P.V.D.; De Belie, N. A service life based global warming potential for high-volume fly ash concrete exposed to carbonation. *Constr. Build. Mater.* **2014**, *55*, 183–193. [CrossRef]
88. Carević, V.; Ignjatović, I.; Dragaš, J. Model for practical carbonation depth prediction for high volume fly ash concrete and recycled aggregate concrete. *Constr. Build. Mater.* **2019**, *213*, 194–208. [CrossRef]

sustainability

MDPI

Review

Application of Waste Tire in Construction: A Road towards Sustainability and Circular Economy

Mohammad R. Hassan [1,*] and Denis Rodrigue [2]

1 Fire Safety, Construction Research Center, National Research Council Canada, 1200 Montreal Road, Ottawa, ON K1A 0R6, Canada
2 Department of Chemical Engineering, Université Laval, 1065 Avenue de la Médecine, Quebec, QC G1V 0A6, Canada
* Correspondence: rokib.hassan@nrc-cnrc.gc.ca

Abstract: The global demand for rubber is on a steady rise, which is driven by the increasing production of automobiles and the growing need for industrial, medical, and household products. This surge in demand has led to a significant increase in rubber waste, posing a major global environmental challenge. End-of-life tire (ELT) is a primary source of rubber waste, having significant environmental hazards due to its massive stockpiles. While landfilling is a low-cost and easy-to-implement solution, it is now largely prohibited due to environmental concerns. Recently, ELT rubber waste has received considerable attention for its potential applications in civil engineering and construction. These applications not only enhance sustainability but also foster a circular economy between ELT rubber waste with the civil engineering and construction sectors. This review article presents a general overview of the recent research progress and challenges in the civil engineering applications of ELT rubber waste. It also discusses commercially available recycled rubber-based construction materials, their properties, testing standards, and certification. To the best of the authors' knowledge, this is the first time such a discussion on commercial products has been presented, especially for civil engineering applications.

Keywords: waste tire; construction materials; circular economy; sustainability

Citation: Hassan, M.R.; Rodrigue, D. Application of Waste Tire in Construction: A Road towards Sustainability and Circular Economy. *Sustainability* **2024**, *16*, 3852. https://doi.org/10.3390/su16093852

Academic Editors: Anibal C. Maury-Ramirez and Nele De Belie

Received: 11 March 2024
Revised: 16 April 2024
Accepted: 27 April 2024
Published: 3 May 2024

1. Introduction

Recently, climate change has emerged as the single biggest challenge in the 21st century [1]. Hence, sustainable development, recycling, and circular economy have become popular research topics within the scientific community worldwide [2]. The global demand for vehicles has been increasing at a significant pace as a result of continued growth in population and social economy [3]. This increase in demand creates a growing concern about generating high levels of ELT wastes, as their disposal causes several environmental issues (Figure 1) [3–26]. According to a recent report by the International Market Analysis Research and Consulting (IMARC) Group, the global tire market size was estimated to be approximately 2.3 billion units in 2022 [27]. The same report also forecasted that the market will reach approximately 2.7 billion units by 2028 with a compound annual growth rate (CAGR) of 2.8% between 2023 and 2028 [27]. Amin et al. have recently reported that approximately 1.5 billion ELTs are generated globally each year [25,28]. This number could potentially reach up to 5 billion ELT by 2030 [6]. In the past, ELT wastes were mostly landfilled, stockpiled, and incinerated [24]. According to a recent study, the global management of ELT wastes now includes recycling (3–15%), reuse (5–23%), and landfilling and stockpiling (20–30%), as well as incineration (25–60%) [25,29].

Figure 1. Stockpiled ELT wastes in a dumpsite.

However, landfilling, stockpiling, and incineration of ELT wastes have severe environmental impacts [24,25]. For example, the landfilling of ELT wastes could result in the leaching of toxic substances and heavy metals into the ecosystem contaminating soil, groundwater, and underground water resources [24,26]. It was also reported that landfilled ELT wastes can trap gases and create punching holes in the landfill cover [23]. The stockpiled ELT wastes can store water as they are mostly empty cavities and impermeable. This trapped water can act as a breeding habitat for mosquitoes, bacteria, mold, and rodents, becoming a health hazard to nearby communities [12,23]. Furthermore, stockpiled ELT wastes create significant fire hazards, as rubber is highly combustible (petroleum-based compounds). The rubber in ELT wastes can serve as a fuel, leading to a prolonged fire event and contributing to greenhouse gas (GHG) emissions (Figure 2) [30–32]. Upon burning, ELT wastes have the potential to generate black smoke, soot, and odor, as well as cause severe air pollution due to the release of toxic gases including dioxin [31,32]. These fires are also challenging to extinguish, since the combustion of a large amount of stockpiled ELT was shown to last for several weeks to months [23,24,33]. For example, a stockpiled ELT fire in Haggerville (Canada) occurred in 1990. This fire lasted for 17 days and forced the evacuation of 1700 people due to severe air pollution and contamination of nearby water wells [33]. Similarly, in 2012, another fire incident in Iowa City (USA) lasted for 18 days and caused severe air pollution [24]. These prolonged fire events resulted from the presence of highly flammable hydrocarbons in ELT wastes and their low thermal conductivity making them difficult to cool down [3,32]. Although the fire can be extinguished from the outside, the tires can still burn from the inside and restart the fire. In addition, the residues generated from burned ELT have the potential to contaminate the soil and groundwater [30]. A study reported that spraying water on an ELT fire caused an increase in pyrolytic oil generation, resulting in the leaching of contaminants off-site [23].

Incineration of ELT wastes could be the cheapest and easiest disposal approach. However, this approach can also have severe environmental impact [3,6,25]. In particular, incineration of ELT wastes is known to release carbon monoxide, sulfur dioxide, nitrogen oxides, hydrogen chloride, butadiene, and other toxic aromatic compounds. It was estimated that the incineration of 1 ton of ELT waste could release about 450 kg of poisonous gases and 270 kg of soot into the atmosphere [25,34]. On the other hand, several different ELT waste recycling processes have been developed, including (i) pyrolysis, (ii) fuel in cement kilns or energy recovery (tire-derived fuel, TDF), (iii) reclamation, (iv) civil engineering, and (v) granulation (ground tire rubber, GTR). However, some recycling processes have their drawbacks. For example, pyrolysis and TDF are expensive and not economically sustainable because these processes generate carbon black (CB) and contribute to GHG emissions. The CB generated from these processes is more expensive and poor quality compared to virgin CB produced from petroleum [6,9]. In addition, pyrolysis requires large processing plants having high operational costs and limited large-scale industrial applications [26]. In contrast, the use of ELT wastes in construction has become very popular in recent years due to their attractive and promising material properties such as long-term durability/stability, good insulation (acoustic and thermal), low density, low earth pressure, high compressibility, and good drainage capability.

Figure 2. A fire incident in an ELT stockpile site generates black smoke and other pollutants.

Managing ELT wastes is highly important given the amount and complexity of these materials. It is estimated that a car contains about 6.7% rubber parts, of which 65.5% are associated with the tires [35]. In general, tires are composed of 7 major parts, namely tread, belts, sidewalls, carcass, inner liner, beads, and bead filler [36]. These parts are made from up to 12 and 20 raw materials designed for passenger cars and trucks, respectively [37]. The typical raw materials used for making these tire parts are listed in Table 1. According to Table 1, it is clear that the main materials to recycle, after their separation, are the rubber in the form of GTR and the reinforcements, including metal and textiles. To this end, several studies have been performed on each type of raw material to find applications to valorize these residues, and a few review articles have been recently published for recycling ELT metals [38–40] and textiles [20,41]. However, the main tire raw material is rubber, which represents about half of a tire's weight. This rubber has been used in the form of GTR either alone or blended with other matrices to produce different compounds and/or products. Some examples are thermoset [42,43] and thermoplastic [44,45] matrices, especially to produce thermoplastic elastomers (TPE) [37,46].

Table 1. List of the main tire raw materials and their concentration (wt.%). The values were compiled from references [20,36,37].

No.	Raw Materials	Content (wt.%)	
		Passenger Car	**Truck**
1.	Rubbers	41–48	41–45
2.	Carbon blacks	21–28	20–28
3.	Metals	13–16	20–27
4.	Textiles	4–6	0–10
5.	Additives	10–12	7–10

Several review articles have been published focusing on the applications of different tire raw materials for their valorization [26,35,45]. This review article focuses on the recent research and development (R&D) progress in civil engineering applications of ELT waste rubber. A discussion on commercially available recycled rubber-based construction materials is also presented. To the best of the authors' knowledge, this is the first time that such a discussion of commercial products is presented with different R&D applications, including asphalt, concrete, sand, and earth/soil. To limit the scope of this review, the subject of functional upcycling [47], including rubber devulcanization [36,48], is not included. The primary objective of this work is to offer a general overview of various applications of ELT and GTR beyond their traditional use as fillers in polymer matrices. Specifically, it highlights a broad spectrum of uses in civil engineering, summarizing current achievements and different possibilities for future research and development.

2. R&D Progress in Civil Engineering Applications

As described above, ELT wastes can be regenerated into different raw materials, which need to be valorized (Table 1). This section focuses on recent research advances in the use of GTR in civil engineering applications, including construction. Since a great deal of literature has previously been published on each subject, the following discussion focuses on the most recent advances to present current R&D trends.

2.1. Asphalt

The addition of GTR into asphalt has been performed for a very long time. These rubberized asphalts were developed to improve the matrix's behavior under different conditions. Recent review articles provide a general overview of the available extensive literature on the subject [49,50].

Different methods (wet, high shear) have been studied to introduce rubber particles into bituminous matrices to improve the service performances of the final blends. The main parameters were temperature, shear intensity, and time. The GTR content (2–50% wt.) and their particle size (0.1–10 mm) were also found to be very important in defining the final blend performance. One important property of GTR is swelling, which can occur in different chemical environments. The swelling results in increasing the GTR particle size, enabling a better interaction with the matrix under a wide range of conditions (pressure, temperature, etc.). Surface roughness must also be accounted for to obtain a complete understanding of all the factors involved, especially for mechanical and rheological (workability) properties.

GTR can also be treated before mixing to obtain better interfacial compatibility. This can be carried out via microwave, plasma, and radiation (UV), as well as chemical modifications (acid, base, solvent, etc.) and grafting (coupling agent). Based on GTR content, a variety of properties can be improved, including ductility, penetration, softening point, toughness, and viscosity. These property improvements can lead to better performance in terms of bending, creep, elastic recovery, fatigue, rutting, thermal cracking, and high/low-temperature storage stability. Nevertheless, the type of rubber and its composition will also affect the overall behavior of the blends, especially for low-temperature applications. The properties of the blends can also be improved by adding a third ingredient such as char (plastics) [51], virgin rubber [52], or natural/synthetic/recycled fibers [53]. GTR can also be added to asphalt–concrete/cement mixtures. Recently, Alsheyab and coworkers reported that the addition of GTR to asphalt–concrete mixtures improved Marshall stability, void mineral aggregate and air voids, water sensitivity, and creep resistance [54]. They conducted a ladder study on GTR content (5–15%) to optimize the performance of asphalt–concrete mixtures. The 10% GTR content in asphalt–concrete mixtures provided the best performance.

2.2. Concrete

Concrete is a highly produced material because of its general application in civil engineering. This is why the material is interesting, because even at low concentrations, there are several possibilities for any replacement. In the past, different types of waste have been investigated, and recycled crumb rubber was one of them [55]. All these materials have been classified as replacements or additives. Due to the wide interest in rubberized concrete, several hundred articles (above 1100 based on the Web of Science, September 2023, combining "recycled rubber" with "concrete") have been published, which can be regrouped into a dozen review articles over the last two years [56–63]. The main results are reported here.

The addition of rubber particles in concrete formulations is mainly to improve the durability of the matrix as the particles are elastic and can easily be deformed under stress. The particles are not hygroscopic and provide better resistance towards water infiltration as well as carbonation and chloride ions to protect structural elements such as rebar (steel). In all cases, extended life is generated for the structures. In most cases, lower sound/thermal conductivity (better sound/thermal insulation) is observed to satisfy ever-increasing building requirements (energy savings). Better durability was also observed in terms of cyclic/dynamic deformation (fatigue and freeze–thaw), but mitigated results have been reported for both increased and decreased drying shrinkage, which might be a function of the particle size distribution. Although workability (viscosity) and most mechanical properties decrease with increasing GTR content, the impact strength usually increases as the elastic rubber particles can deform and absorb the energy before failure. Finally, GTR has a lower density than the neat matrix, leading to weight saving as the content increases. The optimum rubber content is usually around 5–15% wt., but a wide range of particle sizes (0.1–20 mm) have been investigated depending on the property to optimize. Once again, the properties of rubberized concrete can be improved by performing a surface treatment (chemical and thermal) on the rubber particles before mixing [64,65]. Another possibility is the addition of a third ingredient (also of recycled origin), such as thermoplastics [66,67] and fibers [3,28].

2.3. Sand

Rubberized sand has been investigated for several years [68–70]. In the early studies, the effect of the GTR content (5–50%) on the mechanical properties (shear, triaxial, etc.) of different types of sand and their particle size distribution was investigated. Based on the results obtained, several models were proposed for design calculations in terms of geotechnical applications. Depending on the conditions, the addition of GTR (size and shape) mainly changed the internal friction (angle) between the particles and the shear strength under both static and dynamic (damping) conditions. GTR also modifies the ductility, drainage properties, and compressibility of the blends. The optimum performance was achieved with approximately 10% GTR content.

2.4. Earth/Soil

To stabilize the soil for different geotechnical applications, GTR has been added as a low-cost solution to modify properties such as compression, creep, shear, permeability, and drainage (hydraulic properties) [71–73]. Soil properties improvements can be obtained by careful control of the GTR particle size, geometry, and composition. While low GTR content (20%) is used for consolidation, high GTR content (30%) is used for insulation. Since the GTR density is low compared to soil, their mixing provides a low-weight solution to produce backfilling.

Nevertheless, several other recycling approaches have been targeted to use GTR in specific applications. Some examples are railway systems [70,74,75] and geopolymers [76,77]. In all cases, the main objective of producing rubberized composites is to reduce the costs (economics) while reusing waste (environment) for high-volume applications. Furthermore, GTR induces elasticity/toughness in the materials, especially under cyclic deformation. Finally, improved durability and stability (weathering) are observed after optimization of the processing conditions and the composite formulation (concentration of each ingredient). As for any materials, care must be taken while recycling ELT wastes and the residual products. Besides moving away from downcycling and "greenwashing", several factors must be accounted for when working with recycled materials such as GTR for construction applications. The main factors for a complete analysis and development of value-added products for upcycling are economics, environment, health, performance, and social [78]. This is the only way to achieve complete sustainability and develop commercial applications of interest as described next.

3. Commercial Products for Construction Applications

Recently, recycled ELT products have become very popular with builders and designers across all facets of new construction projects. This is because of their excellent durability, impact absorption, safety and comfort, easy installation, low maintenance requirements, and long-term cost-effectiveness. Typical examples include jogging paths, playgrounds, tennis courts, etc. (Figure 3). Other products related to the maintenance and operation of infrastructure, such as traffic-related products, highway crash barriers, etc., were also developed. A recent Transparency Market Research report suggested that the global market for recycled ELT products, including construction and other areas of application, was valued at $5.3 billion in 2021. The report also indicated that this global market is expected to grow to $7.04 billion by 2031 [79].

(a) (b) (c) (d)

Figure 3. Examples of ELT rubber in construction applications: (**a**) playground, (**b**) colored mat, (**c**) tennis court, and (**d**) jogging path.

3.1. Interior and Exterior Construction Products

Table 2 lists commercially available interior and exterior construction products and their applications. The recycled ELT-based interior construction products include floorings, mats, and underlayments. The floorings and mats are used in residential and commercial buildings, sports and fitness centers, and animal farmhouses. The flooring products could be in the form of either rolls or interlocking tiles. Different types of mats are being produced for a wide range of applications, including animal stalls, fitness and sports, anti-fatigue, anti-vibration, etc. These mats are produced by mixing GTR with binders and pigments [80]. Typical GTR size ranges from 0.5 to 3.5 mm. These rubber particles are produced by tire shredding and multi-stage granulating processes followed by separating metals and fibers. Different types of binders are used, but the most important ones are polyurethane, latex, and epoxy binders. However, polyurethane and epoxy binders generate more durable products than latex binders, especially for running tracks [81]. The mats are finally manufactured by hot press molding (compression) and cut into different sizes and shapes based on their application. Different types/geometries are possible including rolls and tiles (flat sheets).

Table 2. List of commercially available construction products and their application.

No.	Product Categories	Product Sub-Categories	Applications
		Interior construction products	
1.	Floorings and mats	Rubber rolls	Residential and commercial buildings, sports, fitness, daycare, agriculture, ice arenas, garages, etc.
		Rubber tiles	
		Garage and warehouse tiles	
		Agricultural stall mats	
		Fitness and sports mats	
		Anti-fatigue mats	
		Anti-vibration mats	
		Arena cover	
2.	Flooring underlayments	Acoustic underlayments	Residential and commercial buildings
		Exterior construction products	
3.	Rooftop walkway tiles	-	Industrial or commercial buildings
4.	Deck and landscape tiles	Interlocking tiles and blocks	Residential and commercial outdoors
5.	Asphalt paving	-	Residential and commercial outdoors, and parks
6.	Mulch	-	Residential and commercial applications
7.	Miscellaneous traffic products	-	Industrial or commercial applications
8.	Noise barrier property fence walls	-	Residential and/or commercial applications

Rubber rolls are produced by skiving (peeling) a hot press-molded rubber cylinder on a computer-controlled and precise cutting system. They are believed to be the least expensive flooring products, which are designed for residential, light commercial, and heavy commercial floors, and come in different thicknesses between 6 mm to 10 mm. While 6 mm rubber rolls are designed for residential floors, 8 mm and 10 mm rubber rolls are designed for light and heavy commercial floors, respectively. These rubber rolls can be 4 feet wide and 25–50 feet long, with a wide variety of colors to satisfy the customer's taste [82]. On the other hand, rubber mats are thicker than rubber rolls. The thickness of rubber mats varies from 9.6 mm to 19 mm depending on their application, and their typical size is 4 ft $\times$ 6 ft. For example, the thickness of multi-purpose rubber mats can be up to 12.7 mm, and their application includes gymnasiums, fitness centers, sports arenas, and complexes, as well as garage and shop floors. Although the thickness of animal stall mats is generally 19 mm, thicker mats up to 25.4 mm are also available. Some companies offer interlocking stall mats, which are also cost-effective and offer easy installation. The stall mats are very durable and are designed to withstand the abuse, harsh weather conditions, and the roughness of farm life. Some companies also offer anti-fatigue mats for workstations and kitchen areas, which have beveled edges to minimize tripping hazards. They offer several attractive features such as easy cleaning, low maintenance, seamless floor surfaces, mold and mildew resistance, shock and sound absorption, and excellent traction. Also, the stall mats can alleviate joint stress for animals. They can be installed over virtually any surface, such as sand, soil, wood, concrete, or asphalt. Besides interior applications, some products are also designed for exterior floor applications. The rubber flooring products also come in tiles, which can be either interlocking or block. The tiles are also manufactured for residential, light commercial, and heavy commercial floors, including both interior and exterior applications. The thickness of the tiles varies from 6 to 38 mm depending on the floor type and application, but 6 mm is typical for residential interior floors. On the other hand, the thicknesses of the tiles for light commercial and heavy commercial floors are 8 mm and 10 mm, respectively. For special applications, such as gym floors and ballistic facilities, the thicknesses of the tiles are 25 mm and 38 mm, respectively. Typical examples are presented in Figure 4.

(a) (b)

Figure 4. Examples of ELT rubber floor applications: (**a**) mats and (**b**) tiles.

The thickness of the tiles for exterior applications including walkways, kids' areas, and patios is 19 mm. All these rubber products are made from 100% recycled rubber, and their compositions consist of up to 92% GTR as the main ingredient. Similar to rubber rolls and mats, the tiles are also very durable, non-slip, and easy to maintain. They also offer excellent features including chemical resistance, low odor, noise reduction, impact absorption, and high traction. The properties and performances of rubber rolls, mats, and tiles are determined following certain test standards depending on the product type and application (Table 3). These products have very low volatile organic compounds (VOC) contents and some are Leadership in Energy and Environmental Design (LEED) certified. Another flooring product is underlayment, which is designed to act as a noise barrier reducing sound transmission from room to room and floor to floor in buildings. The composition of underlayment consists of up to 86% GTR as the main ingredient for applications including commercial, multi-family, education, and healthcare facilities. It can be used with laminate, hardwood, engineered wood, and ceramic tiles. The underlayment comes in rolls that can be 4 ft wide and 25–50 ft long with a wide variety of color options. The thickness of the underlayment varies from 2 mm to 12 mm depending on its applications. It is tested for various properties and performances following relevant test standards as listed in Table 3 [83–109].

Several GTR-based construction products were designed for exterior applications (Table 2). These include rooftop walkway tiles, playground tiles, deck and landscape tiles, rubber paving, mulch, miscellaneous traffic products, and noise barrier property fence walls. The rooftop walkway tiles are designed for industrial or commercial building roofs to minimize slip and/or fall hazards for the workplace crews. They offer (i) exceptional traction even under wet conditions, (ii) UV resistance for long-term durability, and (iii) easy installation. They come in two different tile types, which include standing seam rooftop walkway tiles and flat rooftop walkway tiles, and some brands are LEED-certified. These tiles are made from 100% recycled rubber and are compatible with any modern roof type such as membrane, metal profile, and standing seam roofs. The dimension of these tiles is 24″ wide × 23″ long × 2″ high. At the bottom, these tiles have 0.25″ round standoffs to fully drain water following the roof grade. This prevents any potential for water intrusion into the building, and becoming a breeding ground for mosquitoes.

Table 3. List of tests to determine the properties/performances of interior construction products and related standards.

No.	Properties/Performances	Standards
1.	Chemical Resistance [83,84]	ASTM F925-02 "Standard test method for resistance to chemicals of resilient flooring" ASTM D297-21 "Standard test methods for rubber products—Chemical analysis"
2.	Density [85]	ASTM D729-95 "Standard specification for vinylidene chloride molding compounds"
3.	Tensile Strength [86]	ASTM D412-16(2021) "Standard test methods for vulcanized rubber and thermoplastic elastomers—tension"
4.	Wear Hardness [87]	DIN 53577 "Determination of compression stress value and compression stress-strain characteristic for flexible cellular materials"
5.	Abrasion [88]	DIN 53516 "Testing of rubber and elastomers; determination of abrasion resistance"
6.	Taber Abrasion [89,90]	ASTM C501-21 "Standard test method for relative resistance to wear of unglazed ceramic tile by the Taber abraser" ASTM D4060-19 "Standard test method for abrasion resistance of organic coatings by the Taber abraser"
7.	Fire Resistance [91]	DIN EN 13501-1 "Fire classification of construction products and building elements—Part 1: Classification using data from reaction to fire tests"
8.	Flame Spread and Smoke Development Index [92]	ASTM E84-23d "Standard test method for surface burning characteristics of building materials"
9.	Tear [93]	ASTM D624-00(2020) "Standard test method for tear strength of conventional vulcanized rubber and thermoplastic elastomers"
10.	Compression Set [94,95]	ISO 815-1:2019 "Rubber, vulcanized or thermoplastic—Determination of compression set—Part 1: At ambient or elevated temperatures" ASTM D395-18 "Standard test methods for rubber property—Compression set"
11.	Shore Hardness [96]	ASTM D2240-15(2021) "Standard Test Method for Rubber Property—Durometer Hardness"
12.	Floor Ignition [97]	ASTM D2859-16(2021) "Standard test method for ignition characteristics of finished textile floor covering materials"
13.	Coefficient of Friction [98]	ASTM D1894-14 "Test method for static and kinetic coefficients of friction of plastic film and sheeting"
14.	Static Coefficient of Friction [99]	ASTM D2047-17 "Standard test method for static coefficient of friction of polish-coated flooring surfaces as measured by the James machine"
15.	Use With Wheelchairs [100]	DIN EN 1307:1997-06 "Textile floor coverings—Classification of pile carpets"
16.	Remaining Deformation [101]	DIN EN 433:1994-11 "Resilient floor coverings—Determination of residual indentation after static loading"
17.	Electrostatic Properties [102]	DIN EN 1815:2016 "Resilient and laminate floor coverings—Assessment of static electrical propensity"
18.	Light Fastness [103]	DIN EN ISO 105-B08:2010-02 "Textiles—Tests for colour fastness—Part B08: Quality control of blue wool reference materials 1 to 7"
19.	Sound Absorption (SAA)/Noise Reduction Coefficient (NRC) [104]	ASTM C423-22 "Standard test method for sound absorption and sound absorption coefficients by the reverberation room method"
20.	Oxidation/oil Resistance [105]	ASTM D2440-13(2021) "Standard test method for oxidation stability of mineral insulating oil"
21.	Impact Sound Transmission [106]	ASTM E492-09(2016)e1 "Standard test method for laboratory measurement of impact sound transmission through floor-ceiling assemblies using the tapping machine"

Table 3. *Cont.*

No.	Properties/Performances	Standards
22.	Critical Radiant Flux [107]	ASTM E648-19 "Standard test method for critical radiant flux of floor-covering systems using a radiant heat energy source"
23.	Static Load (1000 lbs) [108]	ASTM F970-17 "Standard test method for measuring recovery properties of floor coverings after static loading"
24.	Acoustics Measurement of Sound Insulation [109]	ISO 10140-3:2021 "Acoustics—Laboratory measurement of sound insulation of building elements—Part 3: Measurement of impact sound insulation"

The playground tiles, which are made from 100% recycled rubber, offer optimum fall safety for kids in the playground and play areas. They are slip-resistant and porous to allow for quick drainage for dry play surfaces. These tiles are fall-height certified and meet Americans with Disabilities Act (ADA) accessibility requirements. The properties and/or performance tests and related standards are listed in Table 4. The size of the tiles is $24'' \times 24''$, and their thickness ranges from 38 mm to 121 mm. The fall height rating is dependent on the tile thickness. For example, the fall height ratings for 38 mm and 57 mm playground tiles are 4 ft and 6 ft, respectively. On the other hand, thicker tiles (121 mm) offer a fall height rating of 10 ft when installed with polyfoam. They are available in black and other pigment colors to meet the end user's taste. They are also available in the interlocking pin system. Besides playground tiles, there are other installation accessories including polyfoam, interlock tubes, ramps, and wedges to improve the fall rating and accessibility. Another exterior construction product is deck and landscape tiles, which are also made from 100% recycled rubber. Similar to playground tiles, the deck and landscape tiles are also slip-resistant, porous supporting quick drainage, and fall-resistant. They also offer high traction even under wet conditions, and long-term durability. The size and thickness of these tiles vary depending on their types (interlocking vs. block) and applications. Similar to other products, they are also tested for different properties and performances to meet any applicable requirements (Table 4) [110–117].

Table 4. List of tests determining the properties/performances of exterior construction products and related standards.

No.	Properties/Performances	Standards
1.	Fall Height [110]	ASTM F1292-22 "Standard specification for impact attenuation of surfacing materials within the use zone of playground equipment"
2.	Freeze-Thaw [111]	ASTM C67/C67M-21 "Standard test methods for sampling and testing brick and structural clay tile"
3.	Static Coefficient of Friction [112]	ASTM C1028-06 "Standard test method for determining the static coefficient of friction of ceramic tile and other like surfaces by the horizontal dynamometer pull-meter method"
4.	High-Temperature Stability [113]	ASTM D573-04(2019) "Standard Test Method for Rubber—Deterioration in an Air Oven"
5.	Critical Radiant Flux [107]	ASTM E648-19 "Standard test method for critical radiant flux of floor-covering systems using a radiant heat energy source"
6.	Mildew Resistance [114]	ASTM G21-15(2021) "Standard practice for determining resistance of synthetic polymeric materials to fungi"
7.	Water Drainage	-
8.	Wind Resistance [115]	UL 1897 "Standard for safety, uplift tests for roof covering systems"
9.	Flame Spread [116]	ASTM E108-20a "Standard test methods for fire tests of roof coverings"
10.	Dimensional Stability [117]	DIN EN 13746-2004 "Surfaces for sports areas—determination of dimensional changes due to the effect of varied water, frost and heat conditions"

Rubber pavement is flexible and porous, made from 100% recycled rubber. It provides a sustainable and environmentally benign alternative solution to concrete pavements. The larger GTR particles allow for faster water drainage and quick drying. This pavement typically comes in block form, and its thickness varies from 38 mm to 51 mm. Similar to other exterior products, it offers high slip resistance, spike resistance, long-term durability, and low maintenance requirements. Some companies also manufacture crumb rubber additives for asphalt applications. These additives help improve asphalt crack and skid resistance, flexibility, and durability of roads as described above.

Rubber mulch is made from 100% recycled rubber and is an eco-friendly alternative to traditional wood mulch. Two different types of rubber mulch are available: nugget mulch and chip mulch. During the manufacturing process, metals are carefully separated from the rubber mulch using powerful magnets in combination with sensitive metal detectors. The main advantage of rubber mulch is that it does not splinter due to its softness compared to wood mulch. It is durable and compression-resistant and can last up to 10 times longer than wood mulch. It also has the potential to prevent wind and water erosion, as well as bug and rodent infestation. Rubber mulch comes in a variety of colors, which are resistant to fading against sunlight, maintaining the original color and beauty of landscaped areas for a long time. It offers fall height ratings up to 16 ft and meets ADA accessibility requirements.

Other miscellaneous traffic products include car parking curbs, speed bumps, shopping cart corral bumps, threshold ramps, pipe and hose ramps, rubber turf infill, delineator bases, sign bases, portable bollard bases, spill containment berms, and engineered trench guards. Another interesting and recent application of ELT wastes is the manufacturing of noise barrier property fence walls. These walls not only provide privacy but also significantly reduce noise improving the quality of living of the building occupants. Currently, a few companies around the world are producing such fence walls to reduce the transmission of highway noise into buildings. These rubber fence walls are produced in panel forms, which are made from 100% recycled rubber, and reinforced with a rigid backbone for stability and good mechanical strength. While the panel length can be up to 16 ft, the thickness can vary from 81 mm to 203 mm. Some companies also manufacture rubber-concrete hybrid noise barrier walls. Besides the sound transmission test, the rubber walls are tested for various properties and/or performances as listed in Table 5 [118–123].

Table 5. List of tests determining the properties/performances of noise barrier property fence walls and related standards.

No.	Properties/Performances	Standards
1.	Road Traffic Noise [118,119]	CEN EN 1793-(1, 2) "Road traffic noise reducing devices—Test method for determining the acoustic performance—Part 1: Intrinsic characteristics of sound absorption under diffuse sound field conditions" CEN EN 1794-(1, 2) "Road traffic noise reducing devices—Non-acoustic performance—Part 2: General safety and environmental requirements"
2.	Sound Absorption [104]	ASTM C423-22 "Standard test method for sound absorption and sound absorption coefficients by the reverberation room method"
3.	Airborne Sound Transmission [120]	ASTM E90-09(2016) "Standard test method for laboratory measurement of airborne sound transmission loss of building partitions and elements"
4.	Flame Spread [92,121]	ASTM E84-23d "Standard test method for surface burning characteristics of building materials" CAN/ULC-S102.2:2018 "Standard method of test for surface burning characteristics of flooring, floor coverings, and miscellaneous materials and assemblies"
5.	Shore Hardness [96]	ASTM D2240-15(2021) "Standard test method for rubber property—Durometer hardness"
6.	Static Coefficient of Friction [112]	ASTM C1028-06 "Standard test method for determining the static coefficient of friction of ceramic tile and other like surfaces by the horizontal dynamometer pull-meter method"

Table 5. *Cont.*

No.	Properties/Performances	Standards
7.	Skid resistance [122]	ASTM E303-22 "Standard test method for measuring surface frictional properties using the British pendulum tester"
8.	Corrosion Resistance [123]	ASTM B117-19 "Standard practice for operating salt spray (fog) apparatus"

3.2. Earth Homes

Recently, Earthship buildings have appeared as an alternative construction practice in many countries around the world [124–127]. Such construction practices are intended to promote locally available recycled, natural, and renewable materials. The sustainability in Earthship buildings is implemented by (i) using the solar system for internal heating and/or cooling, (ii) collecting rainwater as a potable water supply, and (iii) potentially recycling the used water for gardening to produce food [125]. The Earthship buildings are constructed by using recycled aluminum cans, glass bottles, and ELT wastes (Figure 5). The walls of these buildings are constructed with earth-filled ELT wastes, which act as the main load-bearing structure and naturally help regulate indoor temperature [126].

(**a**) (**b**)

Figure 5. Examples of Earthship buildings: (**a**) under construction and (**b**) completed.

4. Conclusions

Based on the information provided in this review, it is clear that ELT wastes are a major environmental issue. This is especially the case as the number of cars and trucks on the roads is still increasing. All these changes will generate a higher number of ELT wastes in the future, but the problem must be addressed now. However, recycling ELT wastes is a complex problem because the tires are highly engineered parts made from different raw materials (additives, metals, fibers, particles, and types of rubber). This is especially the case for the rubber types, which are filled with different additives and made from different origins. There is also substantial variation in the tire composition depending on the manufacturers, types (passenger cars vs. trucks), and seasonal applications (winter, all-season, off-the-road, etc.). The same problems occur for the metal and fiber wastes, which can be of different compositions depending on the tires. A variety of recycled ELT construction products are currently available on the market, which are becoming very popular with builders and designers across all facets of new construction projects. These products offer superior durability and performance making them excellent choices for construction applications. Besides these options, Earthship buildings are also becoming popular by using recycled ELT wastes. Recycling ELT wastes in construction applications will not only help conserve the environment but also support sustainable management of resources. In addition, it will create a more circular economy by limiting the amount of materials going to landfills/incineration.

5. Future Opportunities

To further improve our understanding of recycling ELT wastes, developing new processes, and finding new applications, more investigations are needed from different points of interest (academic, commercial, industrial, and scientific). This will help increase the scope to further implement sustainability in construction and circularity. Here are some key issues that still need further improvement.

Previously, several processes, alone or combined, were developed for reclaiming, and/or regenerating, and/or devulcanizing ELT rubber parts (biological, chemical, mechanical, physical, thermal, etc.). Nevertheless, the relationships between the processing conditions (time, temperature, pressure, velocity, etc.) and the properties of the final ELT rubber raw materials (particle size, geometry, surface state, devulcanization/regeneration level, number of fillers remaining, etc.) are still not well understood. This is why more in-depth scientific investigation to further optimize the processing (lower equipment and processing/energy costs), and reduce the number of residues (gases, wastewater, solvents, etc.) are required. This involves more chemical analyses and a better understanding of the interactions between the components inside the compounds before and during processing/molding.

More work should be carried out on the introduction of GTR into different matrices to improve the overall performance and increase the range of applications. Information on asphalt, concrete, sand, and earth/soil has been presented here, but other materials might be of interest, including ceramics, metal, plastics, and wood. There are good possible opportunities to use GTR not only as fillers but also as functional materials, including impact modifiers (mechanical properties), and durability-improving agents (long-term stability). To achieve this objective, more work is needed regarding the effects of ELT, processing methods/conditions, and final GTR particle sizes and geometry, including the surface state.

On the other hand, much less work has been carried out on recycling ELT waste fibers. Although a large volume of fiber has been generated (about 15% wt. of tires), the complex composition of these fibers (different polymers such as polyesters, polyamides, polyaramids, cellulose and its derivatives, etc.) make their separation and recycling very difficult. Furthermore, the fibers still contain residual rubber particles, creating difficulty in working with them. Also, the fluffy nature of these fibers makes their handling difficult. Hence, there is a need to develop an efficient process to clean and separate the waste fibers before their introduction into a matrix. This is currently under development using different mechanical and physical methods. In addition, the processes must also be optimized to control the fibers' sizes and surface properties to improve their dispersion and adhesion within a variety of matrices. By solving these issues, it will be possible to fully recycle ELT waste fibers and develop new technologies at low cost.

Finally, further investigations are required to find new applications in civil engineering (asphalt, concrete, soil, etc.) and construction. Several factors are impacting the development of Earthship buildings, which include a formal planning process, a lack of vision, and the idea of focusing on the present at the expense of the future. Hence, further studies and cooperation of different stakeholders are required to address these challenges supporting the development of Earthship buildings.

Author Contributions: Conceptualization, M.R.H.; writing—original draft preparation, M.R.H. and D.R.; writing—review and editing, M.R.H. and D.R. All authors have read and agreed to the published version of the manuscript.

Funding: This research received no external funding.

Institutional Review Board Statement: Not applicable.

Informed Consent Statement: Not applicable.

Data Availability Statement: No new data were created or analyzed in this study. Data sharing is not applicable to this article.

Conflicts of Interest: The authors declare no conflicts of interest.

References

1. Wang, T.; Xiao, F.; Zhu, X.; Huang, B.; Wang, J.; Amirkhanian, S. Energy consumption and environmental impact of rubberized asphalt pavement. *J. Clean. Prod.* **2018**, *180*, 139–158. [CrossRef]
2. Barišić, I.; Zvonarić, M.; Grubeša, I.N.; Šurdonja, S. Recycling waste rubber tyres in road construction. *Pol. Acad. Sci.* **2021**, *67*, 499–512. [CrossRef]
3. Islam, M.M.U.; Li, J.; Roychand, R.; Saberian, M.; Chen, F. A comprehensive review on the application of renewable waste tire rubbers and fibers in sustainable concrete. *J. Clean. Prod.* **2022**, *374*, 133998. [CrossRef]
4. André, F.R.; Aboelkheir, M.G. Sustainable approach of applying previous treatment of tire wastes as raw material in cement composites: Review. *Mater. Today Proc.* **2022**, *58*, 1557–1565. [CrossRef]
5. Torgal, F.P.; Ding, Y. 13—Concrete with polymeric wastes. In *Woodhead Publishing Series in Civil and Structural Engineering, Eco-Efficient Concrete*; Pacheco-Torgal, F., Jalali, S., Labrincha, J., John, V.M., Eds.; Woodhead Publishing: Sawston, UK, 2013; pp. 311–339.
6. Thomas, B.S.; Gupta, R.C. A comprehensive review on the applications of waste tire rubber in cement concrete. *Renew. Sust. Energ. Rev.* **2016**, *54*, 1323–1333. [CrossRef]
7. Onuaguluchi, O.; Banthia, N. Value-added reuse of scrap tire polymeric fibers in cement-based structural applications. *J. Clean. Prod.* **2019**, *231*, 543–555. [CrossRef]
8. Chen, M.; Feng, J.; Cao, Y.; Zhang, T. Synergetic effects of hybrid steel and recycled tyre polymer fibres on workability, mechanical strengths and toughness of concrete. *Constr. Build. Mater.* **2023**, *368*, 130421. [CrossRef]
9. Siddique, R.; Naik, T.R. Properties of concrete containing scrap-tire rubber—An overview. *Waste Manag.* **2004**, *24*, 563–569. [CrossRef] [PubMed]
10. Gorde, P.J.; Naktode, P.L. Chemically treated tyre rubber concrete review. *Mater. Today Proc.* **2022**, *60*, 508–512. [CrossRef]
11. Zia, A.; Pu, Z.; Holly, I.; Umar, T.; Tariq, M.A.U.R.; Sufian, M. A comprehensive review of incorporating steel fibers of waste tires in cement composites and its applications. *Materials* **2022**, *15*, 7420. [CrossRef]
12. Assaggaf, R.A.; Ali, M.R.; Al-Dulaijan, S.U.; Maslehuddin, M. Properties of concrete with untreated and treated crumb rubber—A review. *J. Mater. Res. Technol.* **2021**, *11*, 1753–1798. [CrossRef]
13. Valente, M.; Sibai, A. Rubber/crete: Mechanical properties of scrap to reuse tire-derived rubber in concrete: A review. *J. Appl. Biomater. Funct. Mater.* **2019**, *17*, 2280800019835486. [CrossRef] [PubMed]
14. Qin, X.; Kaewunruen, S. Environment-friendly recycled steel fibre reinforced concrete. *Constr. Build. Mater.* **2022**, *327*, 126967. [CrossRef]
15. Kundan, P.; Sharma, S. Rubberized cemented concrete composites: A review. *Mater. Today Proc.* **2021**, *44*, 4838–4842. [CrossRef]
16. Ali, A.S.; Hasan, T.M. Properties of different types of concrete containing waste tires rubber—A review. *IOP Conf. Ser. Mater. Sci. Eng.* **2019**, *584*, 012051. [CrossRef]
17. Reyna, S.L.R.; Hernández, L.S.; Pérez-Gutiérrez, F.G.; Díaz Aguilera, J.H. Mechanical behavior of reinforced concrete with waste-tire particles under an indirect tensile test. *MRS Adv.* **2019**, *4*, 2931–2937.
18. Barbuta, M.; Diaconu, D.; Serbanoiu, A.A.; Burlacu, A.; Timu, A.; Gradinaru, C.M. Effects of tire wastes on the mechanical properties of concrete. *Procedia Eng.* **2017**, *181*, 346–350. [CrossRef]
19. Senin, M.S.; Shahidan, S.; Abdullah, S.R.; Guntor, N.A.; Leman, A.S. A review on the suitability of rubberized concrete for concrete bridge decks. *IOP Conf. Ser. Mater. Sci. Eng.* **2017**, *271*, 012074. [CrossRef]
20. Fazli, A.; Rodrigue, D. Sustainable reuse of waste tire textile fibers (wttf) as reinforcements. *Polymers* **2022**, *14*, 3933. [CrossRef]
21. Singh, J.; Singh, J. Application of waste tyre rubber in construction industry. *Int. J. Civ. Struct. Environ. Infrastruct. Eng. Res. Dev.* **2015**, *5*, 57–64.
22. Mohajerani, A.; Burnett, L.; Smith, J.V.; Markovski, S.; Rodwell, G.; Rahman, M.T.; Kurmus, H.; Mirzababaei, M.; Arulrajah, A.; Horpibulsuk, S. Recycling waste rubber tyres in construction materials and associated environmental considerations: A review. *Resour. Conserv. Recycl.* **2020**, *155*, 104679. [CrossRef]
23. Zornberg, J.G.; Christopher, B.R.; LaRocque, C.J. Applications of tire bales in transportation projects. In *Recycled Materials in Geotechnics*; Aydilek, A.H., Wartman, J., Eds.; American Society of Civil Engineers: Reston, VA, USA, 2004; pp. 42–60.
24. Leong, S.Y.; Lee, S.Y.; Koh, T.Y.; Ang, D.T.C. 4R of rubber waste management: Current and outlook. *J. Mater. Cycles Waste Manag.* **2023**, *25*, 37–51. [CrossRef]
25. Abbas-Abadi, M.S.; Kusenberg, M.; Shirazi, H.M.; Goshayeshi, B.; Geem, K.M.V. Towards full recyclability of end-of-life tires: Challenges and opportunities. *J. Clean. Prod.* **2022**, *374*, 134036. [CrossRef]
26. Fazli, A.; Rodrigue, D. Recycling waste tires into ground tire rubber (gtr)/rubber compounds: A review. *J. Compos. Sci.* **2020**, *4*, 103. [CrossRef]
27. *Tire Market: Global Industry Trends, Share, Size, Growth, Opportunity and Forecast 2023–2028*; Market Research Report; Report ID: SR112023A575; IMARC Group: New York, USA, 2023.

28. Moasas, A.M.; Amin, M.N.; Khan, K.; Ahmad, W.; Al-Hashem, M.N.A.; Deifalla, A.F.; Ahmad, A. A worldwide development in the accumulation of waste tires and its utilization in concrete as a sustainable construction material: A review. *Case Stud. Constr. Mater.* **2022**, *17*, e01677. [CrossRef]
29. Forrest, M.J. 3. Overview of the world rubber recycling market. In *Recycling and Re-Use of Waste Rubber*, 2nd ed.; De Gruyter: Boston, MA, USA; Berlin, Germany, 2019; pp. 13–20. [CrossRef]
30. Talbott, A.F. Tire Fence. U.S. Patent US7387295B2, 17 June 2008.
31. Bekhiti, M.; Trouzine, H.; Asroun, A. Properties of waste tire rubber powder. *Eng. Technol. Appl. Sci. Res.* **2014**, *4*, 669–672. [CrossRef]
32. *Tire Pile Fires: Prevention, Response, Remediation*; Integrated Waste Management Board, Environmental Engineering and Contracting Inc.: Santa Ana, CA, USA, 2002.
33. Pierre, D.K.S. Canadian waste tire practices and their potential in sustainable construction. *Dalhousie J. Interdiscip. Manag.* **2013**, *9*, 1–9.
34. Rumyantseva, A.; Rumyantseva, E.; Berezyuk, M.; Plastinina, J. Waste recycling as an aspect of the transition to a circular economy. *IOP Conf. Ser. Earth Environ. Sci.* **2020**, *534*, 012002. [CrossRef]
35. Czarna-Juszkiewicz, D.; Kunecki, P.; Cader, J.; Wdowin, M. Review in waste tire management-potential applications in mitigating environmental pollution. *Materials* **2023**, *16*, 5771. [CrossRef]
36. Bockstal, L.; Berchem, T.; Schmetz, Q.; Richel, A. Devulcanisation and reclaiming of tires and rubber by physical and chemical processes: A review. *J. Clean. Prod.* **2019**, *236*, 117574. [CrossRef]
37. Ramarad, S.; Khalid, M.; Ratnam, C.T.; Chuah, A.L.; Rashmi, W. Waste tire rubber in polymer blends: A review on the evolution, properties and future. *Prog. Mater. Sci.* **2015**, *72*, 100–140. [CrossRef]
38. Zhang, P.; Wang, C.; Wu, C.; Guo, Y.; Li, Y.; Guo, J. A review on the properties of concrete reinforced with recycled steel fiber from waste tires. *Rev. Adv. Mater. Sci.* **2022**, *61*, 276–291. [CrossRef]
39. Amin, M.N.; Khan, K.; Nazar, S.; Deifalla, A.F. Application of waste recycle tire steel fibers as a construction material in concrete. *Rev. Adv. Mater. Sci.* **2023**, *62*, 20220319. [CrossRef]
40. Modarres, Y.; Ghalehnovi, M. The effect of recycled steel fibers from waste tires on concrete properties. *Civ. Eng. Infrastruct. J.* **2023**, *56*, 1–18.
41. Figueiredoa, F.P.; Shaha, A.H.; Huanga, S.-S.; Angelakopoulosb, H.; Pilakoutasa, K.; Burgessa, I. Fire protection of concrete tunnel linings with waste tyre fibres. *Proc. Eng.* **2017**, *210*, 472–478. [CrossRef]
42. Buss, A.H.; Kovaleski, J.L.; Pagani, R.N.; Silva, V.L.D.; Silva, J.D.M. Proposal to reuse rubber waste from end-of-life tires using thermosetting resin. *Sustainability* **2019**, *11*, 6997. [CrossRef]
43. Hejna, A.; Korol, J.; Przybysz-Romatowska, M.; Zedler, Ł.; Chmielnicki, B.; Formela, K. Waste tire rubber as low-cost and environmentally-friendly modifier in thermoset polymers—A review. *Waste Manag.* **2020**, *108*, 106–118. [CrossRef] [PubMed]
44. Sienkiewicz, M.; Janik, H.; Borzędowska-Labuda, K.; Kucińska-Lipka, J. Environmentally friendly polymer-rubber composites obtained from waste tyres: A review. *J. Clean. Prod.* **2017**, *147*, 560–571. [CrossRef]
45. Taurino, R.; Bondioli, F.; Messori, M. Use of different kinds of waste in the construction of new polymer composites: Review. *Mater. Today Sust.* **2023**, *21*, 100298. [CrossRef]
46. Fazli, A.; Rodrigue, D. Waste rubber recycling: A review on the evolution and properties of thermoplastic elastomers. *Materials* **2020**, *13*, 782. [CrossRef]
47. Guselnikova, O.; Semyonov, O.; Sviridova, E.; Gulyaev, R.; Gorbunova, A.; Kogolev, D.; Trelin, A.; Yamauchi, Y.; Boukherroubf, R.; Postnikov, P. Functional upcycling of polymer waste towards the design of new materials. *Chem. Soc. Rev.* **2023**, *52*, 4755–4832. [CrossRef]
48. Markl, E.; Lackner, M. Devulcanization technologies for recycling of tire-derived rubber: A review. *Materials* **2020**, *13*, 1246. [CrossRef]
49. Li, F.; Zhang, X.; Wang, L.; Zhai, R. The preparation process, service performances and interaction mechanisms of crumb rubber modified asphalt (CRMA) by wet process: A comprehensive review. *Constr. Build. Mater.* **2022**, *354*, 129168. [CrossRef]
50. Duan, K.; Wang, C.; Liu, J.; Song, L.; Chen, Q.; Chen, Y. Research progress and performance evaluation of crumb-rubber-modified asphalts and their mixtures. *Constr. Build. Mater.* **2022**, *361*, 129687. [CrossRef]
51. Lee, S.; Park, Y.-K.; Lee, J. Upcycling of plastic and tire waste toward use as modifier for asphalt binder. *Energy Environ.* **2024**, *35*, 510–524. [CrossRef]
52. Li, H.; Cui, C.; Temitope, A.A.; Feng, Z.; Zhao, G.; Guo, P. Effect of SBS and crumb rubber on asphalt modification: A review of the properties and practical application. *J. Traffic Transp. Eng.* **2022**, *9*, 836–863. [CrossRef]
53. Guo, Y.; Tataranni, P.; Sangiorgi, C. The use of fibres in asphalt mixtures: A state of the art review. *Constr. Build. Mater.* **2023**, *390*, 131754. [CrossRef]
54. Alsheyab, M.A.T.; Khedaywi, T.; Ogiliat, O. Effect of waste tire rubber on properties of asphalt cement and asphalt concrete mixtures: State of the art. *Int. J. Pavement Res. Technol.* **2023**, 1–12. [CrossRef]
55. Jahami, A.; Issa, C.A. Exploring the use of mixed waste materials (MWM) in concrete for sustainable Construction: A review. *Constr. Build. Mater.* **2023**, *398*, 132476. [CrossRef]
56. Bu, C.; Zhu, D.; Lu, X.; Liu, L.; Sun, Y.; Yu, L.; Xiao, T.; Zhang, W. Modification of rubberized concrete: A review. *Buildings* **2022**, *12*, 999. [CrossRef]
57. Li, Y.; Chai, J.; Wang, R.; Zhou, Y.; Tong, X. A review of the durability-related features of waste tyre rubber as a partial substitute for natural aggregate in concrete. *Buildings* **2022**, *12*, 1975. [CrossRef]
58. Surehali, S.; Singh, A.; Biligiri, K.P. A state-of-the-art review on recycling rubber in concrete: Sustainability aspects, specialty mixtures, and treatment methods. *Dev. Built Environ.* **2023**, *14*, 100171. [CrossRef]

59. Mei, J.; Xu, G.; Ahmad, W.; Khan, K.; Amin, M.N.; Aslam, F.; Alaskar, A. Promoting sustainable materials using recycled rubber in concrete: A review. *J. Clean. Prod.* **2022**, *373*, 133927. [CrossRef]
60. Muhammad, S.; Yuan, Q.; Alam, M.; Javed, M.F.; Rehman, M.F.; Mohamed, A. Fresh and hardened properties of waste rubber tires based concrete: A state art of review. *SN Appl. Sci.* **2023**, *5*, 119.
61. Zrar, Y.J.; Younis, K.H. Mechanical and durability properties of self-compacted concrete incorporating waste crumb rubber as sand replacement: A review. *Sustainability* **2022**, *14*, 11301. [CrossRef]
62. He, S.; Jiang, Z.; Chen, H.; Chen, Z.; Ding, J.; Deng, H.; Mosallam, A.S. Mechanical properties, durability, and structural applications of rubber concrete: A state-of-the-art-review. *Sustainability* **2023**, *15*, 8541. [CrossRef]
63. Zhang, P.; Wang, X.; Wang, J.; Zhang, T. Workability and durability of concrete incorporating waste tire rubber: A review. *J. Renew. Mater.* **2023**, *11*, 745. [CrossRef]
64. Tran, T.Q.; Thomas, B.S.; Zhang, W.; Ji, B.; Li, S.; Brand, A.S. A comprehensive review on treatment methods for end-of-life tire rubber used for rubberized cementitious materials. *Constr. Build. Mater.* **2022**, *359*, 129365. [CrossRef]
65. Liu, L.; Wang, C.; Liang, Q.; Chen, F.; Zhou, X. A state-of-the-art review of rubber modified cement-based materials: Cement stabilized base. *J. Clean. Prod.* **2023**, *392*, 136270. [CrossRef]
66. Singh, P.; Singh, D.N.; Debbarma, S. Macro- and micro-mechanisms associated with valorization of waste rubber in cement-based concrete and thermoplastic polymer composites: A critical review. *Constr. Build. Mater.* **2023**, *371*, 130807. [CrossRef]
67. Marinelli, S.; Marinello, S.; Lolli, F.; Gamberini, R.; Coruzzolo, A.M. Waste plastic and rubber in concrete and cement mortar: A tertiary literature review. *Sustainability* **2023**, *15*, 7232. [CrossRef]
68. Cabalar, A.F. Direct shear tests on waste tires–sand mixtures. *Geotech. Geol. Eng.* **2011**, *29*, 411. [CrossRef]
69. Anvari, S.M.; Shooshpasha, I.; Kutanaei, S.S. Effect of granulated rubber on shear strength of fine-grained sand. *J. Rock Mech. Geotech. Eng.* **2017**, *9*, 936–944. [CrossRef]
70. Ding, Y.; Zhang, J.; Chen, X.; Wang, X.; Jia, Y. Experimental investigation on static and dynamic characteristics of granulated rubber-sand mixtures as a new railway subgrade filler. *Constr. Build. Mater.* **2021**, *273*, 121955. [CrossRef]
71. Yang, Z.; Zhang, Q.; Shi, W.; Lv, J.; Lu, Z.; Ling, X. Advances in properties of rubber reinforced soil. *Adv. Civ. Eng.* **2020**, *2020*, 6629757. [CrossRef]
72. Liu, L.; Cai, G.; Zhang, J.; Liu, X.; Liu, K. Evaluation of engineering properties and environmental effect of recycled waste tire-sand/soil in geotechnical engineering: A compressive review. *Renew. Sust. Energ. Rev.* **2020**, *126*, 109831. [CrossRef]
73. Tasalloti, A.; Chiaro, G.; Murali, A.; Banasiak, L. Physical and mechanical properties of granulated rubber mixed with granular soils—A literature review. *Sustainability* **2021**, *13*, 4309. [CrossRef]
74. Farooq, M.A.; Nimbalkar, S.; Fatahi, B. Sustainable applications of tyre-derived aggregates for railway transportation infrastructure. *Sustainability* **2022**, *14*, 11715. [CrossRef]
75. Qiang, W.; Jing, G.; Connolly, D.P.; Aela, P. The use of recycled rubber in ballasted railway tracks: A review. *J. Clean. Prod.* **2023**, *420*, 138339. [CrossRef]
76. Luhar, I.; Luhar, S. Rubberized geopolymer composites: Value-added applications. *J. Compos. Sci.* **2021**, *5*, 312. [CrossRef]
77. Qaidi, S.M.A.; Mohammed, A.S.; Ahmed, H.U.; Faraj, R.H.; Emad, W.; Tayeh, B.A.; Althoey, F.; Zaid, O.; Sor, N.H. Rubberized geopolymer composites: A comprehensive review. *Ceram. Int.* **2022**, *48*, 24234. [CrossRef]
78. Cirino, E.; Curtis, S.; Wallis, J.; Thys, T.; Brown, J.; Rolsky, C.; Erdle, L.M. Assessing benefits and risks of incorporating plastic waste in construction materials. *Front. Built Environ.* **2023**, *9*, 1206474. [CrossRef]
79. *Global Tire Recycling Downstream Products Market*; Market Research Report; Transparency Market Research: Wilmington, DE, USA, 2023.
80. Benjak, P.; Radetić, L.; Tomaš, M.; Brnardić, I.; Radetić, B.; Špada, V.; Grčić, I. Rubber tiles made from secondary raw materials with immobilized titanium dioxide as passive air protection. *Processes* **2023**, *11*, 125. [CrossRef]
81. Hammer, C.; Gray, T.A. *Designing Building Products Made with Recycled Tires*; California Integrated Waste Management Board: Sacramento, CA, USA, 2004.
82. Hart, B. Comparing Rubber Gym Flooring—Rolls, Mats, Tiles. Greatmats. Available online: https://www.greatmats.com/rubber-gym-flooring/comparing-rubber-gym-flooring-rolls-mats-interlocking-tiles.php (accessed on 13 November 2023).
83. *ASTM F925-02*; Standard Test Method for Resistance to Chemicals of Resilient Flooring. ASTM International: West Conshohocken, PA, USA, 2002.
84. *ASTM D297-21*; Standard Test Methods for Rubber Products—Chemical Analysis. ASTM International: West Conshohocken, PA, USA, 1998.
85. *ASTM D729-95*; Standard Specification for Vinylidene Chloride Molding Compounds. ASTM International: West Conshohocken, PA, USA, 1994.
86. *ASTM D412-16(2021)*; Standard Test Methods for Vulcanized Rubber and Thermoplastic Elastomers—Tension. ASTM International: West Conshohocken, PA, USA, 2021.
87. *DIN 53577*; Determination of Compression Stress Value and Compression Stress-Strain Characteristic for Flexible Cellular Materials. Deutsches Institut für Normung: Berlin, Germany, 1988.
88. *DIN 53516*; Testing of Rubber and Elastomers; Determination of Abrasion Resistance. Deutsches Institut für Normung: Berlin, Germany, 1999.
89. *ASTM C501-21*; Standard Test Method for Relative Resistance to Wear of Unglazed Ceramic Tile by the Taber Abraser. ASTM International: West Conshohocken, PA, USA, 2021.

90. *ASTM D4060-19*; Standard Test Method for Abrasion Resistance of Organic Coatings by the Taber Abraser. ASTM International: West Conshohocken, PA, USA, 2019.
91. *DIN EN 13501-1*; Fire Classification of Construction Products and Building Elements—Part 1: Classification Using Data from Reaction to Fire Tests. Deutsches Institut für Normung: Berlin, Germany, 2019.
92. *ASTM E84-23d*; Standard Test Method for Surface Burning Characteristics of Building Materials. ASTM International: West Conshohocken, PA, USA, 2024.
93. *ASTM D624-00(2020)*; Standard Test Method for Tear Strength of Conventional Vulcanized Rubber and Thermoplastic Elastomers. ASTM International: West Conshohocken, PA, USA, 2020.
94. *ISO 815-1:2019*; Rubber, Vulcanized or Thermoplastic—Determination of Compression Set—Part 1: At Ambient or Elevated Temperatures. International Organization for Standardization: Geneva, Switzerland, 2019.
95. *ASTM D395-18*; Standard Test Methods for Rubber Property—Compression Set. ASTM International: West Conshohocken, PA, USA, 2018.
96. *ASTM D2240-15(2021)*; Standard Test Method for Rubber Property—Durometer Hardness. ASTM International: West Conshohocken, PA, USA, 2021.
97. *ASTM D2859-16(2021)*; Standard Test Method for Ignition Characteristics of Finished Textile Floor Covering Materials. ASTM International: West Conshohocken, PA, USA, 2021.
98. *ASTM D1894*; Test Method for Static and Kinetic Coefficients of Friction of Plastic Film and Sheeting. ASTM International: West Conshohocken, PA, USA, 2014.
99. *ASTM D2047-17*; Standard Test Method for Static Coefficient of Friction of Polish-Coated Flooring Surfaces as Measured by the James Machine. ASTM International: West Conshohocken, PA, USA, 2017.
100. *DIN EN 1307:1997-06*; Textile Floor Coverings—Classification of Pile Carpets. Deutsches Institut für Normung: Berlin, Germany, 1997.
101. *DIN EN 433:1994-11*; Resilient Floor Coverings—Determination of Residual Indentation after Static Loading. Deutsches Institut für Normung: Berlin, Germany, 1994.
102. *DIN EN 1815:2016*; Resilient and Laminate Floor Coverings—Assessment of Static Electrical Propensity. Deutsches Institut für Normung: Berlin, Germany, 2016.
103. *DIN EN ISO 105-B08:2010-02*; Textiles—Tests for Colour Fastness—Part B08: Quality Control of Blue Wool Reference Materials 1 to 7. Deutsches Institut für Normung: Berlin, Germany, 2010.
104. *ASTM C423-22*; Standard Test Method for Sound Absorption and Sound Absorption Coefficients by the Reverberation Room Method. ASTM International: West Conshohocken, PA, USA, 2023.
105. *ASTM D2440-13(2021)*; Standard Test Method for Oxidation Stability of Mineral Insulating Oil. ASTM International: West Conshohocken, PA, USA, 2021.
106. *ASTM E492-09(2016)e1*; Standard Test Method for Laboratory Measurement of Impact Sound Transmission through Floor-Ceiling Assemblies Using the Tapping Machine. ASTM International: West Conshohocken, PA, USA, 2016.
107. *ASTM E648-19*; Standard Test Method for Critical Radiant Flux of Floor-Covering Systems Using a Radiant Heat Energy Source. ASTM International: West Conshohocken, PA, USA, 2019.
108. *ASTM F970-17*; Standard Test Method for Measuring Recovery Properties of Floor Coverings after Static Loading. ASTM International: West Conshohocken, PA, USA, 2017.
109. *ISO 10140-3:2021*; Acoustics—Laboratory Measurement of Sound Insulation of Building Elements—Part 3: Measurement of Impact Sound Insulation. International Organization for Standardization: Geneva, Switzerland, 2021.
110. *ASTM F1292-22*; Standard Specification for Impact Attenuation of Surfacing Materials within the Use Zone of Playground Equipment. ASTM International: West Conshohocken, PA, USA, 2022.
111. *ASTM C67/C67M-21*; Standard Test Methods for Sampling and Testing Brick and Structural Clay Tile. ASTM International: West Conshohocken, PA, USA, 2021.
112. *ASTM C1028-06*; Standard Test Method for Determining the Static Coefficient of Friction of Ceramic Tile and Other Like Surfaces by the Horizontal Dynamometer Pull-Meter Method. ASTM International: West Conshohocken, PA, USA, 2006.
113. *ASTM D573-04(2019)*; Standard Test Method for Rubber—Deterioration in an Air Oven. ASTM International: West Conshohocken, PA, USA, 2019.
114. *ASTM G21-15(2021)*; Standard Practice for Determining Resistance of Synthetic Polymeric Materials to Fungi. ASTM International: West Conshohocken, PA, USA, 2021.
115. *UL 1897*; Standard for Safety, Uplift Tests for Roof Covering Systems. Underwriters Laboratories: Northbrook, IL, USA, 2015.
116. *ASTM E108-20a*; Standard Test Methods for Fire Tests of Roof Coverings. ASTM International: West Conshohocken, PA, USA, 2020.
117. *DIN EN 13746-2004*; Surfaces for Sports Areas—Determination of Dimensional Changes Due to the Effect of Varied Water, Frost and Heat Conditions. Deutsches Institut für Normung: Berlin, Germany, 2004.
118. *CEN EN 1793-(1, 2)*; Road Traffic Noise Reducing Devices—Test Method for Determining the Acoustic Performance—Part 1: Intrinsic Characteristics of Sound Absorption under Diffuse Sound Field Conditions. The European Committee for Standardization: Brussels, Belgium, 2017.
119. *CEN EN 1794-(1, 2)*; Road Traffic noise Reducing Devices—Non-Acoustic Performance—Part 2: General Safety and ENVIRONMENTAL Requirements. The European Committee for Standardization: Brussels, Belgium, 2020.

120. *ASTM E90-09(2016)*; Standard Test Method for Laboratory Measurement of Airborne Sound Transmission Loss of Building Partitions and Elements. ASTM International: West Conshohocken, PA, USA, 2016.
121. *CAN/ULC-S102.2:2018*; Standard Method of Test for Surface Burning Characteristics of Flooring, Floor Coverings, and Miscellaneous Materials and Assemblies. Underwriters Laboratories of Canada: Ottawa, ON, Canada, 2018.
122. *ASTM E303-22*; Standard Test Method for Measuring Surface Frictional Properties Using the British Pendulum Tester. ASTM International: West Conshohocken, PA, USA, 2022.
123. *ASTM B117-19*; Standard Practice for Operating Salt Spray (Fog) Apparatus. ASTM International: West Conshohocken, PA, USA, 2019.
124. Kruis, N.J.; Heun, M.K. Analysis of the performance of Earthship housing in various global climates. In Proceedings of the Energy Sustainability Conference, Long Beach, CA, USA, 27–30 June 2007.
125. Booth, C.A.; Rasheed, S.; Mahamadu, A.-M.; Horry, R.; Manu, P.; Awuah, K.G.B.; Aboagye-Nimo, E.; Georgakis, P. Insights into public perceptions of Earthship buildings as alternative homes. *Buildings* **2021**, *11*, 377. [CrossRef]
126. Datla, A.; Pujitha, V.S.; Mahapatra, G.D. Earthship: The reuse of waste materials in construction. *JETIR* **2019**, *6*, 63975.
127. Santic, T.S.; Stanojlovic, D. Earthship—A new habitat on Earth for quality life. In Proceedings of the 1st International Conference on Quality of Life, Kyoto, Japan, 19–21 August 2016; p. 123.

sustainability

Review

A Comprehensive Overview of Recycled Glass as Mineral Admixture for Circular UHPC Solutions

N. Marcela Redondo-Pérez [1], Jesús D. Redondo-Mosquera [2] and Joaquín Abellán-García [1,*]

[1] Department of Civil and Environmental Engineering, Universidad Del Norte, Barranquilla 081007, Colombia; rnurys@uninorte.edu.co
[2] Department of Civil Engineering, Universidad de La Guajira, Riohacha 440002, Colombia; jesusdredondo@uniguajira.edu.co
* Correspondence: jabellan@uninorte.edu.co

Abstract: This review article analyzes the influence of recycled glass (as sand and powder) beyond the durability, rheology and compressive strength of plain UHPC, even exploring flexural and direct tensile performance in fiber-reinforced UHPC. Interactions with other mineral admixtures like limestone powder, rice husk ash, fly ash, FC3R, metakaolin and slags, among others, are analyzed. Synergy with limestone powder improves rheology, reducing superplasticizer usage. Research highlights waste glass–UHPC mixtures with reduced silica fume and cement content by over 50% and nearly 30%, respectively, with compressive strengths exceeding 150 MPa, cutting costs and carbon footprints. Furthermore, with the proper fiber dosage, waste glass–UHPC reported values for strain and energy absorption capacity, albeit lower than those of traditional UHPC formulations with high cement, silica fume and quartz powder content, surpassing requirements for demanding applications such as seismic reinforcement of structures. Moreover, durability remains comparable to that of traditional UHPC. In addition, the reported life cycle analysis found that the utilization of glass powder in UHPC allows a greater reduction of embedded CO_2 than other mineral additions in UHPC without jeopardizing its properties. In general, the review study presented herein underscores recycled glass's potential in UHPC, offering economic and performance advantages in sustainable construction.

Keywords: waste glass; ultra-high-performance concrete (UHPC); life cycle analysis; circular economy; mineral admixtures; compressive strength; rheology; durability; post-cracking behavior; sustainability

Citation: Redondo-Pérez, N.M.; Redondo-Mosquera, J.D.; Abellán-García, J. A Comprehensive Overview of Recycled Glass as Mineral Admixture for Circular UHPC Solutions. *Sustainability* **2024**, *16*, 5077. https://doi.org/10.3390/su16125077

Academic Editors: Anibal C. Maury-Ramirez and Nele De Belie

Received: 13 May 2024
Revised: 7 June 2024
Accepted: 12 June 2024
Published: 14 June 2024

1. Introduction

Over the past few years, notable advancements in concrete technology have occurred, leading to the creation of innovative concretes, like ultra-high-performance concrete (UHPC) [1,2]. With water-to-binder ratios (w/b) ranging from 0.15 to 0.25, UHPC showcases exceptional mechanical and durability properties that outperform standard concrete (SC) and high-performance concrete (HPC) [3,4]. According to the ACI-239 guidelines [5], UHPC features compressive strengths (CS) of over 150 MPa, along with an elastic modulus (MoE) within the range of 40 to 50 GPa. In addition, UHPC has exceptional durability features, which marks it as a promising construction material [6,7]. Despite its inherent brittleness, the practice of incorporating fibers, which has become widespread, successfully provides ductility and toughness to UHPC [8–13].

As observed in previous studies, a typical UHPC formulation encompasses Portland cement, reactive powders (such as silica fume (SF)), polycarboxylate-based superplasticizers, water and fine quartz sand (QS) [3,14,15]. Balancing the arrangement of the constituents is crucial to achieving this specialized cementitious material's outstanding mechanical and durability characteristics [16–18]. For instance, it is important to note that UHPC mixture design greatly depends on particle-packing theories to attain the optimal arrangement of particles [9,19]. The latter is key to achieving a microstructure that exceeds that of typical or even high-performance concrete [14,20].

For a thorough analysis of the unique properties displayed by UHPC, it is essential to consider three crucial factors: (a) the cementitious matrix reinforcement, (b) the porosity reduction and (c) the interfacial transition zone (ITZ) narrowing and densification [1,21,22]. Therefore, the approach of new UHPC formulations, such as those that incorporate recycled glass, must be carried out with care not to harm these three balances previously considered. Moreover, it is important to notice that with a reduced w/b, the cement hydration in UHPC is restricted, resulting in only partial calcium silicate hydrate (C-S-H) and calcium hydroxide (CH) formation [12,14,21]. By utilizing highly reactive pozzolanic mineral admixtures, like SF, the hydrated cement paste can be strengthened through the formation of secondary C-S-H by its reaction with CH [21]. Therefore, it is essential to find the perfect equilibrium between cement quality and content and pozzolanic materials in UHPC to effectively enhance its strength [1]. One important aspect is to decrease the porosity of the concrete by using a low w/b and achieving an enhanced packing density [23]. This way, one must meticulously choose micro- and nano-sized components that can effectively fill the gaps between larger particles, resulting in an improved packing arrangement [22]. It is worth mentioning that certain materials, like SF, quartz powder, slags, fly ash and others, are frequently utilized in this application owing to their fine particle size and pozzolanic features [2,3,6,23–29]. Understanding the achievement of optimal density in concrete relies on the use of particle-packing models, like the modified Andreasen and Andersen (A&Amod) or the Compressive Packing Model (CPM) [19]. The optimal packing density plays a crucial role in lessening interparticle gaps, constraining the space for portlandite crystal growth. In order to obtain the requisite low porosity in UHPC, it is essential to include superplasticizers, which help reduce the w/b to a minimum [20,30]. Due to its unique properties, UHPC has a lower porosity of approximately half of that of the average porosity found in standard concrete [5]. Furthermore, enhancing the strengthening properties of the UHPC matrix can be accomplished by decreasing the maximum aggregate size, encouraging compact particle arrangement and utilizing a low w/b. This improves the ITZ that separates the bulk paste and aggregate [14]. However, the removal of coarse aggregates results in a higher amount of powders, such as cement and SF, being needed. This can have negative effects on cost and sustainability, as will be explored in later sections [25,28].

With its exceptional properties, UHPC proves to be incredibly beneficial in a broad variety of structural and infrastructure uses. The exceptional mechanical properties of this special concrete can be leveraged to create structures with reduced weights by designing reduced cross-sectional areas. Moreover, the exceptional durability of the material extends its lifespan and minimizes the requirement for maintenance, resulting in decreased expenses [21]. Therefore, UHPC is widely used in different areas of the construction industry, such as footbridge construction, retrofitting of existing structures, pavement restoration, tunnel boring machine segments, the ABC bridge construction system, facades and urban furniture [4,12,31–38]. These examples of various applications demonstrate the flexibility and potential of UHPC in addressing the changing requirements of contemporary construction projects.

However, as previously stated, UHPC production requires large amounts of cement, SF and quartz powder, leading to higher costs and a greater impact on the environment [23]. As a result, these limitations have impeded an even more widespread utilization of UHPC around the world [23]. The global research community has been diligently working on addressing this challenge by investigating the potential of alternative components for making UHPC. These components can be employed as cement and SF partial replacements in new formulations. Hence, through extensive research and analysis, most scholars strive to minimize expenses and improve the material life cycle, narrowing the knowledge gap and exploring new avenues for creating UHPC mixtures that are both sustainable and economically viable [12,24,34,39].

For its part, the utilization of glass powder (GP) in UHPC has garnered considerable attention in recent years, yielding substantial advancements in both understanding its properties and practical implementation in construction [40–42]. As a substitute for traditional fillers and pozzolanic materials, glass powder offers unique advantages within concrete mixes [43,44]. Concurrently, the proliferation of waste glass (WG) worldwide has escalated dramatically, resulting in significant environmental pollution [44–46]. This surge in WG underscores the urgent need for sustainable solutions. Furthermore, rising costs and carbon emissions associated with conventional high-strength concrete production have underscored the demand for alternative supplementary cementitious materials [26,47,48].

Chemically, glass comprises inorganic compounds fused at high temperatures and cooled to a solid state, typically consisting of calcium oxide, sodic oxide and non-crystalline silica, though composition varies based on application [40,49–51]. Historically, waste glass has found utility in hydraulic concrete production as a supplementary cementitious material and fine aggregate, owing to its high silica content. Effective recycling strategies are pivotal in preserving the purity and quality of glass [40,49–51]. In this context, GP, obtained by milling WG, emerges as a promising candidate for partially replacing cement in alternative UHPC formulations, leveraging its pozzolanic properties to enhance mechanical characteristics while allowing for substantial cement, silica fume or quartz powder reductions [51–53]. However, careful attention must be directed towards mitigating the potential alkali-silica reaction (ASR) risk introduced by GP particles [54].

This convergence of research underscores the potential of GP as an environmentally friendly solution in UHPC formulations, offering both environmental benefits and economic viability in the construction sector. This review is the first to provide a global perspective, spanning from microstructure to real-world applications. It covers the most relevant mechanical properties, such as the direct tensile behavior of recycled-glass UHPC reinforced with fibers, and examines the effects on costs and carbon footprint. Additionally, it presents a comprehensive overview of UHPC properties containing waste glass in both fresh and hardened states, while also analyzing the advantages, opportunities and risks of using such supplementary cementitious waste material in UHPC formulations. Figure 1 puts forward a scheme of the structure of the review study presented herein.

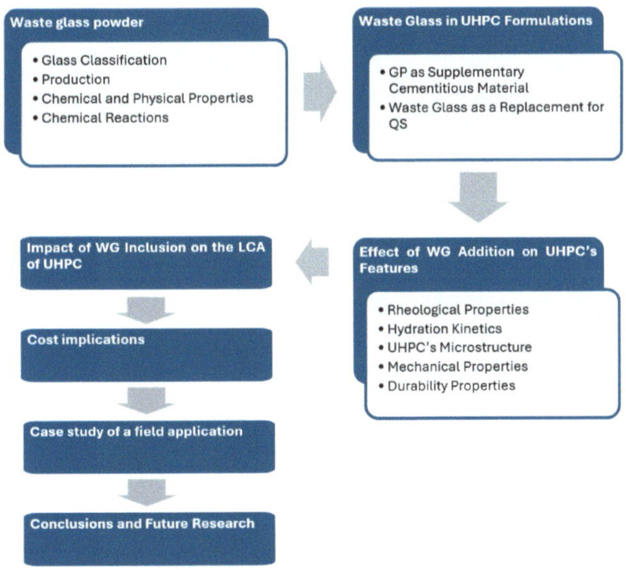

Figure 1. Structure of the review study presented herein.

2. Research Significance

In recent years, there has been a concerted scholarly effort focused on harnessing the potential of GP as a substitute for conventional UHPC-making materials [1,47,55–59].

This approach is driven by the need to address stringent environmental impacts and the escalating costs associated with typical UHPC mixtures, due to their high amount of expensive and/or high carbon footprint raw constituents, like SF, cement or QP. On the one hand, it is imperative to recognize that all materials, including glass, have a finite lifespan and must be recycled or repurposed to mitigate environmental risks [44,60]. Therefore, by incorporating waste glass into concrete production, even UHPC, researchers aim to minimize solid waste, maximize recycling efforts, conserve natural resources and mitigate environmental hazards associated with landfilling [41,44,60]. On the other hand, the employment of WG in concrete offers multifaceted benefits, encompassing the reduction of landfill tipping fees, which typically range from $40 to $100 per ton in the USA [61]. Moreover, the integration of WG as a replacement for typical UHPC-making components in alternative formulations can significantly lower material costs, enhance sustainability and reduce environmental impacts [24,62]. Particularly, by replacing quartz powder, cement and even silica fume with ground glass powder, substantial cost reductions can be achieved in traditional UHPC [63,64]. Waste glass presents a notably lower cost compared to quartz powder, with a price approximately one-third less expensive. Moreover, the utilization of locally sourced waste glass for UHPC production contributes to decreased transportation expenses for materials [41], and also reduces glass waste disposal and, therefore, its associated costs [44]. Furthermore, WG incorporation into concrete production helps reduce CO_2 emissions, energy consumption and air pollution associated with conventional cement clinker production [61]. In addition, incorporating glass powder in UHPC formulations eliminates the need for quartz powder, a known carcinogen, thereby enhancing the safety of the UHPC production process [1,52]. Hence, the strategy of using recycled glass in UHPC production aligns with the broader goal of promoting environmental stewardship and fostering a more sustainable and safer construction industry.

Hence, this review article's purpose is to offer a thorough analysis of the technical, economic and environmental implications, as well as the limitations and possible risks surrounding incorporating recycled waste glass in UHPC production. To that end, the findings of a total of 157 research papers, dissertation theses and proceedings articles including information about rheological, mechanical and durability characteristics, besides costs and carbon footprint implications, were analyzed, organized and presented.

3. Waste Glass Powder

3.1. Glass Classification

This section offers a succinct overview of the primary classifications of waste glass, examining two classification methods while providing detailed insights into their chemical composition and application. On the one hand, concerning chemical composition, WG may be categorized into four main classes: soda-lime glass, lead glass, borosilicate glass and electric glass, also known as E glass [65]. Soda-lime glass finds widespread application in various glass products, including containers and plates. Hence, more than 90% of the glass produced in the European Union corresponds to this type of glass [65]. Even though the chemical compositions across different glass types primarily comprise silica (SiO_2), sodium oxide (Na_2O) (derived from soda ash (Na_2CO_3)) and lime ($CaCO_3$), among others, there are some differences in their proportions [66]. On the other hand, in terms of application, glass can be broadly classified into several categories, each serving distinct purposes. These categories encompass container glass, plate glass (e.g., window glass), continuous filament glass (including roving, mat, textile and optical fiber), domestic glass or tableware, insulation mineral wool and specialty glass (such as high-temperature domestic glass). Each type of glass corresponds to specific applications; for instance, soda-lime glass and plate glass are commonly employed in container and window manufacturing, respectively. Continuous filament glass predominantly consists of electric glass, while domestic glass can vary between soda-lime glass and lead glass. Insulation mineral wool is derived from borosilicate glass, whereas specialty glass typically comprises soda-lime glass or borosilicate glass [65,67]. Table 1 provides a comprehensive breakdown of the chemical compositions associated with various types of glass [67].

Table 1. Major components of the different classes of glass regarding their chemical composition [67].

	Soda-Lime	Lead	Borosilicate	E-Glass
SiO_2	71–75%	54–65%	70–80%	52–56%
Al_2O_3	1–1.5%		7%	12–16%
B_2O_3			7–15%	0–10%
CaO	9–15%			16–25%
PbO		25–30%		
Na_2O+	12–16%	13–15%	4–8%	0–2%
K_2O				

Moreover, glass can be categorized into three types based on color: (i) clear/flint glass, (ii) green glass and (iii) brown/amber glass. These color variations are attributed to distinct chemical compositions, each with specific thresholds for color impurities. The acceptable range for color contamination is 4–6% for clear glass, 5–30% for green glass and 5–15% for amber glass [65]. When it comes to glass aggregate, green glass exhibits superior resistance to ASR reaction due to its high concentration of Cr_2O_3, which effectively mitigates ASR growth [68]. However, due to the particularities of UHPC, other factors such as the particle glass size and the presence and dosage of high pozzolanic supplementary cementitious materials such as SF play more crucial roles in determining ASR formation [50,68,69]. The chemical compositions of different colored glasses are detailed in Table 2 [65].

Table 2. Different color glass chemical compositions reported in the scientific literature [65].

	Glass Color		
	Green	Brown/Amber	Clear/Flint
SiO_2	71.3%	71.9–72.4%	73.2–73.5%
Al_2O_3	2.20%	1.70–1.80%	1.7–1.9%
SO_3	0.05	0.12–0.14%	0.20–0.24%
CaO + MgO	12.20%	11.60%	10.7–10.8%
Fe_2O_3	0.56%	0.30%	0.04–0.05%
$Na_2O + K_2O$	13.10%	13.8–14.4%	13.6–14.1%
Cr_2O_3	0.43%	0.01%	-

3.2. Production

WG is found in municipal solid waste (MSW) as containers, cullets and plate residues. While the majority of the data focus on glass containers, several evaluations also include glass components used in long-lasting items including furniture, appliances and consumer electronics. In 2018, the United States and Europe consumed 12.25 and 16.36 million tons of glass and recycled 3.1 and 12.92 million tons, respectively, for a recycling rate of 31.3% in the case of the USA and 76% in Europe [70,71]. For this part, China generated approximately 20 million tons of glass waste in the same year, with 53% being recycled [72]. In the case of Australia, about 1.1 million tons of WG were produced in 2018, with the recycling rate in the range between 54 and 61% [73]. In contrast, Sweden has achieved a remarkable glass recycling rate exceeding 90% for the same period of time, ranking among the highest in the world [74].

3.3. Chemical and Physical Properties

Table 3 depicts the chemical characterization of the glass in different studies about its use in UHPC formulations carried out previously. As all of the below are soda-lime class, it is observed that the percentages have values close to each other. In other words, although they are different glass powders, their chemical composition is similar. Some interesting findings observed in Table 3 are the high amount of silica oxide (over 79% in all cases) and the alkali content (in the range of 12.4–13.5%). These values have a relevant impact on the concrete's properties and potential damages, as will be seen in future sections of this document.

Table 3. Chemical composition of glass powder reported in UHPC research.

Components	Glass Powder References					
	[29]	[75]	[76]	[77]	[48]	[64]
SiO_2	72.89	75.47	72.2	73.00	71.4	72.76
Al_2O_3	1.67	1.09	1.54	1.5	1.4	1.67
Fe_2O_3	0.81	0.79	0.48	0.4	0.2	0.79
CaO	9.73	9.02	11.42	11.3	10.6	9.74
MgO	2.08	1.97	0.79	1.2	2.5	2.09
SO_3	0.01	-	0.09	-	0.1	0.10
Na_2O	12.54	11.65	12.85	13	12.7	12.56
K_2O	0.76	0.75	0.43	0.5	0.5	0.76
TiO_2	0.04	0.04	-	0.04	-	0.04

Figure 2 presents the X-ray diffraction (XRD) analysis conducted on WG particles utilized in UHPC research [12,49]. As can be observed, the reported XRD analysis of recycled glass particles reveal their amorphous nature, characterized by a minor presence of long-range atomic order. This amorphous phase, typical of soda-lime glass materials [47,49,50,59,78], is a key factor in their pozzolanic activity when used as glass powder in concrete [79,80]. The absence of crystalline structure allows for greater reactivity with calcium hydroxide during cement hydration, resulting in the creation of secondary C-S-H gel. This gel enhances the binding and strengthening properties of concrete, contributing to improved mechanical performance and durability [46,81]. Therefore, confirming the amorphous nature of recycled glass particles through XRD analysis is essential for evaluating their suitability as supplementary cementitious materials and predicting their effectiveness in enhancing concrete properties [50].

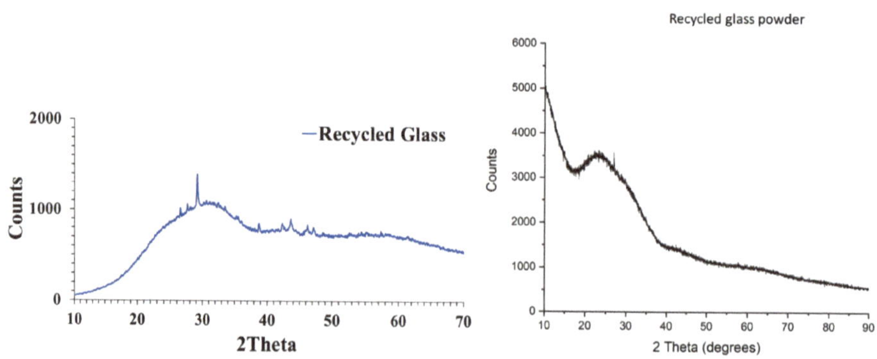

Figure 2. XRD analyses of milled waste glass used in UHCP formulations [12,49].

For its part, Table 4 presents some physical properties of recycled glass particles utilized in UHPC mixtures [24,42,59,64,75].

Table 4. Physical properties reported for milled waste glass particles utilized in UHPC [24,42,59,64,75].

Physical Feature	Value
Specific gravity	2.19–2.60
Water absorption (%)	0.19–0.4

Moreover, Figure 3 puts forward the SEM (Scanning Electron Microscopy) analysis of the particles of recycled waste glass utilized in several UHPC formulations [29,47,82,83]. As can be observed in Figure 3, these particles are characterized by their irregular shapes and angular edges, resulting from the mechanical grinding action applied during production. Their smooth surface and lack of porosity correlate well with the low water absorption depicted in Table 4.

(a) (b)

Figure 3. SEM analysis of milled waste glass used in UHPC formulations: (**a**) recycled glass powder (d_{50}: 7 μm) used as partial cement replacement [82], and (**b**) recycled glass flour (d_{50}: 28 μm) utilized as total QP replacement [29].

3.4. Chemical Reactions

Figure 4 illustrates the reactions of GP within concrete. Initially (Figure 4A), solid amorphous silica is depicted, followed by its dissolution in water (Figure 4B). The silica reactivity is contingent upon its amorphous silica rate and the pH of the pore solution, which dictates the dissolution rate. Equation (1) [84] describes this process.

$$\equiv Si - O - Si \equiv\ + H_2O \rightarrow 2(\equiv Si - OH) \quad (1)$$

When dissolved silica interacts with hydroxide ions OH^- in mixed concrete (see Figure 4C), three types of gel are produced, each contingent upon the environment. The first type is calcium silicate hydrate (Figure 4D), characterized by a pH range of 12–13 and an abundance of calcium ions (Ca^{2+}). The second type is calcium aluminate silicate hydrate (Figure 4E), which shares a similar pH range of 12 or 13 and is rich in both calcium ions (Ca^{2+}) and aluminum ions (Al^{3+}). Lastly, alkali silica gel (Figure 4F) exhibits a higher pH exceeding 13 and is enriched with sodium ions (Na^+) and potassium ions (K^+) [84]. These three types of gel are expounded on in more detail in the next subsections.

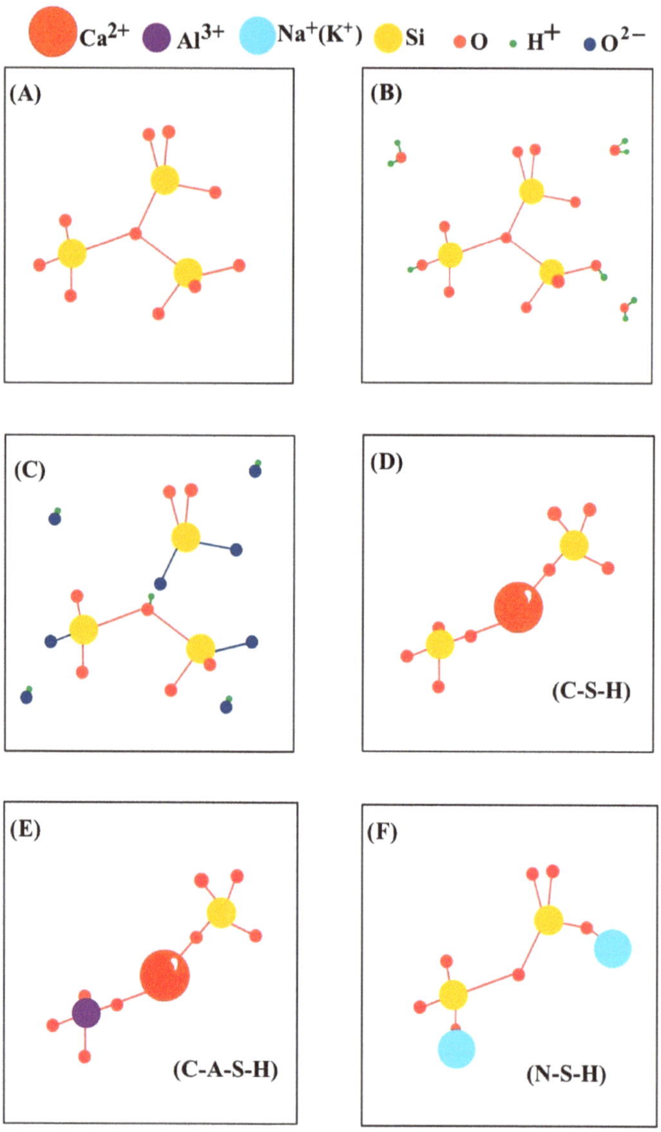

Figure 4. Glass reactions as a UHPC-making component [85]: (**A**) Solid amorphous silica; (**B**) Amorphous silica dissolution in water; (**C**) Dissolved silica interaction with hydroxide ions OH^-; (**D**) Calcium silicate hydrate structure; (**E**) Calcium aluminate silicate hydrate structure; (**F**) Alkali silica gel structure.

3.4.1. Calcium Silicate Hydrate Gel

The main product of Portland cement hydration is C-S-H gel, which plays a crucial role in providing strength. This gel is characterized by its extensive and disordered atomic structure, contributing to the formation of a family of solubility curves within the $CaO - SiO_2 - H_2O$ system. Structurally, most layers resemble tobermorite, while others exhibit imperfections akin to jennite. In addition to cement hydration, this gel can be produced by the pozzolanic reaction between CH and a material with amorphous silica, like GP, in the presence of water [65,86]. Equation (2) [84] illustrates the pozzolanic reaction responsible for C-S-H production, where the coefficients x_1 and x_2 represent variable numbers [84].

$$SiO_2 + Ca^2 + 2OH^- \rightarrow x_1CaO \times SiO_2 \times x_2H_2O \quad (2)$$

Figure 5 represents graphically that the pozzolanic reaction occurs when the SiO_2 reacts with CH and creates C-S-H gel. Investigations showed that the pozzolanic reactivity of WG as a replacement for several percentages of the weight of cement (0, 15, 30, 40 and 60%), and this demonstrated that CS was not reduced by the substitution cement for WG. The pozzolanic reaction between waste glass or glass powder and cement hydration products allows the concrete not to lose strength. In the majority of replacement percentages, the compression resistance is 85% and allows for enhancements in durability owing to the refined microstructures of GP [51].

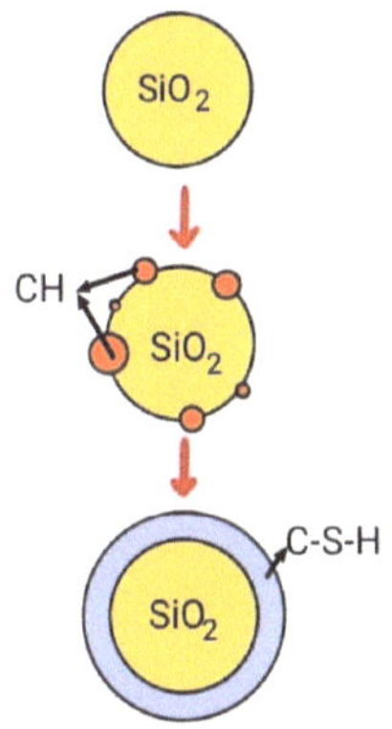

Figure 5. Pozzolanic reactions [85].

3.4.2. Calcium Aluminate Silicate Hydrate

Calcium aluminate silicate hydrate (C-A-S-H) is a calcium silicate hydrate, but this is incorporating aluminum into its structure, as depicted in Figure 4A. Its usual atomic ratio of Al: Si is under 0.25. As can be seen in Figure 4E, the C-S-H adds an aluminum and this gel is shaped during Portland cement hydration in the presence of Al^{3+} ions [87].

To yield C-A-S-H gel the ions OH^- broke the bonds of Al-O and Si-O with a necessary high pH (between 12–13) to form $SiO_2(OH)_3{}^-$, $SiO_2(OH)_2{}^{2-}$ and $Al(OH)_4{}^-$. Equation (3) [88] expresses the situation when the dissolved aluminate and silica react with OH^- and Ca^{2+} to produce C-A-S-H gel.

$$Ca^{2+} + SiO_2(OH)_2{}^{2-}\left(orSiO_2(OH)_3{}^-\right) + Al(OH)_4{}^- \rightarrow C-A-S-H \quad (3)$$

3.4.3. Alkali Silica Gel

The third and last gel, the alkali-silica gel, is the one that is formed during the ASR reaction [79,89,90]. The alkali-silica reaction is considered a concrete disease because it is one of the main risky chemical reactions that is generated between aggregates with some reactive silica content alkalis of the cement (sodium hydroxide and/or potassium sodium) in the concrete. This consists of the generation of an expansive gel that meets water and produces an increase in the matrix volume, which yields the concrete mass to crack and destroy itself. Equation (3) [84] describes how the silica amorphous network becomes more reactive due to high pH (greater than 13) causing the siloxane chemical bond to be broken with the presence of OH^- to form new ions [85].

$$\equiv Si-O-Si\equiv + OH^- \rightarrow \equiv Si-OH + \equiv Si-O^- \quad (4)$$

Once crystalline silicates bonds are formed ($\equiv Si - OH$) they can react with OH^- and produce new ions, as presneted in Equation (5) [85].

$$\equiv Si - OH + OH^- \rightarrow \equiv Si - O^- + H_2O \quad (5)$$

As shown in the Equation (5), the $\equiv Si - O^-$ which in turn reacts with potassium (K^+) and sodium (Na^+) also, with OH^- to produce alkali silica gel (N-S-H).

$$\equiv Si - O^- + 2Na^+ (K^+) + 2OH^- \rightarrow Na_2(K_2)SiO_3{\cdot}H_2O \quad (6)$$

To mitigate the risk of ASR, several factors need careful control, including the water-cement ratio, aggregate size, alkali solubility, type of cement, alkali content, utilization of supplementary cementitious materials and the presence of reactive aggregates [88]. In this regard, the employment of WG powder in concrete composition offers high amounts of amorphous silica (see Table 3 and Figure 2) as well as alkalis (Table 3). Thereby, special care should be taken when incorporating this waste into concrete formulations. However, recent investigations consider that the GP should pass a 325-mesh sieve because the particles smaller than 300 microns do not present a risk for alkali-silica reaction [91].

4. The Use of Waste Glass Powder in UHPC Formulations

4.1. GP as Supplementary Cementitious Material

Various pieces of research have been conducted to define the role of milled WG as a replacement for typical UHPC-making powdered materials like cement, silica fume or quartz powder [3,44,52,75,92]. Results have pointed out how various properties of glass powder, such as particle size, impact the characteristics of UHPC [75,77]. These studies have examined the effects of glass particle size from fine powder (d_{50}: 3.8 μm), powder (d_{50}: 7 μm) or flour (d_{50}: 28 μm) [75,77]. Depending on the particle size, these pieces of research have utilized this waste material for substituting (totally or partially) SF, cement and quartz powder [47,75]. As packaging density and particle-size distribution of components determine UHPC design, Soliman & Tagnit-Hamou [77] employed fine glass powder (d_{50}: 3.8 μm) to partially replace SF in UHPC formulations. The findings of their research showed that substituting 30% and 50% of SF with this fine GP may obtain compressive strengths of 235 and 220 MPa after 2 days of heat curing, in comparison with 204 MPa for the control UHPC formulation. Moreover, the study also depicted that when SF particles were replaced with nonabsorptive glass particles, the fresh UHPC rheology was enhanced [77].

For their part, Abellán et al. [75] achieved CS over 150 MPa with more than 50% of replacement percentages in traditional materials like cement and quartz powder (QP) with glass powder (GP) and glass flour (GF). Their findings revealed that specific combinations led to a reduction in cement content, identifying a mixture comprising 603 kg of cement and over 0.5 kg/m^3 of glass powder (GP) and glass fibers (GF) per kg of cement. This formulation achieved a CS of 152 MPa within 28 days under standard curing.

Tagnit-Hamou et al. [52] studied the implication of the curing temperature and the amount of water in CS of waste glass–UHPC at 2, 28 and 91 days. The tests pointed out that the inclusion of GP as a cement partial replacement increases the spread flow, a factor that allows it to work with a lower w/b [27,75]. Also, with more than 20% of cement replacement with GP, the CS at 28 days was slightly lower than the control [52], but this pattern changes with 91 days of normal curing and with hot curing. This can be considered because of the pozzolanic reaction from the GP particles which need more time to react than SF due to their larger size and, therefore, smaller specific surface area. Other researchers confirm this behavior, pointing out that the key to reaching a faster pozzolanic reaction is the GP particle size [3,75], an outcome that reacts better with CH.

The synergy between milled waste glass powder and other mineral admixtures in UHPC has been extensively explored in various studies. Table 5 summarizes some of these endeavors.

Table 5. Reported research utilizing waste glass combined with other mineral admixtures in UHPC formulations.

Total Powders (kg/m^3)	% Cement	% SF	% WGP	Other Mineral Admixture		Other Mineral Admixture		w/b	Compressive Strength (MPa)	Reference
				Type	%	Type	%			
1180–1727	59.88–90.00	6.83–10.00	0.00–27.03	Rejected-FA	0.00–31.67	-	-	0.15	148–158	[48]
1147–1246	58.99–88.13	0.00–11.87	0.00–33.06	QP	0.00–15.42	-	-	0.15–0.25	153–221 *	[42,64]
1151–1282	47.26–63.38	8.76–17.63	19.03–37.73	-	-	-	-	0.15–0.20	100–177	[1]
1035–1052	78.23–78.31	0.00–21.74	0.00–21.77	-	-	-	-	0.21	100–175	[77]
1131–1147	50.05–58.72	8.76–8.84	7.83–13.92	Limestone	20.37–21.01	FC3R	4.38–4.40	0.16–0.18	132–156	[27]
1158–1210	48.80–56.84	8.26–8.63	25.52–31.66	Limestone	7.60–11.25	-	-	0.14–0.16	101–162	[75]
1287	37.95–48.17	7.70	24.08	Limestone	19.98	MK	0.00–10.21	0.16–0.18	129–159	[3]
1190	56.80	11.93	31.26	-	-	-	-	0.18	155	[15]
1282	46.02	7.80	26.13	Limestone	20.05	-	-	0.15	156	[9,10,18,93]
1165–1195	43.24–59.15	8.36–8.58	3.85–18.04	Limestone	21.79–35.96	EAFS	2.19–14.29	0.14–0.16	100–163	[29]
1205–1280	45.55–57.01	7.65–7.80	26.01–33.02	Limestone	10.15–30.22	-	-	0.14–0.21	142–156	[58]
1259	38.80–49.25	7.94	25.23	Limestone	17.58	RHA	0.00–10.45	0.16–0.18	139–159	[94,95]

* Heat curing conditions.

A multi-criteria optimization technique was carried out to design two eco-friendly UHPCs incorporating micro-limestone powder and two different sizes of GP as cement and SF partial replacements [47,59,75]. The UHPC formulations analyzed, which included a Control dosage without alternative binders, as well as their scheme of pastes' packing density, are presented in Table 6 and Figure 6, respectively. The optimized mixtures exhibited excellent workability and rheological properties, achieving high compressive strength with reduced cement content. Furthermore, new research has corroborated the synergy between glass waste and limestone powders [47,59,96]. Particularly, their joint use demonstrated an improvement in rheology that allowed a reduction in the superplasticizer content and water-to-binder ratios with respect to conventional UHPC dosages. The reductions mentioned carry significant positive implications for the material properties due to two main reasons [94,97]:

(i). Lowering the w/b ratio drives reduced porosity within the material.
(ii). Decreasing the superplasticizer content mitigates the risk of potential incompatibility with cementitious materials.

Table 6. Control, Optimized 1 and Optimized 2 UHPC formulations [47,59].

	Control	Optimized 1	Optimized 2
OPC (kg/m^3)	852	590	603
SF (kg/m^3)	272	100	100
Waste glass powder (kg/m^3)	-	-	169
Waste glass flour (kg/m^3)	-	335	199
LP (kg/m^3)	-	257	118
SS (kg/m^3)	889	778	834
HRWRA (kg/m^3)	26.5	21.5	20.0
w/b	0.218	0.165	0.163
Slump flow (mm)	260	253	258
VPD *	0.83	0.81	0.79

* Virtual Packing Density as per the Sedran & Larrard theory [19,98,99].

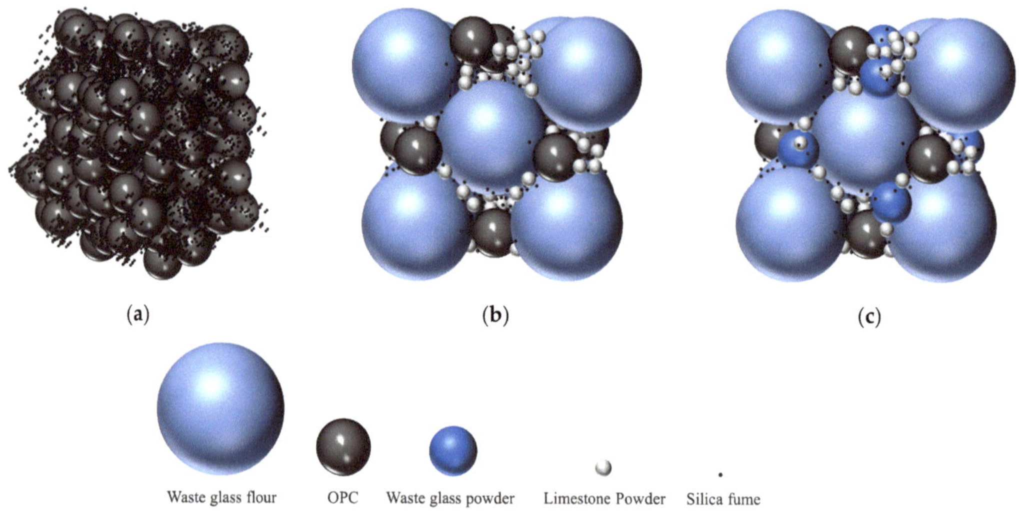

Figure 6. Scheme of paste packing density for three UHPC formulations (**a**) Control mixture without alternative binders; (**b**) Optimized 1 mixture with micro-limestone powder and waste glass flour (d_{50}: 28 μm); (**c**) Optimized 2 mixture with micro-limestone powder, waste glass powder (d_{50}: 7 μm) and waste glass flour (d_{50}: 28 μm) [82].

Similarly, [29] investigated the factorial design of an ultra-high strength mortar containing Electric Arc Slag Furnace (EASF) along with milled waste glass. The study highlighted the importance of incorporating supplementary cementitious materials for enhancing both the mechanical strength and sustainability of UHPC. The effect of high reactive aluminosilicates cementitious materials, like metakaolin (MK) and Fluid Catalytic Cracking Catalyst Residue (FC3R), in waste glass–UHPC was explored [3,27]. Due to the reaction of the high alumina-silicate components of FC3R and MK, yielding ettringite formation, those cementitious materials showed potential for enhancing early-age strength, but their impact on long-term strength and workability was observed to vary based on substitution levels. Hence, the findings of these investigations illustrate the collaborative effect of limestone and glass powder in alleviating rheological issues in mixtures containing FR3C or MK. Thereby, with the optimal dosage of these high alumina materials, the effect of glass and limestone allows the UHPC to exhibit beneficial properties for early strength development, all the while maintaining cost-effectiveness, a proper rheology feature in the fresh state, and reducing the carbon footprint [3,27].

The literature review, as depicted in Table 5, highlights another significant aspect concerning the utilization of GP in these specialized concrete formulations. Specifically, UHPC compositions incorporating GP as a partial or complete substitute for QP have exhibited impressive compressive strengths exceeding 200 MPa [42,64]. This threshold value is crucial for ensuring strong fiber-matrix adherence [100,101], emphasizing the promising potential of glass powder in enhancing the performance of UHPC formulations.

For its part, several investigations have focused on the utilization of rejected fly ash along with recycled glass powder in UHPC formulations [28,48]. On the one hand, the research presented in [28] explored the utilization of local high unburned carbon fly ash as a mineral admixture in waste glass–UHPC. On the other hand, research presented in [48] depicted the utilization of fly ash with particles larger than 45 microns along with recycled glass in alternative formulations of UHPC. Despite challenges in achieving optimal substitutions owing to the elevated content of rejected fly ash, these studies emphasized the potential of utilizing local pozzolans for sustainable UHPC production. Lastly, recent studies [94,95] analyzed the effect of rice husk ash (RHA) as a cement partial replacement for UHPC containing waste glass powder. Despite a decrease in workability, attributed to the high surface area and porous nature of RHA particles which increases water demand [102,103], and a slight reduction in CS, the studies highlighted the potential of RHA for enhancing sustainability and reducing the carbon footprint of UHPC.

Overall, these studies summarized in Tables 5 and 6 underscored the significance of synergistic interactions between waste glass powder and other mineral admixtures in enhancing the sustainability and performance of UHPC formulations. A deeper analysis of the influence of these interactions on each of the material characteristics is presented in the following subsections.

4.2. Waste Glass as a Replacement for QS

UHPC formulations traditionally incorporate quartz sand (QS) as the aggregate, with quartz powder (QP) enhancing mixture density [14,104]. However, QS and QP are known carcinogens, posing health risks to workers and impacting the environment [69,105]. As reported in published research, the replacement of these materials with recycled glass sand (GS) could offer a safer and more sustainable solution [1,49,69,106]. These findings are summarized in Table 7. As can be seen, these studies have explored different replacement ratios and particle size distributions of recycled glass sand, aiming to balance workability and mechanical properties [41,69]. Notably, UHPC formulations with GS have demonstrated exceptional compressive strengths exceeding 200 MPa under thermal curing conditions [1]. These results suggest the potential of GS in UHPC as a viable alternative for even the highest levels of improved microstructure (identified by CS greater than 190 MPa) which indicates a possible exceptional fiber-matrix adhesion [100,101].

Table 7. Results of compressive strength of UHPC with glass sand.

Glass Sand d_{50} (μm)	Glass Sand d_{max} (μm)	Quartz Sand Replacement Percentage (%)	Compressive Strength Normal Curing (MPa)	Compressive Strength Heat Curing (MPa)	Reference
275	630	50	171 ***	196 *	[69]
275	630	100	157 ***	182 *	[69]
350	630	100	128 ***	153 *	[69]
225	630	100	127 ***	164 *	[69]
300	600	100	124 **	-	[50]
-	800	25	175 ***	200 *	[1]
-	800	50	165 ***	190 *	[1]
-	800	50	162 ***	175 *	[1]
275	630	50	140 ***	150 *	[106]

* After 2 days of heat curing; ** after 28 days of normal curing; *** after 91 days of normal curing.

Additionally, research on alternative glass UHPC formulations revealed that varying combinations of glass components, such as GP, fine glass powder (FGP) and GS, significantly influenced workability and mechanical performance, with CS values of up to 171 MPa observed even under normal curing conditions [77].

Moreover, the study conducted by [49,50] showcased the highest proportion of recycled glass utilization in UHPC formulations. These formulations replaced 100% of quartz sand with GS and incorporated various sizes of GP, resulting in up to 52% of the mixture by weight being comprised of recycled glass (both sand and powder). Finally, it is important to note the risk of ASR associated with glass sand [91], which will be addressed in a specific section of the review article.

5. Effect of WG Addition on UHPC's Features

5.1. Rheological Properties

Concrete's rheological properties are a measure of its ability to flow and resist deformation. In the case of UHPC, rheological properties are especially important due to the high particle density and low w/b [97,107]. The utilization of GP as a supplementary cementitious agent can significantly improve these properties of UHPC [40,62,96]. In this sense, the research conducted by [75] revealed that replacing some of the cement with GP in UHPC formulations leads to a higher spread flow due to the glass powder's minimal water absorption and the consequently higher alkaline content resulting from the recycled glass's elevated Na_2O levels [75]. The high-level alkali essence and the low water absorption of glass particles are well known, and they are depicted in Tables 2–4 of the present document. On the one hand, utilizing GP in cement partial replacement capitalizes on its significantly lower water absorption [1,69], thereby liberating more free water to effectively contribute to the material's rheological behavior in its fresh state [58,75]. On the other hand, the alkalinity provided by GP can enhance the performance of superplasticizers in cement pastes by facilitating the dispersion of cement particles and reducing their agglomeration. This improved dispersion allows the superplasticizer molecules to effectively adsorb onto the cement particles, leading to better fluidity and workability of the paste [108].

The impact of incorporating GP into the mixture on UHPC's rheology is evident in Figure 7 [12]. On the left, the slump flow of a blend devoid of recycled glass is depicted, whereas the image on the right illustrates the outcome following the substitution of 13.5% of the cement with GP. It is notable that the mixture lacking GP exhibits a slump flow of 210 mm, while the addition of GP results in a higher slump flow of 250 mm, despite maintaining the same w/b ratio and dosage of high-range water-reducing polycarboxylate-based superplasticizer (HRWR) (w/b = 0.165 and HRWR dosage of 23 kg/m^3).

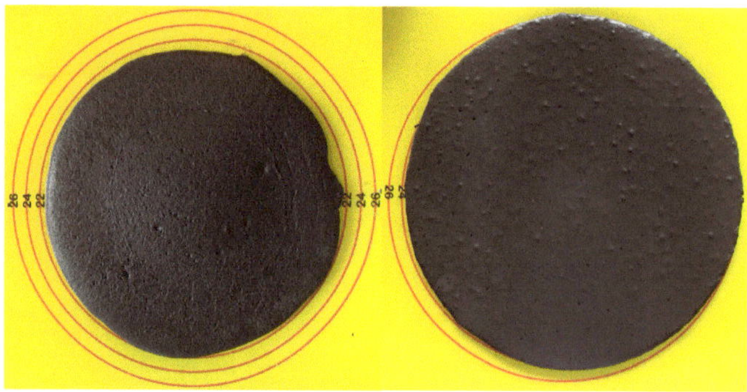

Figure 7. Impact of GP on UHPC's rheology. The left image shows the slump flow of a GP-free mixture, contrasting with the right image depicting the effect of replacing 13.5% of cement with GP [12].

Other studies have assessed the UHPC mechanical and rheological features incorporating varying proportions of glass powder as a pozzolanic mineral admixture [1,41,63]. Their findings indicated that the GP addition enhanced workability and flowability without compromising strength and durability. Notably, the incorporation of glass powder mitigated exudation and bleeding, thereby improving bonding capacity and resistance to segregation. Finally, the positive influence of recycled glass together with limestone is observed in Table 6, which shows that thanks to the incorporation of these components in cement and SF partial replacement, the optimized dosages achieve the same value of slump flow as the Control mix, but with a lower w/b value and lower superplasticizer content [47,59].

5.2. Hydration Kinetics

Comprehending the hydration is vital for understanding the behavior of cementitious materials, particularly in the context of advancing UHPC [41]. The study delved into the cement hydration kinetics, focusing on the influence of GP addition, using isothermal calorimetry to examine the pozzolanic reaction during the cement hydration early stages [41]. The investigation scrutinized the rate at which UHPC emits heat during hydration and the cumulative heat curves of hydration for varying levels of glass powder replacement (0%, 20%, 40% and 50%) within the initial 48 h of water-cementitious material contact, normalized to the total binder weight in the mixture. It became apparent that both the maximum heat flow and total heat decreased with increasing GP replacement ratios. Moreover, the study revealed that the maximum value of the second exothermic peak decreased by 35%, 32% and 50% for the mixtures containing 20%, 40% and 50% glass powder, respectively, in comparison with a typical UHPC. The decrease in heat release was attributed, on the one hand, to the dilution effect caused by the incorporation of glass powder, which, in the reported research, possessed a lower specific surface area than cement [41]. On the other hand, the GP pozzolanic reaction produces less heat than ordinary Portland cement due to its similarity to a C2S reaction [109]. This lower heat of hydration aided in minimizing cracking as a consequence of the elevated temperatures [41].

Moreover, when GP replaced 20% of the cement, the end of the induction period and the acceleration period ascribed to the major peak were shorter [41]. This acceleration was explained by the ability of fine GP to expedite the hydration of cement by adsorbing calcium ions and acting as nucleation sites for hydrate formation. Additionally, the high alkali content in GP was probed to catalyze the early formation of C-S-H [41]. Conversely, with GP replacement of over 20%, the periods of induction and acceleration were delayed. For instance, at the end of the induction period for typical UHPC formulation, and formulations with 40 and 50% cement replacement by GP were about 9.1, 9.7 and 11.2 h, respectively. This delay was attributed to increased levels of GP replacement, resulting in higher water-to-cement ratios and superplasticizer dosages, as is common in UHPC manufacturing [41]. This, according to Jansen et al. [110], could lead to the complexation of Ca^{2+} ions by the superplasticizer, impeding the early hydration of cement and delaying the pozzolanic process owing to the insufficient CH presence in the concrete.

5.3. *UHPC's Microstructure*

The cement-based materials' mechanical properties are influenced by various factors, such as their chemical characterization, microstructure, aggregate features and the ITZs characteristics [21,111]. Understanding the microstructure of these matrices is key to unraveling their mechanical behavior. Although UHPC presents a highly heterogeneous and intricate microstructure, efforts to create realistic models remain challenging [112–114]. Nevertheless, delving into the microstructure of UHPC allows for the determination of optimal mix compositions, which in turn yield desirable fresh and mechanical properties and enhanced durability, thereby reducing production costs and related CO_2 emissions [56,113,115].

Therefore, microstructure analysis procedures like X-ray diffraction (XRD), scanning electron microscopy/energy-dispersive spectroscopy (SEM/EDS), mercury intrusion porosimetry (MIP), ^{29}Si magic-angle sample-spinning nuclear magnetic resonance (^{29}Si MAS NMR) analysis and thermogravimetric/differential thermal analysis (TG/DTA) have been extensively reported in the literature as methodologies to gain insights into the microstructure of UHPC [41,64,116]. In this sense, studies on UHPC microstructure indicate that incorporating cement partial replacements like limestone and granulated blast furnace slag not only reduces unhydrated cement content and the C2S/C3S ratio but also diminishes CH levels while promoting the creation of C-S-H, resulting in a more compact and homogeneous cementitious matrix. Among the various pozzolans assessed, silica fume stands out for its ability to decrease porosity, lower the calcium/silicon (Ca/Si) ratio and maintain minimal CH levels, all contributing to defining optimal mechanical specifications for UHPC. Furthermore, the inclusion of blast furnace slag, electric arc furnace slag and limestone powder as mineral admixtures aids in further reducing UHPC matrix porosity and allows for a reduction in cement consumption. While each substitute offers specific advantages, their optimized use strengthens the mechanical or maintains (depending on the case) the final properties of UHPC while supporting sustainable waste management practices [21,29,56,117].

Additionally, integrating nano-silica as a cement replacement presents a promising avenue to enhance the mechanical characteristics of UHPC further. Nano-silica fosters the formation of a denser matrix with reduced porosity and increased C-S-H content [117]. Hence, considering the properties of recycled glass discussed in earlier sections (e.g., chemical composition and crystallography), it is reasonable to anticipate that the addition of GP in UHPC formulations would yield a beneficial influence akin to that of reported mineral admixtures. Soliman & Tagnit-Hamou [41] conducted some research to analyze the influence of the cement and/or quartz powder partial replacements with glass powder- with d_{50} equal to 50 μm—on the UHPC's microstructure by utilizing SEM approaches. Three samples were evaluated: (i) a control one, which represented a typical UHPC formulation; (ii) 80C/20GP formulation in which the 20% of the weight of cement was replaced by GP; and (iii) 0QP/100GP in which QP was totally replaced by GP. Their findings using backscattered electron (BSE) images revealed that the ITZ in the control mix was thin, and the addition of GP did not significantly affect it. Moreover, owing to the low w/b value, a considerable amount of unreacted cement, QP particles and GP particles were observed in the matrix. The study also revealed that the absence of portlandite (CH) indicated its consumption by the SF plus GP pozzolanic effects. Another conclusion that can be drawn from their BSE research is that the design of the UHPC promoted a dense and homogeneous matrix with reduced porosity, with the presence of spherical air voids being a consequence of the high superplasticizer content. Moreover, BSE images after thermal treatment revealed the formation of a C-S-H hydration layer around the cement and GP particles, demonstrating the pozzolanic reaction of GP and its contribution to the improved microstructure [118–120]. Regarding the characteristics of GP particles, it was observed that their appropriate fineness prevented the formation of microcracks associated with the ASR reaction. Their study proved that the incorporation of GP yielded the formation of more C-S-H, thereby improving the microstructure of UHPC. In addition, their findings also revealed that the presence of GP particles did not adversely affect the ITZ or increase porosity, suggesting adequate compatibility between GP and the cement matrix. This highlights that GP can be an effective additive for enhancing the properties of UHPC without compromising its internal structure [41].

For their part, Vaitkevicius et al. [64] carried out a study to investigate the effects of GP as a complete replacement for QP and SF in UHPC formulations. Through experimental analysis using techniques such as MIP, XRD and ^{29}Si MAS NMR, the study elucidated the beneficial reactions of GP with cement phases, leading to the formation of C-S-H and subsequent enhancements in mechanical strength and microstructure. Notably, the research highlighted the elimination of macroporosity, increased cement dissolution rate, and the role of fast soluble alkali from the surface of GP in accelerating cement hydration.

5.4. Mechanical Properties

Building on the insights gleaned from earlier sections regarding its effects on rheology, kinetics and microstructure, the incorporation of glass powder into UHPC is anticipated to yield enhancements or minimal reductions in its mechanical features. Nevertheless, the following sections of this review article will analyze the impact of the addition of GP in the UHPC formulations on their CS, elasticity modulus, ultrasonic pulse velocity, bending performance and direct tensile behavior.

5.4.1. Compressive Strength

The effectiveness of WG powder in enhancing the CS of concrete, particularly in UHPC mixtures, has been extensively explored [55,58,117]. Some of these results are presented in Table 8. The role of GP on the CS of concrete is a consequence of different mechanisms. For instance, the implementation of a particle size of GP finer than cement particles reduces mixture porosity due to the filler effect as a replacement for cementitious materials, leading to a trend of increasing CS. Specially, these fine particles of glass also promote the cement hydration in a UHPC mixture by increasing the hydration surface-volume ratio and the filler effect, thus producing more hydration products owing to the pozzolanic reactions of fine glass and increasing the creation speed of hydrates that improve the concrete's CS [40,41,51]. In addition, this pozzolanic reaction of glass is due to the amorphous silica composition (as seen in Table 3 and Figure 2) that reacts with CH producing secondary C-S-H gel that mostly provides this mechanical property [120]. Hence, researchers have investigated various sizes and replacement percentages of GP, revealing notable effects on concrete strength [40,41,77]. For instance, studies have applied GP with a d_{50} particle diameter of 10 μm, replacing 25% of the cement, leading to UHPC exhibiting compressive strength values ranging from 150 to 200 MPa after 91 days of standard curing conditions [40]. Similarly, investigations utilizing GP with a particle size of 12 μm as a partial cement replacement in UHPC with a water-to-binder ratio of 0.189 demonstrated a significant increase in CS, reaching up to 204 MPa at 91 days under normal curing conditions [41].

Table 8. Results of WGP application in UHPC compressive strength from 28 to 91 days of SC.

Replaced Material	Percentage of Replacement (%)	GP Fineness d_{50} (μm)	Concrete Strength Trend	Reference
Cement	20	12	Increase of 13%	[41]
Cement	50	12	Decrease of 9%	[41]
Cement	25	10	Increase of 11%	[40]
Silica Fume	30	3.8	Increase of 8%	[77]
Silica Fume	50	3.8	Increase of 1%	[77]
Silica Fume	100	3.8	Decrease of 16%	[77]
Quartz Powder	100	12	Increase of 17%	[41]
Quartz Powder	50	12	Increase of 11%	[41]
Quartz Powder	100	10	Increase of 19%	[40]

However, as shown in Table 8, substitution ratios of cement with GP exceeding 25% were found to be ineffective, leading to a reduction in the CS [40,41,51]. This phenomenon can be attributed to the delicate balance between two key factors in the UHPC matrix: the creation of portlandite during cement hydration and the pozzolanic activity of glass powder [83]. This way, when a relevant cement portion is replaced with GP, the amount of portlandite generated may decrease due to the reduced cement content, which can lead to several of the pozzolanic materials remaining unreacted [75]. For its part, the lowering of cement also leads to a decrease in the primary C-S-H gel obtained by cement hydration [61,62]. Therefore, the decrease in compressive strength observed at higher replacement ratios of cement with GP suggests that the pozzolanic activity of the glass powder may not fully compensate for the decrease in C-S-H from cement hydration for the reduction in portlandite content.

Furthermore, studies have explored the use of GP as a partial or complete substitution for SF and quartz powder QP in UHPC mixtures. When it comes to the case of silica fume, fine GP, with d_{50} values of 3.8 μm, was applied as a partial and complete replacement for SF in a UHPC formulation [74]. The findings for UHPC compressive strength are provided in Figure 8. According to the reported study results, and as can be seen in Table 8, replacements of over 50% lead to a decrease in the CS value. The reason for that could be explained by the following: (i) the improvement of the packing density provided by the reduced size of SF, with d_{50} values of 0.15 μm, can not be attained with glass particles. A proof of the latter can be observed in Table 6, where the VPD of the Control formulation is higher than those of the optimized values [47,59]; (ii) the improved pozzolanic nature of SF than GP, due to its reduced size and higher amorphous silicon oxide [59,63,121].

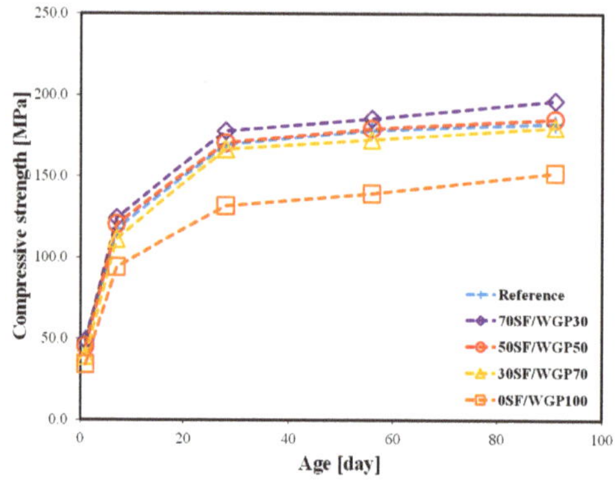

Figure 8. Results of UHPC CS of silica fume replacement with fine d_{50} 3.8 GP [77].

Moreover, in another study, glass powder (GP) was introduced as a substitute for QP in a UHPC mixture with a w/b value of 0.189 [41]. The complete replacement of QP with GP led to an enhancement in the compressive strength of the concrete, ranging from 30 MPa to 35 MPa after 91 days of standard curing (SC). The pozzolanic reaction between waste glass powder and cement-hydrated products facilitated an accelerated hydration process, consequently augmenting the strength of the UHPC [40,41].

5.4.2. MoE

The modulus of elasticity is a measure of the stiffness or ability of a material to resist deformation under applied loads. In the case of UHPC, according to the ACI-239R, its value evaluated after 28 days is commonly within the range of 40 to 50 GPa [122]. UHPC's MoE can vary depending on factors like material composition, particle size and distribution, density and porosity [123]. However, in general, the modulus of elasticity of UHPC is much higher than that of other types of concrete, due to its stiffness and the fact that it can withstand larger loads before permanently deforming [123,124].

Jaramillo-Murcia et al. [58] analyzed the modulus of elasticity of two alternative UHPC formulations that incorporated waste glass in their composition. To establish comparisons, a typical UHPC formulation was also tested. The information on the considered formulations can be observed in Table 6 while and scheme of their paste's structure is presented in Figure 6. The findings of this study, measured at different ages, are put forward in Figure 9.

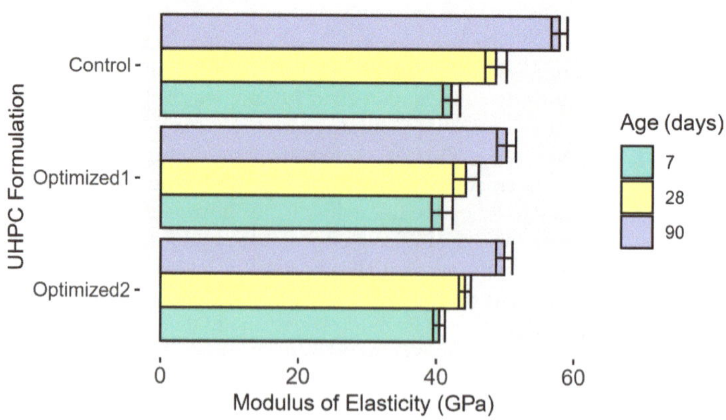

Figure 9. MoE of control and optimized UHPC formulations evaluated at 7, 18 and 90 days [58].

Throughout their study, it was noticed that the Control formulation consistently exhibited a higher Modulus of Elasticity (MoE) compared to the waste glass formulations at each stage. Nevertheless, all examined UHPC formulations met the Modulus of Elasticity criteria outlined by ACI 239 [5]. The rate of MoE increase over curing time varied among formulations containing different supplementary cementitious materials [125]. Incorporating pozzolans with higher activity, such as SF, resulted in accelerated enhancement of MoE during the initial stages due to their smaller particle size and faster hydration kinetics. Additionally, the improved packing density of the Control mix contributed to its improved MoE. Nevertheless, differences observed in younger specimens were slightly less marked in comparison with the older ones. This could be attributed to the influence of limestone powder and GP, which, as suggested by previous studies [12,75], speeds up the hydration process in the early stages, driving denser particle structures and enhanced mechanical properties. Moreover, the GP addition was found to expedite the cement dissolution rate, thereby facilitating a faster hydration process, as stated in previous sections [40,41,51].

5.4.3. Ultrasonic Pulse Velocity

Upon analyzing the ultrasonic pulse velocity data, on the same three UHPC formulations presented in the previous section, Jaramillo-Murcia et al. [58] reported a consistent upward trend in values over time. Figure 10 depicts the results obtained by [58] in relation to this property. Notably, as Figure 10 reveals, the Control mixture exhibits the highest response, followed by the Optimized 2 mixture. The authors attributed this increase in velocity to the VPD of the UHPC's formulations (see Table 6), resulting in increased stiffness of the material and accelerated pulse wave propagation within the UHPC cylinder. Consequently, as per their findings, the ranking of UHPC types as per their VPD values corresponds to the dosages determined by their ultrasonic pulse velocity at any given age. Moreover, this research also suggests a significant correlation between velocity enhancement and the reactive powders' hydration and pozzolanic activities [58].

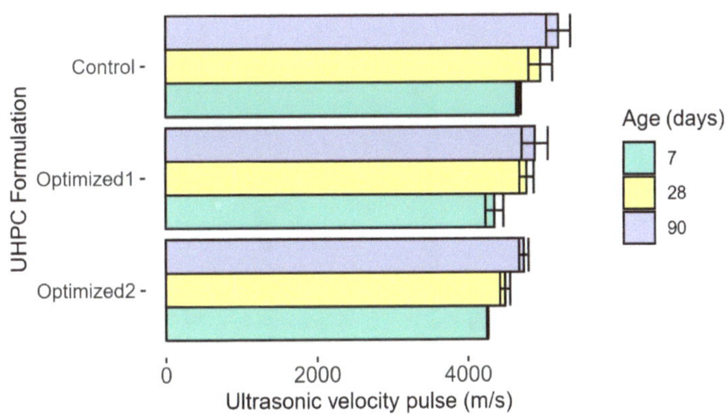

Figure 10. Ultrasonic pulse velocity findings of control and optimized UHPC formulations evaluated at 7, 18 and 90 days [58].

The study also established that concrete of superior quality typically exhibits an ultrasonic pulse velocity exceeding 4575 m/s. As illustrated in Figure 10, the Control dosage reaches the threshold value after seven days of curing, while Optimized 1 requires 28 days, and Optimized 2 necessitates 90 days to achieve a similar threshold value. It is worth mentioning that the mix in Optimized 2 at 28 days achieves a UPV value of 4491 m/s [58].

5.4.4. Flexural Behavior

The literature review shows that substituting some of the Portland cement with GP diminishes the resistance to bending in plain UHPC formulations [15,62]. However, while it is notable that the flexural strength lowers with increasing substitution of GP in the case of plain UHPC, it is important to recognize that in fiber-reinforced UHPC, the primary determinant of performance is the fiber reinforcement system [126–128]. Neira-Medina et al. [15] researched to analyze the impact of different fiber-reinforced systems on the flexural behavior of UHPC utilizing a GP as a cementitious material. The UHPC cementitious matrix was characterized by a w/b value of 0.25, and glass powder with a d_{50} value of 10 µm was utilized as a cement replacement at a rate of substitution by weight of 45%. The mix design was formulated based on fractal-based particle-packing theories [15]. The experimental campaign included four types of commercial synthetic fibers, both macro and micro, along with high-strength steel fibers, for evaluation purposes. In addition, the fiber reinforcement system encompassed mono and hybrid systems with volume ratios of reinforcement from 1 to 2%, whose results are depicted in Figures 11 and 12, respectively. Notably, only the UHPC specimens reinforced with 2% fibers demonstrated ductile behavior, with the exception of the beam reinforced with 1% steel microfibers [15].

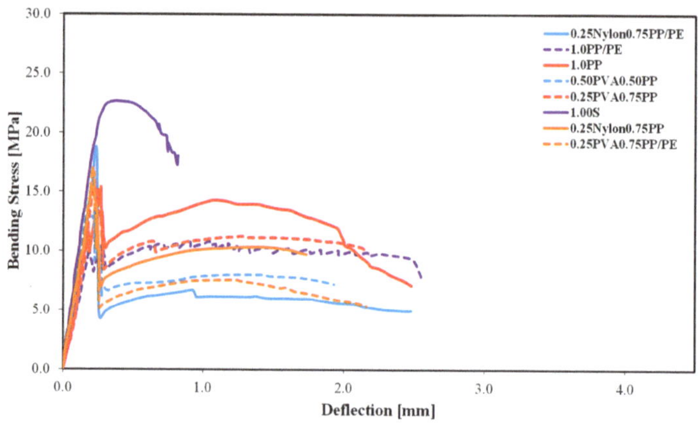

Figure 11. Effect of 1% of fibers on the waste glass–UHPC flexural behavior [15].

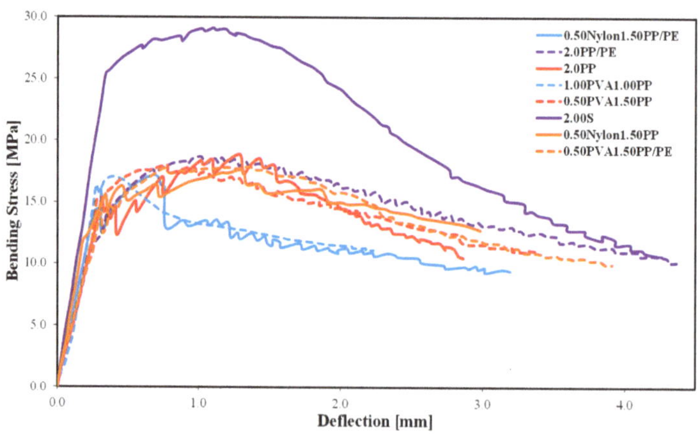

Figure 12. Effect of 2% of fibers on the waste glass–UHPC flexural behavior [15].

Moreover, it is important to note that the performance of series reinforced with 1% and 2% high-strength steel fibers (represented as 1.00S and 2.00S in Figures 11 and 12) achieved results comparable to those reported for the same reinforcement in traditional UHPC matrices without recycled glass. Among these pieces of research with typical UHPC matrices and similar limits of proportionality and modulus of rupture when using 1% or 2% of OL 13/0.20 fiber, the following references can be consulted [129–131].

5.4.5. Tensile Behavior

Mousa et al. [106] studied the tensile behavior of UHPC through the utilization of GP and GS as replacements for cement and fine aggregates, respectively. This study compares two distinct UHPC compositions: one incorporating GP as a cement substitute, and the other integrating both GP and GS. Both mixtures featured a 2% volume fraction of OL 13/02 steel microfibers with an aspect ratio of 65. In addition, the research explored the influence of two curing regimes, hot curing and normal curing, on the tensile behavior of the recycled glass UHPC samples at the ages of 28 and 91 days. A significant experimental factor considered was the alignment of the steel microfibers (either parallel or perpendicular to the fracture plane). Fiber orientation was induced during the pouring procedure. The trials were conducted following the technique known as the Double Edge Wedge Splitting Test (DEWS) [132]. Figure 13 shows the graphs with the results obtained in the specimens with fibers aligned perpendicular to the fracture [106]. It is noteworthy that these results are the only ones that could be corroborated with those obtained from dogbone specimens, as the use of dogbone specimens, owing to their gauge dimensions, promotes a favorable orientation of the fibers [101,133].

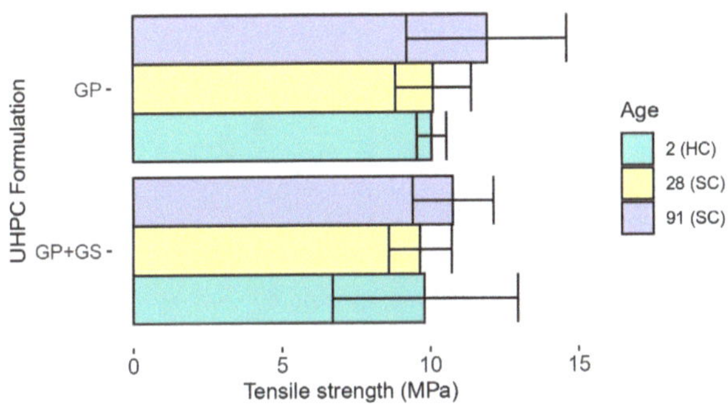

Figure 13. Maximum tensile stresses of UHPC mixtures with only glass powder (GP) or glass powder and glass sand (GP+GS) after 2 days of heat curing, and 28 and 91 days of standard curing (SC) [106].

As can be observed in Figure 13, both cementitious matrices achieved excellent tensile strength [106], in the same range as those achieved with standard UHPC reinforced with a 2% OL 13/0.2 fiber [123,133–135]. Another conclusion that can be drawn from this figure is the improvement in the CS between 28 and 90 days of standard curing (SC) which may be attributed to the slow pozzolanic reaction of glass powder [61,92]. Finally, it is relevant to denote that, the results obtained in the specimens with fibers aligned parallel to the fracture were about half of those presented in Figure 13 [106].

Nevertheless, the most relevant feature of UHPC is not its maximum tensile strength, but its direct tensile performance [10,100,136]. The introduction of fibers not only augments the material's tensile strength but, with the proper dosage of adequate fibers, also facilitates efficient force transfer even in the presence of cracks, thereby impeding or decelerating their propagation [137,138]. Fibers within the fissures serve to convey a portion of the matrix's tensile strength, enabling the composite to endure greater strains under favorable circumstances [11,138]. This state is achieved through the strategic alignment of a sufficient number of fibers possessing proper properties within the cementitious matrix [139].

In strain-hardening UHPC, the fiber-reinforced composite demonstrates the capacity to withstand stress increments beyond its cracking stress (σ_{cc}) until it reaches its maximum peak stress (σ_{pc}) [18,140]. Figure 14 depicts the various phases of the stress-strain curve for a strain-hardening UHPC under uniaxial tensile loading [106]. These stages encompass the elastic phase, multicracking phase, crack-straining phase and strain-softening phase. Initially, the material experiences elastic deformation until it reaches the point of tension, known as σ_{cc}, and attains a strain of ε_{cc}. Beyond this point, the material enters the inelastic strain zone characterized by repeated micro-cracking, surpassing the σ_{cc} tension, with stress levels remaining relatively stable [11,106]. Subsequently, the crack-straining phase ensues, marked by a significant increase in crack opening as the fiber reinforcement undergoes debonding and elastic straining, contingent upon the fiber type [101]. This branch extends until it reaches the maximum stress σ_{pc} and strain ε_{pc} [139].

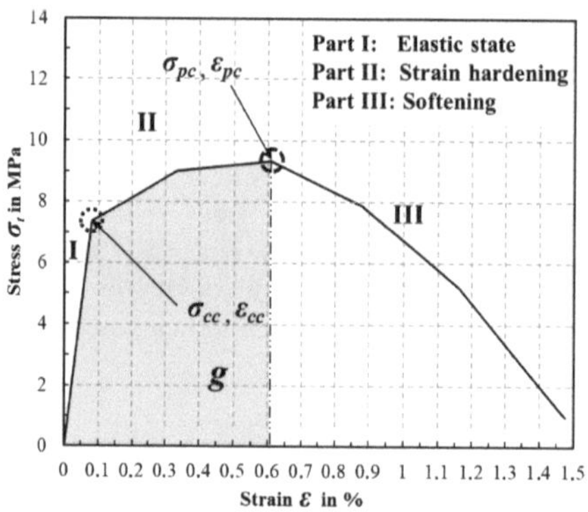

Figure 14. Phases of the uniaxial stress-strain curve for a strain-hardening UHPC [93].

In Figure 14, the energy absorption capacity (g) of the material can be computed as the area below the strain-stress curve until reaching the stress level denoted by σ_{pc} and ε_{pc} [18]. Subsequently, in the third step, a notable decline in the finer-reinforced UHPC strength develops as the strain increases, a phenomenon known as the strain-softening branch. During this phase, specimen failure is attributed to the slippage and/or rupture of fibers within critical fissures [20,106].

Abellán-García [12,93] investigated the ductility performance of uniaxial tensile behavior across 52 different series of recycled glass UHPC tests. His research focused on a recycled-glass cementitious matrix, particularly examining its suitability for seismic retrofitting applications. The experimentation encompassed various fiber types, including both steel and synthetic options, whose descriptions are presented in Table 9. It is also worth noticing that the S_1 fiber utilized in these studies corresponds to the smooth and high-strength steel OL 13/02 fiber.

Table 9. Detailed description of the fibers reported by [93].

Notation	Form	d_f/l_f	Material	Strength (MPa)
S_1	Smooth	65	Steel	≈2600
S_2	Smooth	30	Steel	≈2600
H_1	Hooked	70	Steel	≈2000
H_2	Hooked	80	Steel	≈1600
PP	Twisted	26	Steel	≈1700
PE	Corrugated	75	Polypropylene	≈650
PVA	Fibrillated	67	Polyethylene	≈550

Figure 15 presents the multicracking pattern in a strain-hardening glass–UHPC specimen reported in reference [12]. Some of the strain stress curves of strain hardening glass UHPC developed in [12,93] are presented in Figure 16.

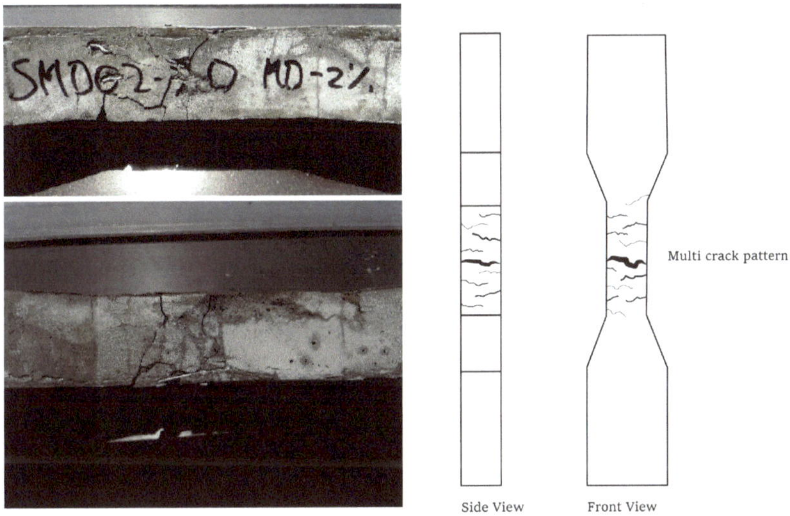

Figure 15. Recycled glass UHPC specimens with strain hardening behavior after failure. Detail of muli crack pattern [12].

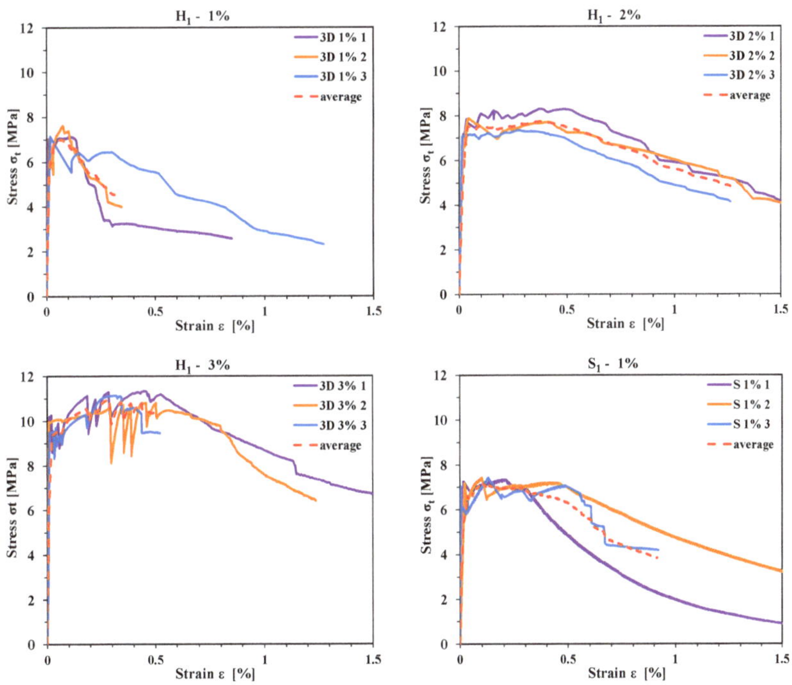

Figure 16. *Cont.*

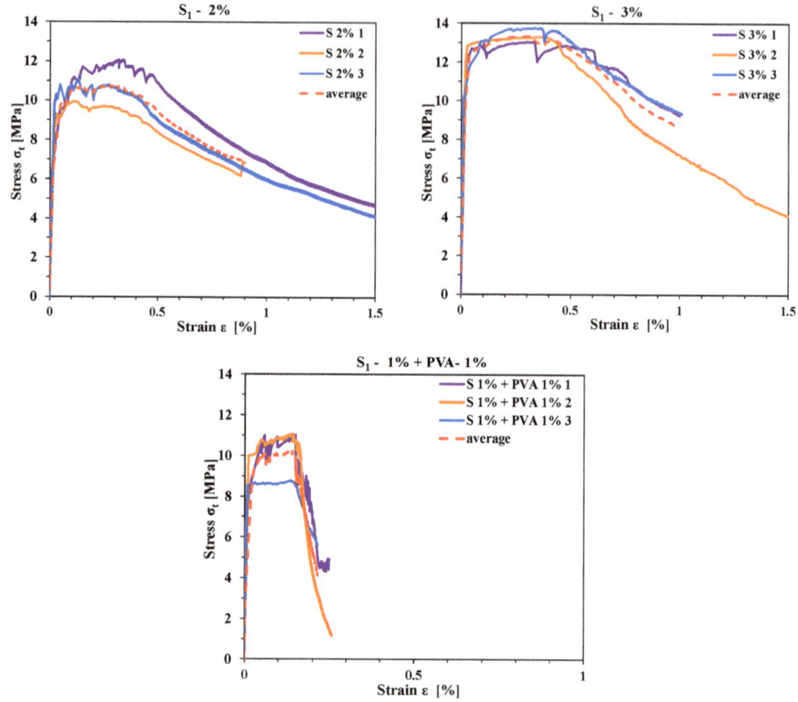

Figure 16. Direct traction stress-strain graphs [93]. H_1 represents hooked steel fiber; S_1 represents smooth steel fiber; and PVA represents polyvinyl alcohol fiber.

Among the conclusions drawn from these studies, it is notable that the ductility parameters achieved with the glass cementitious matrix closely approximate those reported in studies employing conventional UHPC matrices [93,96]. For instance, the waste glass–UHPC reinforced with a 2.0% volume fraction of OL 13/0.2 fibers exhibited a peak post-cracking strength of 11.03 MPa, only marginally lower than the 11.30 MPa reported in previous research, with a strain of 0.20% compared to 0.19%. The performance achieved, despite reducing cement content by approximately 35% and silica fume by 50%, can be attributed to the chemical and physical properties of micro-limestone and recycled glass powder [10,93,96]. These materials facilitated substantial substitution of the typical UHPC-making constituents without significantly compromising the chemical balance and packing density of the matrix.

5.5. Durability Properties

This review investigates the effect of GP on UHPC's durability by analyzing several key properties reported in existing literature. These properties include voids in hardened concrete, chloride penetration, initial surface absorption, freeze-thaw performance, ASR, mechanical abrasion resistance, drying shrinkage and resistance to deicing salt scaling.

5.5.1. Voids in Hardened Concrete

In their study, Jaramillo-Murcia et al. [58] presented compelling findings regarding void measurements in hardened concrete, following the ASTM C642 procedure, comparing the results of a Control UHPC mixture, without any alternative SCMs, with two optimized mixtures that incorporate recycled glass as mineral admixtures. More information about these dosages can be found in Table 6 and Figure 6. The reported results, measured at different ages, are illustrated in Figure 17. Their investigation revealed that all UHPC mixes consistently exhibit low void content throughout the curing process. This way, by the seventh day of curing, samples demonstrate void content below 5%, indicating a promising trend that persists with advancing treatment time.

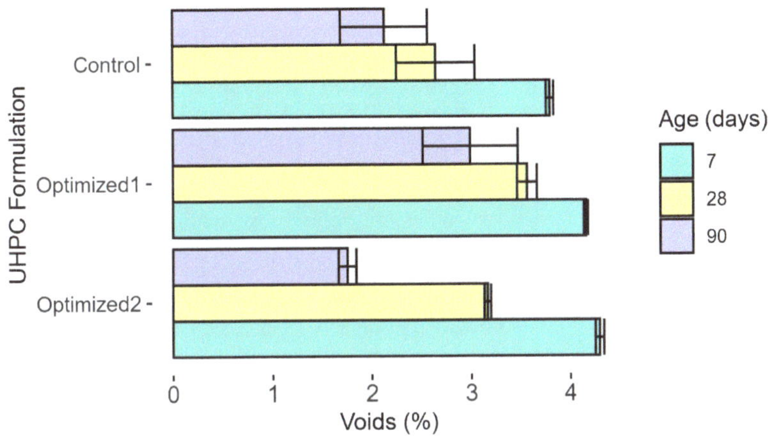

Figure 17. Findings in voids in hardened concrete measured by Jaramillo-Murcia et al. [58] at different ages.

Of particular interest is the exceptional performance of the Optimized 2 mixture, surpassing even the Control mixture by the 90-day mark. This noteworthy discovery highlights the positive impacts of the investigated admixtures, attributed to the development of C-S-H gel facilitated by the interaction of CH crystals with SF and GP particles [58]. Furthermore, Jaramillo-Murcia et al. [58] shed light on the implications of silica fume particle content in the Control mixture, which necessitates a higher water dosage, potentially contributing to increased internal fracturing. Their findings suggest a correlation between reduced silica fume content and mitigated drying shrinkage, thereby minimizing the risk of microcracking. These findings are also aligned with [12,141]. The study observes that the Optimized 2 mixture requires the lowest water dosage, a phenomenon attributed to the elevated alkali content in this formulation, synergistically enhanced by both types of glass powders. The effect of alkali content in rheology was previously explained and has been widely reported by [75,108,142], whose results underscore the intricate relationship between material composition and superplasticizer efficacy.

5.5.2. Chloride Penetration

As per Nancy Soliman et al. [40], in the realm of UHPC, the incorporation of recycled GP introduces a notable enhancement in resistance to chloride-ion penetration. As per these authors, the dense matrix inherent to UHPC serves as a critical barrier against the ingress of detrimental materials, effectively sealing the structure and bolstering its durability. Therefore, Soliman's findings reveal that UHPC formulations containing recycled glass powder exhibit remarkably low levels of chloride-ion penetration, with average total Coulombs passed values of 5.0 and 3.0 at 28 and 91 days, respectively. These findings align with the "negligible" classification as per ASTM C1202 standards, underscoring the effectiveness of the dense matrix in mitigating chloride ingress [40]. By comparison, traditional UHPC formulations subjected to standard heat treatment demonstrate significantly higher total charge passed values, reaching 18 Coulombs for heat-treated samples and a staggering 360 Coulombs for untreated samples. Such stark contrasts highlight the superior chloride-ion penetration resistance offered by UHPC formulations incorporating recycled glass powder, positioning them as promising candidates for improving the durability and longevity of concrete structures [40].

In the same line, Jaramillo-Murcia et al. [58] also compared the two recycled glass UHPC formulations with the control one but using just standard curing for the specimens. Their results are represented in Figure 18. Notably, the findings reveal a consistent enhancement in durability across different UHPC mixtures and ages. Initially, discernible differences are observed among the mixtures, particularly at the seven-day mark, attributable to variations in packaging density. The Control mixture exhibited Coulombs values of 1161, 80 and 49 at 7, 28 and 90 days, respectively, while Optimized 1 displayed values of 1211, 488 and 75 for the same time intervals. Optimized 2, on the other hand, showed Coulombs values of 2178, 531 and 90 at 7, 28 and 90 days, respectively. However, as the glass content matures within the mixtures, these disparities diminish due to the activation of its pozzolanic properties. Specifically, at 28 days, the penetration of chloride ions for Control and Optimized 1 dosages is deemed negligible, while that of Optimized 2 dosage is exceptionally low. Remarkably, at 90 days, all blend outcomes demonstrate negligible chloride ion penetration, underscoring the exceptional efficacy of the mineral admixtures. These findings not only corroborate the potential application of the investigated UHPCs in infrastructure elements facing harsh environmental conditions, such as coastal environments but also resonate with previous studies conducted on various UHPC dosages by multiple researchers, including Soliman's research. The remarkable performance exhibited by the mineral admixtures further fortifies their suitability for deployment in demanding construction scenarios [58].

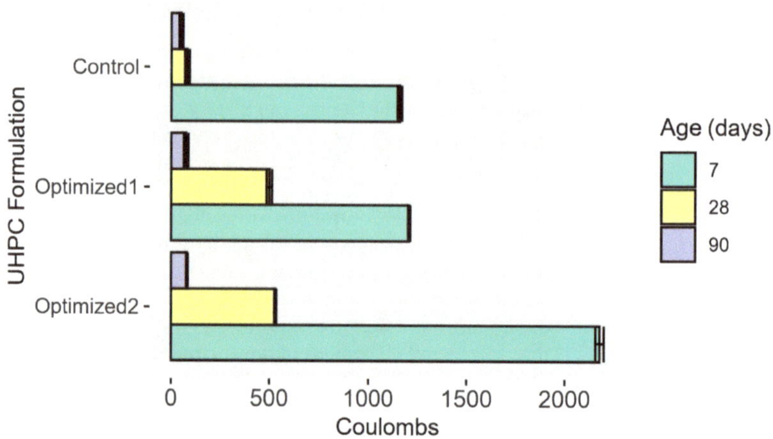

Figure 18. Chloride in penetration results of control and optimized UHPC formulations assessed at 7, 18 and 90 days [58].

5.5.3. Initial Surface Absorption

The Initial Surface Absorption Test (ISAT) for concrete assesses the rate at which water is absorbed by the concrete's specimen's surface over a specific duration, providing insights into its permeability and potential durability. Figures 19 and 20 present the findings of the ISAT measured during a period of 10 and 30 min respectively, conducted by [12] on the three UHPC dosages depicted in Table 6 and whose particle packing is presented in Figure 6. The data reveal consistently low absorption levels across all samples, with a gradual decline as curing progresses. Notably, there is a marked reduction in absorption values over time for the Control dosage, suggesting potential densification of the structure, likely attributed to pore filling [58,82]. The findings from density, absorption and void testing on hardened concrete support this observation. Figure 19 indicates that the use of recycled glass and limestone powders as a partial cement replacement yields comparable initial absorption rates, as evidenced at the 10-min interval. Similarly, Figure 20 depicts data for the initial absorption rate at 30 min and across all test ages. Analysis indicates a direct correlation between exposure duration and water absorption, with prolonged exposure resulting in decreased absorption due to surface pore saturation. Further examination reveals the most notable discrepancy between the control dosage and optimized dosages at 7 days of age, with diminishing disparities as curing time progresses. This suggests a gradual convergence in performance as the curing duration extends [58,82].

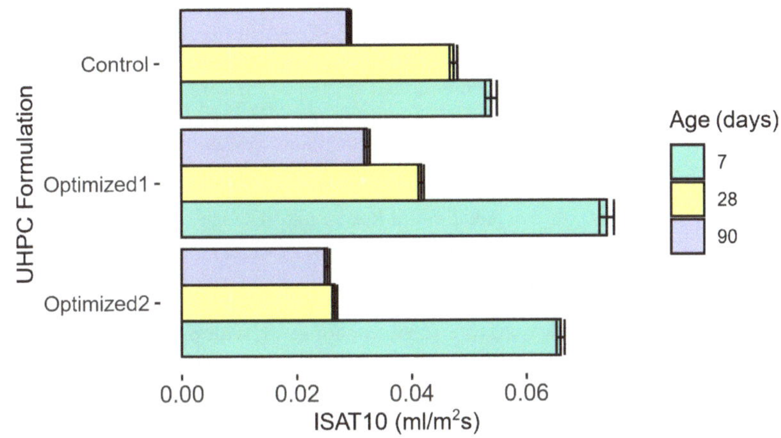

Figure 19. ISAT-10 results of control and optimized UHPC formulations assessed at 7, 18 and 90 days [12].

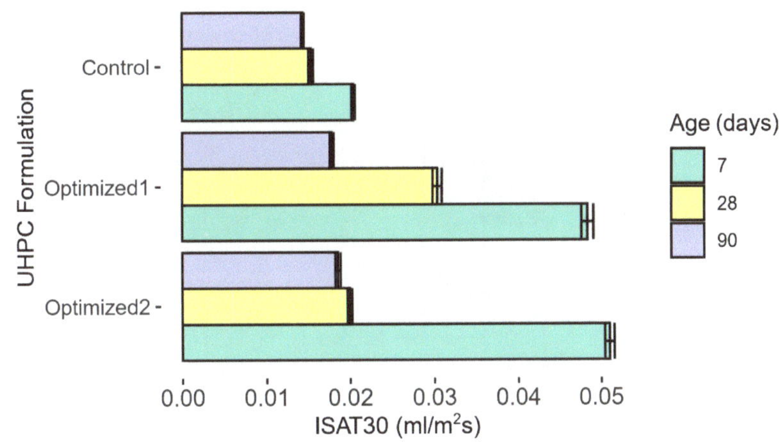

Figure 20. ISAT-30 results of control and optimized UHPC formulations assessed at 7, 18 and 90 days [12].

In accordance with the Brithishstandard [143], concrete exhibiting absorption rates under 0.25 mL/m^2s for ISAT-10 and 0.17 mL/m^2s for ISAT-30 is defined as low-absorption concrete. As depicted in Figures 19 and 20, the recorded values across all dosages and ages significantly surpass these thresholds. This discrepancy highlights the enhanced microstructure of all UHPC formulations considered [12].

5.5.4. Freeze-Thaw Performance

Freeze-thaw performance represents the ability of a material to withstand the adverse effects of repeated cycles of freezing and thawing. In the context of construction materials, such as UHPC, resistance to freezing and thawing is a key feature to ensure durability and structural integrity over time, especially in regions with cold climates or subject to extreme seasonal changes [112,144]. A recent study, based on ASTM C 666 specification, evaluated the freeze-thaw resistance of UHPC modified with glass particles [40]. In that research, a portion of the cement was replaced by glass particles in the mixture, resulting in a refined microstructure that enhanced its resistance to the freeze-thaw cycle. During 1000 cycles of freezing and thawing, the recycled glass added UHPC showed no significant signs of deterioration, suggesting excellent freeze-thaw performance [40].

Moreover, the resistance to freeze-thaw cycles, evaluated as per ASTM C666 standard, the aforementioned study put forward a mean dynamic MoE of 101% after 1000 cycles for glass UHPC, with no sign of jeopardizing or cracking at the end of the experiment. In comparison, minimal degradation was observed in traditional UHPC after 600 to 800 cycles, while the relative dynamic modulus decreased to 90% after 1000 freeze-thaw cycles [40].

5.5.5. Alkali-Silica Reaction Resistance

When GP is utilized as a cementitious agent, as in the case of glass powder, it does not represent a hazard of ASR expansion in the UHPC, as the pozzolanic action of the glass is activated before the ASR [54,85,145]. The situation could change when recycled glass is used as a substitute for sand, with particle diameters greater than when it is used as cementitious material [91]. In a research conducted by Redondo-Mosquera et al. [49] 100% of the sand was replaced by GS with an average diameter of particles of 300 μm in a UHPC with a w/b less than 0.2. Three mixtures were designed with different amounts of SF (from 100 to 210 kg/m^3) to evaluate the ASR expansion according to ASTM C 1260 [146]. With an amount of 100 kg/m^3 of SF the volume expansion of the UHPC was 0.28%, which is higher than the 0.1% recommended in ASTM C 1260, however when using 155 kg/m^3 and 210 kg/m^3 of SF the results are satisfactory (less than 0.1%). The trend shown in Figure 21 indicated that at such a high glass replacement level it is appropriate to add mineral admixtures with high pozzolanic reactivity to the UHPC in order to mitigate the ASR expansion. The latter is in concordance with what was exposed in [65]. Consistent results were reported for a UHPC with w/cm of 0.189 designed with glass (d_{50}: 275 mm) used as a sand replacement. The UHPC volume expansion was no more than 0.05% [69].

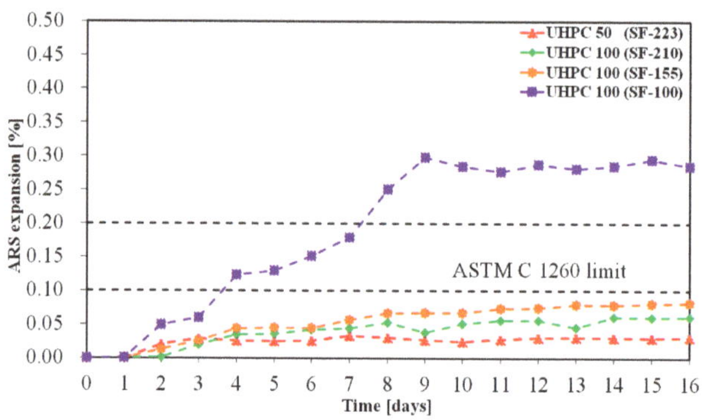

Figure 21. Rapid ASR expansion test of UHPC [49,69]. UHPC 50 (SF-223): 50% sand replacement and 223 kg/m^3 of SF; UHPC 100 (SF-210): 100% sand replacement and 210 kg/m^3 of SF; UHPC 100 (SF-155): 100% sand replacement and 155 kg/m^3 of SF; UHPC 100 (SF-100): 100% sand replacement and 100 kg/m^3 of SF.

5.5.6. Resistance to Mechanical Abrasion

Measuring mechanical abrasion resistance is particularly useful for evaluating the durability of materials used in buildings, like concrete, masonry and cementitious products [40,147]. It helps assess how well these materials can withstand abrasive forces encountered in real-world conditions, such as foot traffic, vehicular traffic and other mechanical actions. Soliman et al. analyzed the influence of GP on UHPC's mechanical abrasion [1,40]. According to the ASTM C944 standard, this resistance is determined by the relative volume loss index [148]. In the case of waste glass–UHPC under study, this index was measured at 1.35 mm after 28 days of standard curing. In comparison, for conventional UHPC, this index varied between 1.1 and 1.7 mm. It is important to note that the maximum specified threshold in ASTM C944 is 3.0 mm [40].

5.5.7. Drying Shrinkage

The study presented by [96] delved into the impact of limestone and milled glass powders on UHPC drying shrinkage, by considering the three dosages depicted in Table 6 and Figure 6, with and without fiber reinforcement. As previously stated, three different dosages were examined: two with partial substitution of cement and SF, and one serving as a reference without any substitution [47,59]. By integrating 2% (vol.) of steel microfiber into these three UHPC dosages, fiber-reinforced UHPC samples were successfully produced. Figure 22 depicts the findings of the tests conducted in the aforementioned research. In that figure, 2%Control corresponds to the Control dosage depicted in Table 6 reinforced with 2% of fiber OL 13/0.2, and 2%Optimzed 1 and 2%Optimized 2 represented the same fiber reinforcement for Optimized 1 and Optimized 2, respectively. The study demonstrated that the inclusion of limestone and GP enhances the rheological characteristics of concrete, thereby reducing dependence on chemical admixtures and enhancing cost-effectiveness. Thereby, as per this study's findings, partial substitution of cement and SF had been found to effectively mitigate drying shrinkage in both UHPC and UHPFRC compared to Control dosage [96]. The reported experimentation also revealed that incorporating microfibers into reinforcement and adjusting cement and SF proportions significantly reduced UHPC drying shrinkage by up to 40% compared to the Control UHPC formulation. The addition of 2% straight steel fibers led to notable reductions in Control dosage shrinkage, with reductions of 10.8%, 18.1%, 12.1% and 12.2% observed at 5, 15, 20 and 25 days, respectively [96].

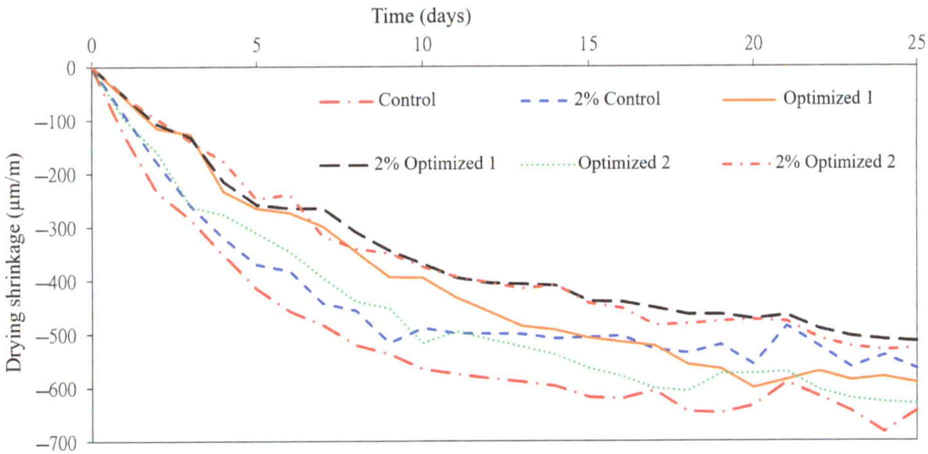

Figure 22. Reported drying shrinkage of UHPC formulations depicted in Table 6 with and without 2% volume fiber reinforcement [96].

Significantly, findings also depicted that the incorporation of a substantial volume of fibers into the mix of limestone and GP powders (i.e., into Optimized 1 or Optimized 2) resulted in considerable shrinkage reductions of 40.4%, 28.3%, 25.0% and 18.1% at the same ages [96]. Similar results in the same orders of magnitude were reported by PP when analyzing mixtures with and without UHPC fibers with recycled glass and calcium carbonate. Similar results in the same orders of magnitude were reported by [12,47,59] when analyzing mixtures with and without UHPC fibers with recycled glass and calcium carbonate.

5.5.8. Resistance to Deicing Salt Scaling

The resistance to deicing salt scaling is a critical factor for assessing concrete durability, especially in structures subjected to saltwater or in which de-icing salts are used, like pavements and bridge decks. In the study conducted by [1,40], this resistance was evaluated on a UHPC mixture proportion that utilized recycled glass powder for total substitution of quartz powder. The UHPC formulation is put forward in Table 10.

Table 10. Glass UHPC formulation reported by [40] in kg/m^3.

Cement	SF	Glass Powder	Quartz Sand	PVA Fiber	Water
549	204	403	888	32.5	224

The test was conducted by measuring mass loss on the concrete surface after 56 freeze-thaw cycles, resulting in a scaled mass of approximately 12 g/m^2, which compares favorably with the ASTM C672 specified limit of 1000 g/m^2. This measured value falls within the range reported in the literature for traditional UHPC, which varies from 8 to 60 g/m^2 after 28 to 50 freeze-thaw cycles. As per the authors, these results suggested excellent durability of the concrete evaluated in this study under freeze-thaw cycle exposure conditions, which is promising for its application in structures exposed to marine environments or subject to de-icing salt usage [1,40].

6. Impact of Recycled Glass Inclusion on the Life Cycle Assessment of UHPC

Life Cycle Assessment (LCA) is a pivotal methodology employed to comprehensively assess the environmental impact of a product, material or technology by considering its entire life cycle—from raw material extraction to manufacturing, application and disposal. When it comes to UHPC incorporating alternative mineral admixtures, such as recycled glass, LCA serves as a robust tool to quantify and analyze the environmental implications of this innovative construction material [41,52,94]. The LCA analysis considers various stages in the life cycle of UHPC, including the extraction and transportation of raw materials, the manufacturing process, construction and end-of-life scenarios. This holistic approach allows researchers and practitioners to assess the overall environmental footprint, providing insights into potential environmental benefits and areas for improvement [94]. Key aspects taken into account during the LCA analysis include [41,52,94]:

1. Raw Material Extraction and Transportation: Examining the environmental impact associated with the extraction of raw materials such as cement, recycled glass and other supplementary materials, as well as the energy-intensive transportation of these materials to the production site.
2. Manufacturing Process: Assessing the environmental consequences of the UHPC production process, including energy consumption, emissions and waste generation. This stage is critical for understanding the influence of incorporating recycled glass on factors like carbon footprint and energy efficiency.
3. Construction Phase: Considering the environmental impact during the construction phase, which includes transportation of UHPC to the construction site, energy use during placement and potential implications for construction-related activities.
4. Service Life: Evaluating the durability and performance of UHPC over its service life, as these factors can significantly influence the overall environmental sustainability of the material.
5. End-of-Life Scenarios: Investigating the environmental implications of various end-of-life scenarios, such as recycling, reuse or disposal in landfills. This aspect is particularly relevant for assessing the potential environmental benefits of recycling glass in UHPC compared to conventional end-of-life options.

Therefore, the LCA analysis provides a quantitative and systematic approach to understanding the environmental trade-offs associated with UHPC incorporating recycled glass. By considering these various stages in the life cycle, researchers can identify opportunities for optimization and innovation, ultimately contributing to the development of more sustainable construction materials and practices [41,52,94]. In this sense, a comprehensive study conducted by Tagnit-Hamou et al. [52] demonstrated the noteworthy environmental benefits of incorporating recycled glass into concrete, including those of normal, high and ultra-high strength, compared to conventional end-of-life scenarios such as landfilling. In addition, Abellan-García and colleagues have analyzed the carbon footprint of UHPC dosages which incorporates other mineral admixtures such as LP and RHA in addition to glass powder [47,59,94].

Figure 23 synthesizes the findings from various research works that analyzed the use of alternative mineral admixtures for lowering the carbon footprint of the cement-based material, focusing on the use of waste glass, along with several typical UHPC dosages with higher dosages of cement in their composition. This way, Figure 23 presents a comprehensive view of UHPC dosages, showcasing their coordinates of compressive strength in MPa and their corresponding CO_2-equivalent emissions. Notably, the increase in CS of analyzed UHPCs corresponded to higher embedded CO_2 emissions and environmental impact. However, as Figure 23 depicts, the compressive strength reached by the UHPC which utilizes waste glass in their composition as alternative SCM outperforms the compressive strength of those with other alternative mineral admixtures, such as LP, GGBFS and FA, for the same range of embedded CO_2 emissions. The reason behind the latter has been reported to lie in the optimized particle-packing model, coupled with cement replacement by waste glass, which emerged as a strategic approach to achieve both 'greener' concrete and higher mechanical properties [24,52,63,117]. This design strategy not only reduces CO_2 emissions but also facilitates the reuse of glass, minimizing the need for landfilling and conserving natural resources [50].

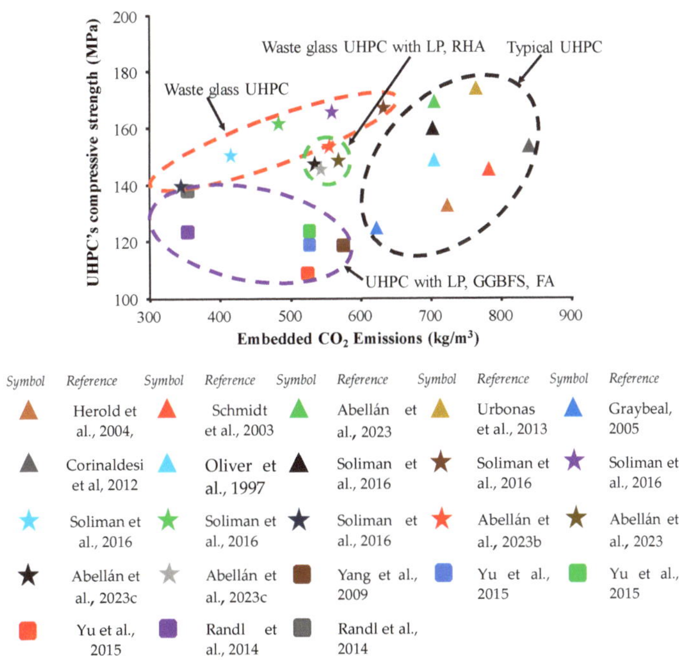

Figure 23. Correlation between the amount of embedded CO_2 emissions in kg/m^3 and the UHPC's CS in MPa for different formulations [41,47,59,94,130,149–156].

7. Cost Implications

Over the past decade, considerable data have been acquired on UHPC developments and uses, which has demonstrated that despite its superior qualities, the higher production cost of UHPC—largely due to increased cement, quartz powder and silica fume requirements—is among the causes that have limited its widespread adoption in the construction sector [157]. Therefore, substituting typical UHPC-making ingredients, such as QP, SF, cement and even quartz sand, with ground waste glass has the potential to substantially reduce the expenses associated with traditional UHPC formulations [76,77]. In addition, due to the chemical features depicted in Table 3, the low water absorption (see Table 4) and the lack of porosity of milled waste glass particles, the utilization of GP and GS allows for reducing the necessary amount of HRWR [47,59,82], which has a significant positive impact on the final cost of UHPC [12]. Moreover, the utilization of locally available WG for UHPC manufacturing has the added advantage of reducing material transportation expenses [41].

To illustrate, as per studies conducted in Canada by [41] the cost of quartz powder exceeds that of GP by threefold. For its part, research carried out in Colombia by [12] found that the cost of recycled glass powder after the milling process was about half the price of cement and a tenth of the price of quartz powder. For its part, the cost of glass sand was also reported to be half of that of UHPC typical quartz sand [50]. Moreover, as per [12], the utilization of waste glass and limestone powders as alternative mineral admixtures in Optimized 1 and 2 UHPC formulations (see Table 6) represents a decrease in the final prize of about 40% with respect to the Control dosage.

Furthermore, beyond the direct cost savings, incorporating waste glass into UHPC has several indirect economic benefits. For instance, the reduction in cement and silica fume content not only lowers material costs but also reduces mixing times [12,47,59,82], thereby decreasing the overall cost of mixing in plant-recycled-glass–UHPC compared to typical UHPC formulations. This efficiency can further drive down production costs [157].

For its part, the high processing costs, as well as the energy implications of UHPC, are partly due to the extensive use of cement, which functions as an expensive filler [157]. A more effective method of enhancing strength involves improving binder phase packing using cement substitute materials and fine fillers, rather than simply increasing cement content. Recent methodologies for designing economical and efficient UHPC binders incorporate recycled glass achieving high microstructural packing and desirable rheology for flowable concrete, which could lessen the pouring process and formwork costs [47,75,157].

In any case, practical applications of UHPC reveal that its higher initial costs can be offset by savings in construction supplies, labor, transportation expenses and reduced use of lifting and moving equipment on construction sites. The smaller quantity of concrete required, reduced reinforcement, less formwork and improved floor space utilization due to thinner sections all contribute to lowering overall costs [41,157]. In this sense, the properties both in the fresh state, as well as mechanical and durability reported in the present review study, as well as the real case study in Section 8 of this document, support the idea that waste glass–UHPC follows the trend described above. This way, the reduction in material costs depicted before can make UHPC more competitive compared to conventional concretes, potentially expanding its market reach. Moreover, the environmental benefits of using recycled materials such as waste glass could be a hook for attracting projects with sustainability goals and enhancing the marketability of UHPC products.

8. Case Study of a Field Application

The waste glass–UHPC has not only remained in laboratory tests. In addition, real-field applications have been reported. Among them, the construction of two similar footbridges, replacing the University of Sherbrooke campus's damaged wooden structures, stands out. The bridge utilized the recycled glass UHPC formulation depicted in Table 10, which was manufactured in the University of Sherbrooke's laboratory [40]. More information about the waste glass–UHPC formulation and mixing procedure can be found in the references [1,40,52].

As per reference [40], the construction of the two pedestrian bridges met the university's architectural and structural requirements for pedestrian applications, as well as aligned with its sustainability policy. The waste glass–UHPC's mechanical properties facilitated the creation of spans with minimal cross-sections, with each bridge weighing approximately four tons. In addition, the concrete exhibited remarkable workability and rheological properties [1]. This may be attributed to GP's particles' completely non-absorbent nature and the meticulously optimized packing density across the concrete matrix [75]. Moreover, the extensive investigation of this material boasted robustness, exceptional abrasion and impact resistance [1,40,52]. The bridge exhibits the following properties: the arch slab has a length of 4.91 m, a width of 2.5 m and a thickness of 0.075 m. It is supported by longitudinal ribs of different heights and a uniform width of 0.13 m [1]. The mid-height of the arch slab was reinforced with welded-wire reinforcement (M10 at 300 mm in both directions), and the bottom of each rib was strengthened with a single M20 reinforcing bar [1].

9. Conclusions

After a careful and deep review of 157 documents from the scientific literature, this paper discusses the advantages and disadvantages of using milled WG in the form of GP and/or GS in the manufacture of UHPC mixes, based on past and present research. The following conclusions are drawn:

1. Environmental and Economic Benefits: Using WG in UHPC helps preserve the environment and reduces costs by substituting traditional materials such as cement, silica fume, quartz powder and quartz sand. Incorporating waste glass and limestone powders in UHPC formulations substantially lowers production costs, with savings of up to 40% compared to traditional formulations.
2. Supplementary Cementitious Material: Due to its high silicon oxide content and amorphous nature, GP, when properly milled, effectively acts as a supplementary cementitious material, enhancing the formation of C-S-H and improving UHPC properties.
3. Enhanced Hydration Kinetics: The inclusion of GP improves hydration kinetics, reduces heat of hydration and mitigates microcrack formation.
4. Compressive Strength: Substituting up to 25% of cement with GP can enhance compressive strength through pozzolanic reactions. However, cement replacement ratios over 30% may reduce it.
5. Material Efficiency: The incorporation of GP allows for significant reductions in cement and silica fume content (up to 30% and 50%, respectively), while maintaining critical compressive strength thresholds above 150 MPa and achieving strengths exceeding 200 MPa in some formulations.
6. Synergistic Effects with Limestone Powder: The combined use of GP and limestone powder improves rheology, reduces superplasticizer demand and lowers water-to-binder ratios, enhancing both mechanical properties and sustainability.
7. Early Strength Development: The utilization of glass and limestone powder, along with FC3R and MK, shows promising advancements in early strength development. Through the synergy between glass and limestone, the negative effect on the rheology of the aluminum silicates in these components can be mitigated.
8. Fiber-reinforced UHPC mechanical Performance: Waste glass–UHPC exhibits bending and direct tensile performances comparable to traditional UHPC with similar fiber reinforcement.
9. Durability: The inclusion of milled GP in UHPC has been thoroughly analyzed, focusing on properties such as voids in hardened concrete, chloride penetration, initial surface absorption, freeze-thaw performance, alkali-silica reaction resistance, mechanical abrasion resistance, drying shrinkage and resistance to deicing salt scaling. While the addition of GP generally produces favorable results, slightly lower than traditional UHPC without recycled glass, it demonstrates superior performance in specific areas such as freeze-thaw resilience and drying shrinkage.

10. Environmental Impact: Incorporating GP into UHPC formulations offers significant environmental benefits, as evidenced by life cycle analysis studies. Revised studies have shown that UHPC incorporating recycled glass exhibits lower carbon footprints compared to conventional end-of-life scenarios like landfilling, contributing to sustainability in construction. Furthermore, research indicates that UHPC formulations with GP achieve higher compressive strengths at lower CO_2 emissions, in comparison with other supplementary cementitious materials.
11. Scalable Application: The successful use of WG in UHPC for constructing the footbridge at the University of Sherbrooke demonstrates its potential for scalable and practical applications.

10. Future Research Directions

Further research is essential to explore the long-term performance and durability of UHPC incorporating various types and proportions of waste glass under diverse environmental conditions. Investigating the impact of glass particle size distribution on UHPC properties and the combined effects of waste glass with other supplementary materials could further enhance UHPC's performance and sustainability. Large-scale field applications and pilot projects are necessary to validate laboratory findings and assess practical implications. Expanded life cycle analyses and cost-benefit studies across different scenarios and locations will provide a comprehensive understanding of the benefits. Additionally, innovative recycling techniques and advanced processing methods for waste glass should be explored to improve its quality and performance in UHPC formulations.

Funding: This research received no external funding.

Acknowledgments: Special thanks go to the Department of Civil and Environmental Engineering at the Universidad del Norte for their invaluable support and assistance throughout the research process. Additionally, the authors also extend their appreciation to the Karl C. Parrish Jr. Library at the same university for their generous assistance in providing access to essential documents for this review article. Their contributions have been instrumental in the completion of this work.

Conflicts of Interest: The authors declare no conflict of interest.

References

1. Tagnit-Hamou, A.; Soliman, N.A.; Omran, A. Green Ultra-High-Performance Glass Concrete. In Proceedings of the International Interactive Symposium on Ultra-High Performance Concrete 2016, Des Moines, IA, USA, 18–20 July 2016.
2. Abellán-García, J. K-Fold Validation Neural Network Approach for Predicting the One-Day Compressive Strength of UHPC. *Adv. Civ. Eng. Mater.* **2021**, *10*, 223–243. [CrossRef]
3. Abellan-Garcia, J.; Santofimo-Vargas, M.A.; Torres-Castellanos, N. Analysis of Metakaolin as Partial Substitution of Ordinary Portland Cement in Reactive Powder Concrete. *Adv. Civ. Eng. Mater.* **2020**, *9*, 368–386. [CrossRef]
4. Nehdi, M.; Abbas, S.; Soliman, A. Exploratory Study of Ultra-High Performance Fiber Reinforced Concrete Tunnel Lining Segments with Varying Steel Fiber Lengths and Dosages. *Eng. Struct.* **2015**, *101*, 733–742. [CrossRef]
5. *ACI Committe 239R*; ACI Committee 239 ACI—239 Committee in Ultra-High Performance Concrete. ACI: Toronto, ON, Canada, 2018.
6. Ahmad, S.; Hakeem, I.; Maslehuddin, M. Development of UHPC Mixtures Utilizing Natural and Industrial Waste Materials as Partial Replacements of Silica Fume and Sand. *Eur. J. Environ. Civ. Eng.* **2014**, *2014*, 1106–1126. [CrossRef] [PubMed]
7. Habel, K.; Charron, J.-P.; Braike, S.; Hooton, R.D.; Gauvreau, P.; Massicotte, B. Ultra-High Performance Fibre Reinforced Concrete Mix Design in Central Canada. *Can. J. Civ. Eng.* **2008**, *35*, 217–224. [CrossRef]
8. Abellán-García, J.; Fernández-Gómez, J.A.; Torres-Castellanos, N.; Núñez-López, A.M. Machine Learning Prediction of Flexural Behavior of UHPFRC. In *Fibre Reinforced Concrete: Improvements and Innovations, BEFIB 2020*; Serna, P., Llano-Torre, A., Martí-Vargas, J.R., Navarro-Gregori, J., Eds.; RILEM Bookseries: Valencia, Spain, 2020; Volume 4, pp. 570–583. ISBN 978-3-030-58482-5.
9. Abellán-García, J.; Guzmán-Guzmán, J.S. Random Forest-Based Optimization of UHPFRC under Ductility Requirements for Seismic Retrofitting Applications. *Constr. Build. Mater.* **2021**, *285*, 122869. [CrossRef]
10. Abellán-García, J.; Fernández-Gómez, J.; Torres-Castellanos, N.; Núñez-López, A. Tensile Behavior of Normal Strength Steel Fiber Green UHPFRC. *ACI Mater. J.* **2021**, *118*, 127–138. [CrossRef]
11. Abellán-García, J.; Guzmán-Guzmán, J.S.; Sánchez-Díaz, J.A.; Rojas-Grillo, J. Experimental Validation of Artificial Intelligence Model for the Energy Absorption Capacity of UHPFRC. *Dyna* **2021**, *88*, 150–159. [CrossRef]
12. Abellán-García, J. *Dosage Optimization and Seismic Retrofitting Applications of Ultra-High-Performance Fiber Reinforced Concrete (UHPFRC)*; Polytechnic University of Madrid: Madrid, Spain, 2020.

13. Abellán-García, J.; Ortega-Guzmán, J.J.; Chaparro-Ruiz, D.A.; García-Castaño, E. A Comparative Study of LASSO and ANN Regressions for the Prediction of the Direct Tensile Behavior of UHPFRC. *Adv. Civ. Eng. Mater.* **2022**, *11*, 235–262. [CrossRef]
14. Shi, C.; Wu, Z.; Xiao, J.; Wang, D.; Huang, Z.; Fang, Z. A Review on Ultra High Performance Concrete: Part I. *Raw Materials and Mixture Design. Constr. Build. Mater.* **2015**, *101*, 741–751. [CrossRef]
15. Neira-Medina, A.; Abellan-Garcia, J.; Torres-Castellanos, N. Flexural Behavior of Environmentally Friendly Ultra-High-Performance Concrete with Locally Available Low-Cost Synthetic Fibers. *Eur. J. Environ. Civ. Eng.* **2021**, *26*, 6281–6304. [CrossRef]
16. Lowke, D.; Stengel, T.; Schießl, P.; Gehlen, C. Control of Rheology, Strength and Fibre Bond of UHPC with Adittions—Effect of Packing Density and Addittion Type. In Proceedings of the 3rd International Symposium on UHPC and Nanotechnology for High Performance Construction Materials, Kassel, Germany, 7–9 March 2012; University of Kassel: Kassel, Germany, 2012; pp. 215–224, ISBN 9783862192649.
17. Nguyen, N.-H.; Abellán-García, J.; Lee, S.; Garcia-Castano, E.; Vo, T.P. Efficient Estimating Compressive Strength of Ultra-High Performance Concrete Using XGBoost Model. *J. Build. Eng.* **2022**, *52*, 104302. [CrossRef]
18. Abellán-Garcia, J.; Sánchez-Díaz, J.; Ospina-Becerra, V. Neural Network-Based Optimization of Fibers for Seismic Retrofitting Applications of UHPFRC. *Eur. J. Environ. Civ. Eng.* **2021**, *26*, 6305–6333. [CrossRef]
19. de Larrard, F.; Sedran, T.; Larrard, F. Optimization of Ultra-High-Performance Concrete by the Use of a Packing Model. *Cem. Concr. Res.* **1994**, *24*, 997–1009. [CrossRef]
20. Aghdasi, P.; Heid, A.E.; Chao, S.H. Developing Ultra-High-Performance Fiber-Reinforced Concrete for Large-Scale Structural Applications. *ACI Mater. J.* **2016**, *113*, 559–569. [CrossRef]
21. Shi, C.; Wu, Z.; Xiao, J.; Wang, D.; Huang, Z.; Fang, Z. A Review on Ultra High Performance Concrete: Part II. Hydration, Microstructure and Properties. *Constr. Build. Mater.* **2015**, *96*, 368–377. [CrossRef]
22. Wang, D.; Shi, C.; Wu, Z.; Xiao, J.; Huang, Z.; Fang, Z. Durability of an Ultra High Performance Fiber Reinforced Concrete (UHPFRC) under Progressive Aging. *Cem. Concr. Res.* **2015**, *55*, 1–13. [CrossRef]
23. Abellán-García, J.; Fernández-Gómez, J.; Torres-Castellanos, N. Properties Prediction of Environmentally Friendly Ultra-High-Performance Concrete Using Artificial Neural Networks. *Eur. J. Environ. Civ. Eng.* **2020**, *26*, 2319–2343. [CrossRef]
24. Abellán, J.; Fernández, J.; Torres, N.; Núñez, A. Development of Cost-Efficient UHPC with Local Materials in Colombia. In Proceedings of the Hipermat 2020—5th International Symposium on UHPC and Nanotechnology for Construction Materials, Kessel, Germany, 11–13 March 2020; Middendorf, B., Fehling, E., Wetzel, A., Eds.; University of Kassel: Kassel, Germany, 2020; pp. 97–98.
25. Abellán-García, J. Four-Layer Perceptron Approach for Strength Prediction of UHPC. *Constr. Build. Mater.* **2020**, *256*, 119465. [CrossRef]
26. Aghdasi, P.; Ostertag, C.P. Green Ultra-High Performance Fiber-Reinforced Concrete (G-UHP-FRC). Constr. *Build. Mater.* **2018**, *190*, 246–254. [CrossRef]
27. Abellán-García, J.; Núñez-López, A.; Torres-Castellanos, N.; Fernández-Gómez, J. Effect of FC3R on the Properties of Ultra-High-Performance Concrete with Recycled Glass. *Dyna* **2019**, *86*, 84–92. [CrossRef]
28. Abellán-García, J.; Torres-Castellanos, N.; Fernández-Gómez, J.A.; Núñez-López, A.M. Ultra-High-Performance Concrete with Local High Unburned Carbon Fly Ash. *Dyna* **2021**, *88*, 38–47. [CrossRef]
29. Abellán-García, J.; Núñez-López, A.; Torres-Castellanos, N.; Fernández-Gómez, J. Factorial Design of Reactive Powder Concrete Containing Electric Arc Slag Furnace and Recycled Glass Powder. *Dyna* **2020**, *87*, 42–51. [CrossRef]
30. Mishra, O.; Singh, S.P. An Overview of Microstructural and Material Properties of Ultra-High-Performance Concrete. *J. Sustain. Cem. Mater.* **2019**, *8*, 97–143. [CrossRef]
31. Tayeh, B.A.; Abu Bakar, B.H.; Megat Johari, M.A.; Voo, Y.L. Utilization of Ultra-High Performance Fibre Concrete (UHPFC) for Rehabilitation—A Review. *Procedia Eng.* **2013**, *54*, 525–538. [CrossRef]
32. Kalny, M.; Kvasnicka, V.; Komanec, J. First Practical Applications of UHPC in the Czech Republic. In Proceedings of the Proceedings of Hipermat 2016—4th International Symposium on UHPC and Nanotechnology for Construction Materials, Kassel, Germany, 9–11 March 2016; Fehling, E., Middendorf, B., Thiemicke, J., Eds.; University of Kassel: Kassel, Germany, 2016; pp. 147–148.
33. Abellán, J.; Núñez, A.; Arango, S. Pedestrian Bridge of UNAL in Manizales: A New UPHFRC Application in the Colombian Building Market. In Proceedings of the Hipermat 2020—5th International Symposium on UHPC and Nanotechnology for Construction Materials, Kessel, Germany, 11–13 March 2020; Middendorf, B., Fehling, E., Wetzel, A., Eds.; Kassel University Press: Kassel, Germany, 2020; pp. 43–44.
34. Abellán-García, J. An Overview of the Development and Applications of UHPFRC in Colombia. In *Concrete Plant International Worldwide*; Conrete Technology: Cape Town, South Africa, 2020; pp. 48–52. ISBN 1437-9023.
35. Abellán-García, J.; Nuñez-Lopez, A.; Arango-Campo, S. Pedestrian Bridge over Las Vegas Avenue in Medellín. First Latin American Infrastructure in UHPFRC. In *BEFIB 2020*; Serna, P., Llano-Torre, A., Martí-Vargas, J.R., Navarro-Gregori, J., Eds.; RILEM Bookseries: Valencia, Spain, 2020; pp. 864–872. ISBN 978-3-030-58482-5.
36. Dagenais, M.A.; Massicotte, B.; Boucher-Proulx, G. Seismic Retrofitting of Rectangular Bridge Piers with Deficient Lap Splices Using Ultrahigh-Performance Fiber-Reinforced Concrete. *J. Bridg. Eng.* **2018**, *23*, 04017129. [CrossRef]

37. Rai, B.; Wille, K. Development and Testing of High/Ultra-High Early Strength Concrete. In Proceedings of the 5th International Symposium on Fracture Mechanics of Concrete and Concrete Structures, Vail, CO, USA, 12–16 April 2004; Middendorf, B., Fehling, E., Wetzel, A., Eds.; University of Kassel: Kassel, Germany, 2020; pp. 7–8.
38. Pham, H.D.; Khuc, T.; Nguyen, T.V.; Cu, H.V.; Le, D.B.; Trinh, T.P. Investigation of Flexural Behavior of a Prestressed Girder for Bridges Using Nonproprietary UHPC. *Adv. Concr. Constr.* **2020**, *10*, 71–79. [CrossRef]
39. Abellán-García, J. Artificial Neural Network Model for Strength Prediction of Ultra-High-Performance Concrete. *ACI Mater. J.* **2021**, *118*, 3–14. [CrossRef]
40. Soliman, N.A.; Omran, A.F.; Tagnit-Hamou, A. Laboratory Characterization and Field Application of Novel Ultra-High-Performance Glass Concrete. *ACI Mater. J.* **2016**, *113*, 307–316. [CrossRef]
41. Soliman, N.A.; Tagnit-Hamou, A. Development of Ultra-High-Performance Concrete Using Glass Powder—Towards Ecofriendly Concrete. *Constr. Build. Mater.* **2016**, *125*, 600–612. [CrossRef]
42. Šerelis, E.; Vaitkevičius, V.; Kerševičius, V. Mechanical Properties and Microstructural Investigation of Ultra-High Performance Glass Powder Concrete. *J. Sustain. Archit. Civ. Eng.* **2016**, *1*, 5–11. [CrossRef]
43. Shi, C.; Wu, Y.; Riefler, C.; Wang, H. Characteristics and Pozzolanic Reactivity of Glass Powders. *Cem. Concr. Res.* **2005**, *35*, 987–993. [CrossRef]
44. Hamada, H.; Alattar, A.; Tayeh, B.; Yahaya, F.; Thomas, B. Effect of Recycled Waste Glass on the Properties of High-Performance Concrete: A Critical Review. *Case Stud. Constr. Mater.* **2022**, *17*, e01149. [CrossRef]
45. Rashad, A.M. Recycled Waste Glass as Fine Aggregate Replacement in Cementitious Materials Based on Portland Cement. *Constr. Build. Mater.* **2014**, *72*, 340–357. [CrossRef]
46. Rajabipour, F.; Maraghechi, H.; Fischer, G. Investigating the Alkali-Silica Reaction of Recycled Glass Aggregates in Concrete Materials. *J. Mater. Civ. Eng.* **2010**, *22*, 1201–1208. [CrossRef]
47. Abellan-Garcia, J.; Molinares, M.; Daza, N.; Abbas, Y.M.; Iqbal Khan, M. Formulation of Inexpensive and Green Reactive Powder Concrete by Using Milled-Waste-Glass and Micro Calcium-Carbonate—A Multi-Criteria Optimization Approach. *Constr. Build. Mater.* **2023**, *409*, 134167. [CrossRef]
48. Kou, S.C.; Xing, F. The Effect of Recycled Glass Powder and Reject Fly Ash on the Mechanical Properties of Fibre-Reinforced Ultrahigh Performance Concrete. *Hindawi Publ. Corp. Adv. Mater. Sci. Eng.* **2012**, *2012*, 1–8. [CrossRef]
49. Redondo-Mosquera, J.D.; Sánchez-Angarita, D.; Redondo-Pérez, M.; Gómez-Espitia, J.C.; Abellán-García, J. Development of High-Volume Recycled Glass Ultra-High-Performance Concrete with High C3A Cement. Case Stud. *Constr. Mater.* **2023**, *18*, e01906. [CrossRef]
50. Abellan-Garcia, J.; Mosquera- Redondo, J.; Khan, M.I.; Abbas, Y.M.; Castro, A. Development of a Novel 124 MPa Strength Green Reactive Powder Concrete Employing Waste Glass and Locally Available Cement. *Arch. Civ. Mech. Eng.* **2023**, *7*, 1–17. [CrossRef]
51. Du, H.; Tan, K.H. Waste Glass Powder as Cement Replacement in Concrete. *J. Adv. Concr. Technol.* **2014**, *12*, 468–477. [CrossRef]
52. Tagnit-Hamou, A.; Zidol, A.; Soliman, N.; Deschamps, J.; Omran, A. Ground Glass Pozzolan in Conventional, High, and Ultra-High Performance Concrete. In Proceedings of the 2nd International Congress on Materials & Structural Stability (CMSS-2017), Rabat, Morocco, 22–25 November 2017.
53. Amin, M.; Zeyad, A.M.; Tayeh, B.A.; Agwa, I.S. Effect of Glass Powder on High-Strength Self-Compacting Concrete Durability. *Key Eng. Mater.* **2023**, *495*, 117–127. [CrossRef]
54. Dhir, R.K.; Dyer, T.D.; Tang, M.C. Alkali-Silica Reaction in Concrete Containing Glass. *Mater. Struct. Constr.* **2009**, *42*, 1451–1462. [CrossRef]
55. Shaaban, M.; Ahmed, S. Development of Ultra-High Performance Concrete Jointed Precast Decks and Concrete Piles in Integral Abutment Bridges. In Proceedings of the The First International Symposium on Jointless & Sustainable Bridges, Fuzhou, China, 20–22 November 2016.
56. Amin, M.; Agwa, I.S.; Mashaan, N.; Mahmood, S.; Abd-Elrahman, M.H. Investigation of the Physical Mechanical Properties and Durability of Sustainable Ultra-High Performance Concrete with Recycled Waste Glass. *Sustainability* **2023**, *15*, 3085. [CrossRef]
57. Xu, J.; Zhan, P.; Zhou, W.; Zuo, J.; Shah, S.P.; He, Z. Design and Assessment of Eco-Friendly Ultra-High Performance Concrete with Steel Slag Powder and Recycled Glass Powder. *Powder Technol.* **2023**, *419*, 118356. [CrossRef]
58. Jaramillo-Murcia, D.C.; Abellán-Garcia, J.; Torres-Castellanos, N.; García-Castaño, E. Properties Analysis of UHPC with Recycled Glass and Limestone Powders. *ACI Mater. J.* **2022**, *119*, 153–164. [CrossRef]
59. Abellán-García, J.; Daza, N.; Molinares, M.; Abbas, Y.M.; Khan, M.I. Multi-Criteria Optimization of Cost-Effective and Environmentally Friendly Reactive Powder Concrete Incorporating Waste Glass and Micro Calcium Carbonate. *Materials* **2023**, *16*, 6434. [CrossRef]
60. Wang, H.Y. A Study of the Effects of LCD Glass Sand on the Properties of Concrete. *Waste Manag.* **2009**, *29*, 335–341. [CrossRef] [PubMed]
61. Shi, C.; Zheng, K. A Review on the Use of Waste Glasses in the Production of Cement and Concrete. *Resour. Conserv. Recycl.* **2007**, *52*, 234–247. [CrossRef]
62. Esmaeili, J.; Oudah Al-Mwanes, A. A Review: Properties of Eco-Friendly Ultra-High-Performance Concrete Incorporated with Waste Glass as a Partial Replacement for Cement. *Mater. Today Proc.* **2021**, *42*, 1958–1965. [CrossRef]
63. Soliman, N.A.; Tagnit-Hamou, A. Using Particle Packing and Statistical Approach to Optimize Eco-Efficient Ultra-High-Performance Concrete. *ACI Mater. J.* **2017**, *114*, 847–858. [CrossRef]

64. Vaitkevicius, V.; Šerelis, E.; Hilbig, H. The Effect of Glass Powder on the Microstructure of Ultra High Performance Concrete. *Constr. Build. Mater.* **2014**, *68*, 102–109. [CrossRef]
65. Guo, P.; Meng, W.; Nassif, H.; Gou, H.; Bao, Y. New Perspectives on Recycling Waste Glass in Manufacturing Concrete for Sustainable Civil Infrastructure. *Constr. Build. Mater.* **2020**, *257*, 119579. [CrossRef]
66. Siddique, R. Waste Glass. In *Waste Materials and By-Products in Concrete*; Springer: Berlin/Heidelberg, Germany, 2008; pp. 147–175. ISBN 978-3-540-74294-4.
67. Villanueva, A.; Eder, P. *End-of-Waste Criteria for Waste Paper: Technical Proposals*; European Commission: Luxembourg, 2011.
68. Weihua Jin and Stephen Baxter, C.M. "Glascrete"—Concrete with Glass Aggregate. *ACI Mater. J.* **2000**, *97*, 208–213. [CrossRef]
69. Soliman, N.A.; Tagnit-Hamou, A. Using Glass Sand as an Alternative for Quartz Sand in UHPC. *ACI Mater. J.* **2017**, *145*, 847–858. [CrossRef]
70. FEVE. Containrer Glass Recycling Un Europe. Available online: https://feve.org/glass_recycling_stats_2018/ (accessed on 1 August 2023).
71. United States Environmental Protection Agency. *Advancing Sustainable Materials Management, 2017 Fact Sheet*; United States Environmental Protection Agency: Washington, DC, USA, 2017.
72. Jiao, Y.; Zhang, Y.; Guo, M.; Zhang, L.; Ning, H.; Liu, S. Mechanical and Fracture Properties of Ultra-High Performance Concrete (UHPC) Containing Waste Glass Sand as Partial Replacement Material. *J. Clean. Prod.* **2020**, *277*, 123501. [CrossRef]
73. Sood, T.; Gurmu, A. Reusing and Repurposing of Glass Waste: A Literature Review. In Proceedings of the 10th World Construction Symposium, Colombo, Sri Lanka, 24–26 June 2022.
74. Al-Awabdeh, F.W.; Al-Kheetan, M.J.; Jweihan, Y.S.; Al-Hamaiedeh, H.; Ghaffar, S.H. Comprehensive Investigation of Recycled Waste Glass in Concrete Using Silane Treatment for Performance Improvement. *Results Eng.* **2022**, *16*, 100790. [CrossRef]
75. Abellán, J.; Fernández, J.; Torres, N.; Núñez, A. Statistical Optimization of Ultra-High-Performance Glass Concrete. *ACI Mater J.* **2020**, *117*, 243–254. [CrossRef]
76. Esmaeili, J.; AL-Mwanes, A.O. Performance Evaluation of Eco-Friendly Ultra-High-Performance Concrete Incorporated with Waste Glass-A Review. IOP Conf. *Ser. Mater. Sci. Eng.* **2021**, *1094*, 012030. [CrossRef]
77. Soliman, N.A.; Tagnit-Hamou, A. Partial Substitution of Silica Fume with Fine Glass Powder in UHPC: Filling the Micro Gap. *Constr. Build. Mater.* **2017**, *139*, 374–383. [CrossRef]
78. Jain, K.L.; Sancheti, G.; Gupta, L.K. Durability Performance of Waste Granite and Glass Powder Added Concrete. *Constr. Build. Mater.* **2020**, *252*, 119075. [CrossRef]
79. Taha, B.; Nounu, G. Using Lithium Nitrate and Pozzolanic Glass Powder in Concrete as ASR Suppressors. *Cem. Concr. Compos.* **2008**, *30*, 497–505. [CrossRef]
80. Helmy, S.H.; Tahwia, A.M.; Mahdy, M.G.; Elrahman, M.A. Development and Characterization of Sustainable Concrete Incorporating a High Volume of Industrial Waste Materials. *Constr. Build. Mater.* **2023**, *365*, 130160. [CrossRef]
81. Hasan, S.; Nayyef, D. Investigation of Using Waste Glass Powder as a Supplementary Cementitious Material in Reactive Powder Concrete. *Proc. Int. Struct. Eng. Constr.* **2020**, *7*, 1–6. [CrossRef]
82. Abellan-Garcia, J.; Abbas, Y.M.; Khan, M.I.; Pellicer-Martínez, F. ANOVA-Guided Assessment of Waste Glass and Limestone Powder Influence on Ultra-High-Performance Concrete Properties. *Case Stud. Constr. Mater.* **2024**, *20*, e03231. [CrossRef]
83. Abellán-García, J.; García-Castaño, E. Development and Research on Ultra-High-Performance Concrete Dosages in Colombia: A Review. *ACI Mater. J.* **2022**, *119*, 209–221. [CrossRef]
84. Niibori, Y.; Chida, T.; Tochiyama, O. Dissolution Rates of Amorphous Silica in Highly Alkaline Solution. *J. Nucl. Sci. Technol.* **2000**, *37*, 349–357. [CrossRef]
85. Guo, P.; Bao, Y.; Meng, W. Review of Using Glass in High-Performance Fiber-Reinforced Cementitious Composites. *Cem. Concr. Compos.* **2021**, *120*, 104032. [CrossRef]
86. Islam, G.M.S.; Rahman, M.H.; Kazi, N. Waste Glass Powder as Partial Replacement of Cement for Sustainable Concrete Practice. *Int. J. Sustain. Built Environ.* **2017**, *6*, 37–44. [CrossRef]
87. Kunhi Mohamed, A.; Moutzouri, P.; Berruyer, P.; Walder, B.J.; Siramanont, J.; Siramanont, J.; Harris, M.; Negroni, M.; Galmarini, S.C.; Parker, S.C.; et al. The Atomic-Level Structure of Cementitious Calcium Aluminate Silicate Hydrate. *J. Am. Chem. Soc.* **2020**, *142*, 11060–11071. [CrossRef]
88. Kupwade-Patil, K.; Allouche, E.N. Impact of Alkali Silica Reaction on Fly Ash-Based Geopolymer Concrete. *J. Mater. Civ. Eng.* **2013**, *25*, 131–139. [CrossRef]
89. Shayan, A.; Xu, A. Value-Added Utilisation of Waste Glass in Concrete. *Cem. Concr. Res.* **2004**, *34*, 81–89. [CrossRef]
90. Idir, R.; Cyr, M.; Tagnit-Hamou, A. Use of Waste Glass in Cement-Based Materials. In Proceedings of the SBEIDCO—1st International Conference on Sustainable Built Environement Infrastructures in Developing Countries, Orán, Algeria, 12–14 October 2009; pp. 109–116.
91. Kaminsky, A.; Krstic, M.; Rangaraju, P.; Tagnit-Hamou, A.; Thomas, M.D.A. Ground-Glass Pozzolan for Use in Concrete. *Concr. Int.* **2020**, *42*, 24–32.
92. Shao, Y.; Lefort, T.; Moras, S.; Rodriguez, D. Studies on Concrete Containing Ground Waste Glass. *Cem. Concr. Res.* **2000**, *30*, 91–100. [CrossRef]
93. Abellan-Garcia, J. Tensile Behavior of Recycled-Glass-UHPC under Direct Tensile Loading. *Case Stud. Constr. Mater.* **2022**, *17*, 1–16. [CrossRef]

94. Abellan-garcia, J.; Martinez, D.M.; Khan, M.I.; Abbas, Y.M.; Pellicer-martı, F. Environmentally Friendly Use of Rice Husk Ash and Recycled Glass Waste to Produce Ultra-High- Performance Concrete. *J. Mater. Res. Technol.* **2023**, *25*, 1869–1881. [CrossRef]
95. Abellan-Garcia, J. Effect of Rice Husk Ash as Partial Replacement of Ordinary Portland Cement in Ultra-High-Performance Glass Concrete. *Eur. J. Environ. Civ. Eng.* **2023**, *2*, 396–401. [CrossRef]
96. Abellan-Garcia, J.; Iqbal Khan, M.; Abbas, Y.M.; Martínez-Lirón, V.; Carvajal-Muñoz, J.S. The Drying Shrinkage Response of Recycled-Waste-Glass-Powder-and Calcium-Carbonate-Based Ultrahigh-Performance Concrete. *Constr. Build. Mater.* **2023**, *379*, 131163. [CrossRef]
97. Abellán-garcía, J.; Iqbal-khan, M.; Abbas, Y.M.; Pellicer-martínez, F. Predicting the Flowability of UHPC and Identifying Its Significant Influencing Factors Using an Accurate ANN Model. *Dyna* **2024**, *91*, 27–36. [CrossRef]
98. De Larrard, F. *Concrete Mixture Proportioning: A Scientific Approach*; Modern Concrete Technology Series; CRC Press: Boca Raton, FL, USA, 1999.
99. De Larrard, F.; Sedran, T. Mixture-Proportioning of High-Performance Concrete. *Cem. Concr. Res.* **2002**, *32*, 1699–1704. [CrossRef]
100. Nguyen, N.H.; Abellán-García, J.; Lee, S.; Nguyen, T.K.; Vo, T.P. Simultaneous Prediction the Strain and Energy Absorption Capacity of Ultra-High Performance Fiber Reinforced Concretes by Using Multi-Output Regression Model. *Constr. Build. Mater.* **2023**, *384*, 131418. [CrossRef]
101. Abellan-Garcia, J.; Fernández, J.; Khan, M.I.; Abbas, Y.M.; Carrillo, J. Uniaxial Tensile Ductility Behavior of Ultrahigh-Performance Concrete Based on the Mixture Design—Partial Dependence Approach. *Cem. Concr. Compos.* **2023**, *140*, 105060. [CrossRef]
102. Huang, H.; Gao, X.; Wang, H.; Ye, H. Influence of Rice Husk Ash on Strength and Permeability of Ultra-High Performance Concrete. *Constr. Build. Mater.* **2017**, *149*, 621–628. [CrossRef]
103. Chindaprasirt, P.; Homwuttiwong, S.; Jaturapitakkul, C. Strength and Water Permeability of Concrete Containing Palm Oil Fuel Ash and Rice Husk-Bark Ash. *Constr. Build. Mater.* **2007**, *21*, 1492–1499. [CrossRef]
104. Yang, R.; Yu, R.; Shui, Z.; Guo, C.; Wu, S.; Gao, X.; Peng, S. The Physical and Chemical Impact of Manufactured Sand as a Partial Replacement Material in Ultra-High Performance Concrete (UHPC). *Cem. Concr. Compos.* **2019**, *99*, 203–213. [CrossRef]
105. National Cancer Institute Crystalline Silica. Available online: https://www.cancer.gov/about-cancer/causes-prevention/risk/substances/crystalline-silica (accessed on 1 August 2023).
106. Mousa, M.; Cuenca, E.; Ferrara, L.; Roy, N.; Tagnit-Hamou, A. Tensile Characterization of an "Eco-Friendly" UHPFRC with Waste Glass Powder and Glass Sand. In Proceedings of the Strain-Hardening Cement-Based Composites. SHCC 2017, Dresden, Germany, 18–20 September 2017; Mechtcherine, V., Slowik, V., Kabele, P., Eds.; RILEM Bookseries: Valencia, Spain, 2017; pp. 238–248.
107. Azmee, N.M.; Shafiq, N. Ultra-High Performance Concrete: From Fundamental to Applications. *Case Stud. Constr. Mater.* **2018**, *9*, e00197. [CrossRef]
108. Pedrajas, C.; Rahhal, V.; Talero, R. Determination of Characteristic Rheological Parameters in Portland Cement Pastes. *Constr. Build. Mater.* **2014**, *51*, 484–491. [CrossRef]
109. Mirzahosseini, M.; Riding, K.A. Effect of Curing Temperature and Glass Type on the Pozzolanic Reactivity of Glass Powder. *Cem. Concr. Res.* **2014**, *58*, 103–111. [CrossRef]
110. Jansen, D.; Neubauer, J.; Goetz-Neunhoeffer, F.; Haerzschel, R.; Hergeth, W.D. Change in Reaction Kinetics of a Portland Cement Caused by a Superplasticizer—Calculation of Heat Flow Curves from XRD Data. *Cem. Concr. Res.* **2012**, *42*, 327–332. [CrossRef]
111. Li, W.; Huang, Z.; Zu, T.; Shi, C.; Duan, W.H.; Shah, S.P. Influence of Nanolimestone on the Hydration, Mechanical Strength, and Autogenous Shrinkage of Ultrahigh-Performance Concrete. *J. Mater. Civ. Eng.* **2016**, *28*, 1–9. [CrossRef]
112. Li, J.; Wu, Z.; Shi, C.; Yuan, Q.; Zhang, Z. Durability of Ultra-High Performance Concrete—A Review. *Constr. Build. Mater.* **2020**, *255*, 119296. [CrossRef]
113. Lee, N.K.; Koh, K.T.; Park, S.H.; Ryu, G.S. Microstructural Investigation of Calcium Aluminate Cement-Based Ultra-High Performance Concrete (UHPC) Exposed to High Temperatures. *Cem. Concr. Res.* **2017**, *102*, 109–118. [CrossRef]
114. Ge, W.; Zhang, Z.; Ashour, A.; Li, W.; Jiang, H.; Hu, Y.; Shuai, H.; Sun, C.; Li, S.; Liu, Y.; et al. Hydration Characteristics, Hydration Products and Microstructure of Reactive Powder Concrete. *J. Build. Eng.* **2023**, *69*, 106306. [CrossRef]
115. Amin, M.; Zeyad, A.M.; Tayeh, B.A.; Saad Agwa, I. Effects of Nano Cotton Stalk and Palm Leaf Ashes on Ultrahigh-Performance Concrete Properties Incorporating Recycled Concrete Aggregates. *Constr. Build. Mater.* **2021**, *302*, 124196. [CrossRef]
116. Bahmani, H.; Mostofinejad, D. Microstructure of Ultra-High-Performance Concrete (UHPC)—A Review Study. *J. Build. Eng.* **2022**, *50*, 104118. [CrossRef]
117. Faried, A.S.; Mostafa, S.A.; Tayeh, B.A.; Tawfik, T.A. Mechanical and Durability Properties of Ultra-High Performance Concrete Incorporated with Various Nano Waste Materials under Different Curing Conditions. *J. Build. Eng.* **2021**, *43*, 102569. [CrossRef]
118. Cheyrezy, M.; Maret, V.; Frouin, L. Microstructural Analysis of RPC (Reactive Powder Concrete). *Cem. Concr. Res.* **1995**, *25*, 1491–1500. [CrossRef]
119. Maso, J.C. (Ed.) *Interfacial Transition Zone in Concrete*; CRC Press: Boca Raton, FL, USA, 1996.
120. Reda, M.M.; Shrive, N.G.; Gillott, J.E. Microstructural Investigation of Innovative UHPC. *Cem. Concr. Res.* **1999**, *29*, 323–329. [CrossRef]
121. Demiss, B.A.; Oyawa, W.O.; Shitote, S.M. Mechanical and Microstructural Properties of Recycled Reactive Powder Concrete Containing Waste Glass Powder and Fly Ash at Standard Curing. *Cogent Eng.* **2018**, *5*, 1464877. [CrossRef]
122. ACI. *ACI Committe 239R Ultra-High-Performance Concrete: An Emerging Technology Report*; ACI: Toronto, ON, Canada, 2018.

123. Hassan, A.M.T.; Jones, S.W.; Mahmud, G.H. Experimental Test Methods to Determine the Uniaxial Tensile and Compressive Behaviour of Ultra High Performance Fibre Reinforced Concrete (UHPFRC). *Constr. Build. Mater.* **2012**, *37*, 874–882. [CrossRef]
124. Shafieifar, M.; Farzad, M.; Azizinamini, A. Experimental and Numerical Study on Mechanical Properties of Ultra High Performance Concrete (UHPC). *Constr. Build. Mater.* **2017**, *156*, 402–411. [CrossRef]
125. Nassif, H.; Najm, H.; Sukawang, N. Effect of Pozzolanic Materials and Curing Methods on the Elastic Modulus of HPC. *Cem. Concr. Compos.* **2005**, *27*, 661–670. [CrossRef]
126. Yoo, D.Y.; Kim, M.J.; Kim, S.W.; Park, J.J. Development of Cost Effective Ultra-High-Performance Fiber-Reinforced Concrete Using Single and Hybrid Steel Fibers. *Constr. Build. Mater.* **2017**, *150*, 383–394. [CrossRef]
127. Kim, D.J.; Park, S.H.; Ryu, G.S.; Koh, K.T. Comparative Flexural Behavior of Hybrid Ultra High Performance Fiber Reinforced Concrete with Different Macro Fibers. *Constr. Build. Mater.* **2014**, *25*, 4144–4155. [CrossRef]
128. Abellán-García, J. Comparison of Artificial Intelligence and Multivariate Regression in Modeling the Flexural Behavior of UHPFRC. *Dyna* **2020**, *87*, 239–248. [CrossRef]
129. Yu, R.; Spiesz, P.; Brouwers, H.J.H. Mix Design and Properties Assessment of Ultra-High Performance Fibre Reinforced Concrete (UHPFRC). *Cem. Concr. Res.* **2014**, *56*, 29–39. [CrossRef]
130. Yu, R.; Spiesz, P.; Brouwers, H.J.H. Development of an Eco-Friendly Ultra-High Performance Concrete (UHPC) with Efficient Cement and Mineral Admixtures Uses. *Cem. Concr. Compos.* **2015**, *55*, 383–394. [CrossRef]
131. Wu, Z.; Shi, C.; He, W.; Wu, L. Effects of Steel Fiber Content and Shape on Mechanical Properties of Ultra High Performance Concrete. *Constr. Build. Mater.* **2016**, *103*, 8–14. [CrossRef]
132. Monte, R.; Pereira, L.d.S.; Figueiredo, A.D.d.; Blanco, A.; Bitencourt Júnior, L.A.G. Simplified DEWS Test for Steel Fibre-Reinforced Concrete Characterisation. *Rev. IBRACON Estruturas Mater.* **2023**, *16*, e16604. [CrossRef]
133. Maya Duque, L.F.; Graybeal, B. Fiber Orientation Distribution and Tensile Mechanical Response in UHPFRC. *Mater. Struct. Constr.* **2017**, *50*, 55. [CrossRef]
134. Wille, K.; Kim, D.; Naaman, A.E. Strain Hardening UHP-FRC with Low Fiber Contents. *Mater. Struct.* **2011**, *44*, 538–598. [CrossRef]
135. Graybeal, B.A.; Baby, F. Development of Direct Tension Test Method for Ultra-High-Performance Fiber-Reinforced Concrete. *ACI Mater. J.* **2013**, *110*, 177–186. [CrossRef]
136. Abellán-García, J. Artificial Neural Network-Based Methodology for Optimization of Low-Cost Green UHPFRC Under Ductility Requirements. In *Numerical Modeling Strategies for Sustainable Concrete Structures*; Rossi, P., Tailhan, J.-L., Eds.; Springer: Cham, Switzerland, 2023; pp. 1–11. ISBN 9783031077463.
137. Kim, J.J.; Jang, Y.S.; Yoo, D.Y. Enhancing the Tensile Performance of Ultra-High-Performance Concrete through Novel Curvilinear Steel Fibers. *J. Mater. Res. Technol.* **2020**, *9*, 7570–7582. [CrossRef]
138. Meng, W.; Khayat, K.H. Effect of Hybrid Fibers on Fresh Properties, Mechanical Properties, and Autogenous Shrinkage of Cost-Effective UHPC. *J. Mater. Civ. Eng.* **2018**, *30*, 04018030. [CrossRef]
139. Kim, D.J.; Wille, K.; Naaman, A.E.; El-Tawil, S. Strength Dependent Tensile Behavior of Strain Hardening Fiber Reinforced Concrete. In *High Performance Fiber Reinforced Cement Composites 6*; Springer: Berlin/Heidelberg, Germany, 2011; pp. 3–10.
140. Shaikh, F.U.A.; Nishiwaki, T.; Kwon, S. Effect of Fly Ash on Tensile Properties of Ultra-High Performance Fiber Reinforced Cementitious Composites (UHP-FRCC). *J. Sustain. Cem. Mater.* **2018**, *7*, 357–371. [CrossRef]
141. Ghafari, E.; Costa, H.; Júlio, E.; Portugal, A.; Durães, L. Enhanced Durability of Ultra High Performance Concrete by Incorporating Supplementary Cementitious Materials. In Proceedings of the Second International Conference Microdurability, Amsterdam, The Netherlands, 11–13 April 2012; pp. 11–13.
142. Abellan-Garcia, J.; Khan, M.I.; Abbas, Y.M.; Castro-Cabeza, A.; Carrillo, J. Multi-Criterion Optimization of Low-Cost, Self-Compacted and Eco-Friendly Micro-Calcium-Carbonate- and Waste-Glass-Flour-Based Ultrahigh-Performance Concrete. *Constr. Build. Mater.* **2023**, *371*, 130793. [CrossRef]
143. *BS1881-208*; Recommendations for the Determination of the Initial Surface Absorption of Concrete. British Standards: London, UK, 1996.
144. Liu, Z.; El-Tawil, S.; Hansen, W.; Wang, F. Effect of Slag Cement on the Properties of Ultra-High Performance Concrete. *Constr. Build. Mater.* **2018**, *190*, 830–837. [CrossRef]
145. Taha, B.; Nounu, G. Utilizing Waste Recycled Glass as Sand/Cement Replacement in Concrete. *J. Mater. Civ. Eng.* **2009**, *21*, 709–721. [CrossRef]
146. Graybeal, B.; Tanesi, J. Durability of an Ultrahigh-Performance Concrete. *J. Mater. Civ. Eng.* **2007**, *19*, 848–854. [CrossRef]
147. Sadek, M.M.; Hassan, A.A.A. Abrasion and Scaling Resistance of Lightweight Self Consolidating Concrete Containing Expanded Slate Aggregate. *ACI Mater. J.* **2021**, *118*, 31–42. [CrossRef]
148. Adresi, M.; Lacidogna, G. Investigating the Micro/Macro-Texture Performance of Roller-Compacted Concrete Pavement under Simulated Traffic Abrasion. *Appl. Sci.* **2021**, *11*, 5704. [CrossRef]
149. Bonneau, O.; Lachemi, M.; Dallaire, E.; Dugat, J.; Aitcin, P.C. Mechanical Properties and Durability of Two Industrial Reactive Powder Concretes. *ACI Mater. J.* **1997**, *94*, 286–290. [CrossRef] [PubMed]
150. Yang, S.L.; Millard, S.G.; Soutsos, M.N.; Barnet, S.J.; Le, T.T. Influence of Aggregate and Curing Regime on the Mechanical Properties of Ultra-High Performance Fibre Reinforced Concrete (UHPFRC). *Constr. Build. Mater.* **2009**, *23*, 2291–2298. [CrossRef]

151. Randl, N.; Steiner, T.; Ofner, S.; Baumgartner, E.; Mészöly, T. Development of UHPC Mixtures from an Ecological Point of View. *Constr. Build. Mater.* **2014**, *67*, 373–378. [CrossRef]
152. Herold, G.; Müller, H.S. Measurement of Porosity of Ultra High Strength Fibre Reinforced Concrete. In Proceedings of the International Symposium on Ultra-High Performance Concrete, Kassel, Germany, 13–15 September 2004; Kassel University Press: Kassel, Germany, 2004; pp. 685–694.
153. Schmidt, M.; Fehling, E.; Teichmann, T.; Kai, B.; Roland, B. Ultra-High Performance Concrete: Perspective for the Precast Concrete Industry. *Betonw. Fert. Precast. Plant Technol.* **2003**, *69*, 16–29.
154. Urbonas, L.; Heinz, D.; Gerlicher, T. Ultra-High Performance Concrete Mixes with Reduced Portland Cement Content. *J. Sustain. Archit. Civ. Eng.* **2013**, *3*, 47–51. [CrossRef]
155. Graybeal, B.A. *Characterization of the Behaviour of Ultra-High Performance Concrete*; University of Maryland: College Park, MD, USA, 2005.
156. Corinaldesi, V.; Moriconi, G. Mechanical and Thermal Evaluation of Ultra High Performance Fiber Reinforced Concretes for Engineering Applications. *Constr. Build. Mater.* **2012**, *26*, 289–294. [CrossRef]
157. Akeed, M.H.; Qaidi, S.; Ahmed, H.U.; Faraj, R.H.; Mohammed, A.S.; Emad, W.; Tayeh, B.A.; Azevedo, A.R.G. Ultra-High-Performance Fiber-Reinforced Concrete. Part IV: Durability Properties, Cost Assessment, Applications, and Challenges. *Case Stud. Constr. Mater.* **2022**, *17*, e01271. [CrossRef]

sustainability

MDPI

Review

Exposed Clay Bricks Made with Waste: An Analysis of Research and Technological Trends

Ingrid Silva Assis Santana [1], Mariana da Penha Novaes [2], Ryan Carvalho Chagas de Araújo [2] and Luara Batalha-Vieira [2,*]

[1] Department of Creative Economy, Ruy Wyden Institute of Higher Education, Salvador 41730-101, Brazil
[2] Department of Civil Construction, SENAI CIMATEC Institute of Higher Education, Salvador 41650-010, Brazil; ryan.araujo@ba.estudante.senai.br (R.C.C.d.A.)
* Correspondence: luara.batalha@fieb.org.br

Abstract: Properly disposal of industrial waste is a recurring issue due to its large volume and environmental impact. In turn, civil construction has shown itself to be a potential consumer of waste and can contribute to expanding the circular economy. Clay matrix materials have been a focus of interest for absorbing waste, with the possibility of varying their aesthetics, depending on the waste, as an exposed clay brick. Therefore, it is necessary to understand the research and technological trends on the topic to truly meet the demands of the market and society, in an innovative and sustainable way. To this end, a bibliometric review was carried out considering articles published in journals and an analysis of patent trends was carried out. The use of industrial waste was considerably influential in the growth of research on clay bricks. However, while the scientific community focuses on understanding the impact of industrial waste on clay brick properties, inventors focus on processes and methods for synthesizing clay particles associated with contaminants. The existence of gaps to be explored was identified, such as the aesthetic evaluation of clay bricks. The need to further study the properties of bricks made with waste, optimizing production processes and evaluating the life cycle of these materials are some of the challenges for future research.

Keywords: industrial waste; clay; exposed bricks; review

Citation: Santana, I.S.A.; Novaes, M.d.P.; Araújo, R.C.C.d.; Batalha-Vieira, L. Exposed Clay Bricks Made with Waste: An Analysis of Research and Technological Trends. *Sustainability* **2024**, *16*, 11274. https://doi.org/10.3390/su162411274

Academic Editors: Anibal C. Maury-Ramirez and Nele De Belie

Received: 18 November 2024
Revised: 13 December 2024
Accepted: 18 December 2024
Published: 23 December 2024

1. Introduction

The incorporation of waste into the construction supply chain is aligned with environmental sustainability. It promotes the reduction in consumption of natural raw materials, allows the recycling and reuse of waste and ensures its proper disposal [1–5]. Some additional benefits are the commercial valorization of discarded materials, reduction in energy expenditure in transformation processes, reduction in the volume of materials sent to landfill areas and compensation for the imbalance in local sustainability [1,2,6].

The generation of solid waste is understood as a serious problem in the modern world. A study conducted by the United Nations Environment Program (UNEP) and the International Solid Waste Association (ISWA) in 2024 highlights the need to increase waste management capacity for both public and private entities. This study presents data on municipal solid waste generation, in addition to the amount of uncontrolled or unmanaged waste for the year 2020. Control, in this context, is associated with appropriate waste disposal.

South America had a controlled waste generation rate of approximately 100,000 thousand tons/year. Uncontrolled generation was estimated at 75,000 thousand tons/year. North America, with greater economic power and a higher level of industrialization, has a controlled waste generation rate of about 310,000 thousand tons/year. No data are available regarding waste without management processes. Oceania, with its main representatives Australia and New Zealand, has the lowest waste generation rates. The rate of managed waste is less than 50 thousand tons/year [7].

Regarding waste quantification in Brazil, data mainly originates from business associations, the National Sanitation Information System (SNIS), and the National Solid Waste Management Information System (SINIR) established by the National Solid Waste Policy, linked to the federal government. The SNIS and SINIR databases do not use the concept of waste generation, but rather per capita collected mass for the urban population, which includes the amount collected through selective collection by all executing agents (public agents, private agents, waste pickers' associations or cooperatives, and other executing agents). However, the use of the term "generation" is not entirely inaccurate, since Brazil has a high waste collection rate (98.8%).

It is estimated that waste collection in 2019, in Brazil, reached an annual amount of 65.11 million tons, equivalent to 178.4 thousand tons per day [8,9]. The year 2019 was chosen due to the complexity of the studies conducted up to that year. It is believed that data collection and processing in subsequent years were impacted by the COVID-19 pandemic.

Brazil still falls short of its potential for reuse or recycling, according to the data presented in Table 1. The value of 1.67% corresponds to the Waste Recovery Index or IRR, quantified by the sum of reused, recycled, and energy recovery waste divided by the total waste generated in the country. This value exposes the challenges that must be faced in Brazil regarding this issue [8].

Table 1. Waste recycling index and absolute value of waste recycled in Brazil.

Year	IRR %	Recycled Waste (Thousand Tons/Year)
2019	1.67	928.96
2018	1.75	923.2
2017	1.70	890.08
2016	1.65	842.5
2015	1.90	1020
2014	1.60	937.28

Source: [8,9].

Specifically, regarding industrial waste, Table 2 presents data on generated/collected waste, categorized by type of industrial activity and hazardousness to human health and the environment, available on the SINIR platform. The sectors classified as industrial mainly encompass waste originating from filter media, scrap metal and slag, scrap/used tires, fly ash from coal and zinc combustion, sugarcane bagasse, and waste containing hydrocarbons. Among mining waste, dust and particulate matter, waste contaminated with hydrocarbons, packaging, gravel, and rock fragments are noteworthy. The column labeled "Other Activities" encompasses various activities; however, the most frequently cited waste materials include ash, slag, and boiler dust, wood waste, sludge from urban effluent treatment, and caustic waste.

Table 2. Mass of industrial waste collected in Brazil.

Type	Industry (Thousand Tons)		Mining (Thousand Tons)		Other Activities (Thousand Tons)	
Year	Hazardous	Non-Hazardous	Hazardous	Non-Hazardous	Hazardous	Non-Hazardous
2019	28.60	2200.00	4.00	898.00	99.70	4400.00
2018	16.30	157.00	3.40	366.40	135.30	940.60
2017	26.40	118.00	3.90	24.50	187.10	1100.00
2016	39.50	60.00	3.50	33.20	854.90	2800.00
2015	23.10	2200.00	10.30	12.80	70.30	4100.00
2014	15.80	40.90	44.90	134.00	53.00	3700.00

Source: [8,9].

Civil construction presents itself as a consumer of waste from other industries, contributing to the circular economy. One possibility that has been studied is the use of waste in the production of ceramic pieces, such as exposed brick, which is widely used in masonry systems. Exposed brick is a parallelepiped solid piece, made of clay, and obtained by extrusion. It is dried and burned under high temperatures to obtain important properties, such as resistance and durability. Its sides are full of material and can have different finishing characteristics on both or one of the sides, making it visible without the need for layers of coating. This brick forms a unique aesthetic expression related to the Brazilian architectural tradition. Notable examples include the Convent of São Francisco in São Paulo and the Church of São Francisco de Assis in Belo Horizonte, where the exposed brick is not only a structural material but also a decorative element [8,9]

Among the various waste materials studied as additions or substitutes for clay raw materials, those rich in aluminum stand out due to the high recyclability and control of aluminum-based products in Brazil, and the affinity of aluminum oxides with the clay mixture. Aluminum industry filter dust was used as a raw material for the manufacture of clay bricks, at a content of 20% by mass, producing bricks with higher bulk density and compressive strength, and water absorption similar to reference bricks [10]. An aluminum-rich sludge, produced from anodizing or surface coating processes, was utilized for the fabrication of refractory ceramic bodies with high thermal inertia and mechanical strength [11]. An abundant residue in Brazil that has also been studied as an addition in the manufacture of ceramic bricks is sugarcane bagasse. Studies show that the mechanical and physical properties of bricks made with sugarcane bagasse are better than conventional ones [12].

Therefore, considering the need to conserve traditional construction materials and expand their possibilities of use, there is an effort to search for alternative raw materials. Given the different possibilities for incorporating industrial waste into ceramic products, this study sought to identify the existing gaps on the subject, focusing on the preservation of the raw material (natural clay), possibilities for reducing energy consumption for production and the need to improve the properties of mechanical resistance, water absorption, porosity, linear shrinkage and aesthetic appearance [6,13]. To this end, a bibliometric review was carried out on the recycling of waste in the production of exposed bricks, covering the main patents and published articles. It is expected to contribute to the advances in the compatibility of clay with mining, petrochemical and sewage treatment industry waste, due to its importance in the Brazilian scenario [14–16].

2. Materials and Methods

The ProKnow-C (Knowledge Development) method, a systematic approach for preparing review articles, was used, which aims to ensure the careful selection of data. This method is structured into four main steps (Figure 1).

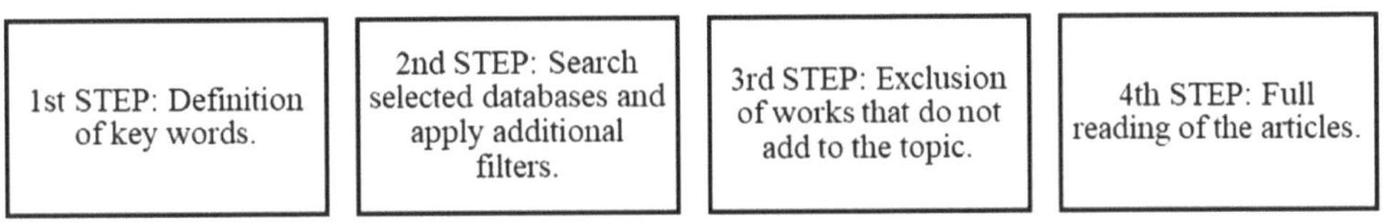

Figure 1. Description of the ProKnow-C method used to prepare the systematic review.

The Scopus database was used to search for scientific articles, while the Lens.org database was used to search for patents. The keywords used in each database were different, as indicated by the ProKnow-C method, to obtain as many works related to the topic as possible (Figure 2). Among the key words, the Boolean operator "AND" was used. Additionally, the Boolean operators "NOT" or "ANDNOT" were used for the word cement, aiming to exclude works on bricks with Portland cement.

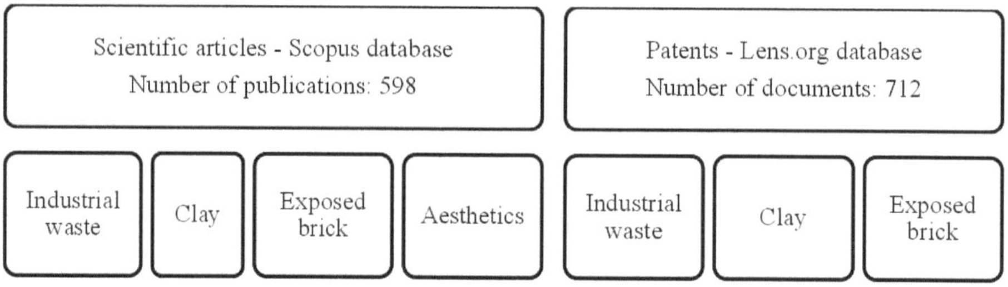

Figure 2. Keywords selected to apply the data collection method.

For scientific articles, the period between 2013 and 2023 and the types of publications (only articles published in journals) were used as filters. Also, Biblioshiny and Bibliometrix (software package for the R language, version RStudio 4.4.2) were used for analyzing the data.

Biblioshiny, the graphical interface for the Bibliometrix package in the R language, offers a comprehensive suite of tools for bibliometric analysis. These tools facilitate complex analyses of bibliographic data, even for users without extensive R programming expertise. The interactive interface allows for dynamic exploration of results, simplifying data interpretation and the identification of patterns and trends. The combined use of these tools enables the extraction of metadata from renowned databases such as Scopus and Web of Science, the generation of bibliometric indicators including H-indices, co-authorship networks, and bibliographic coupling, and data visualization through network graphs and word clouds.

For the global analysis of patents, the tool available on the Lens.org platform was used. To analyze the most found words in patents, the VOSViewer (2023) tool, version 1.6.20, was used because the Bibliometrix tool proved to be ineffective in analyzing the data generated through the Lens.org platform.

Thereby, a robust bibliometric analysis was carried out, highlighting information regarding the number of publications per year, authors, countries and institutions. Through a complete reading of reference articles and patents, the main industrial waste and their interference with the properties of exposed brick were discussed. In this way, it was possible to identify gaps in studies and topics of greatest interest to society related to exposed bricks.

3. Results and Discussion

3.1. Bibliometric Review Based on Articles Published in Journals

The number of articles on clay bricks with industrial waste showed an increasing trend between 2017 and 2019, reaching a maximum of 93 publications (Figure 3) in that period. Between 2019 and 2021 there was again a new interest in the topic with 148 publications. Since then, a decrease has been observed, with 56 productions recorded in 2023. However, in recent years the number of publications has effectively grown, which indicates a search for compatibility between the construction materials industry and the waste generators and possibilities to validate the recycling of their waste.

The 598 publications were separated into conference articles (7.78%), review articles (16.68%) and research articles (75.54%). This result shows that there is still space for more research on clay bricks with industrial waste. It also justifies the focus of this study and of debates and publications at conferences, fundamental for the dissemination and debate of the topic.

There was a total of 225 authors who published about incorporating industrial waste into the composition of clay bricks. The 10 researchers with the most publications are presented in Figure 4. The author considered first in the volume of publications was Aeslina Abdul Kadir, with 25 articles. The focus of her research is the different burning temperatures of clay bricks with waste in their composition, as well the use of waste with the presence of heavy metals, bodywork mill sludge and quarry dust.

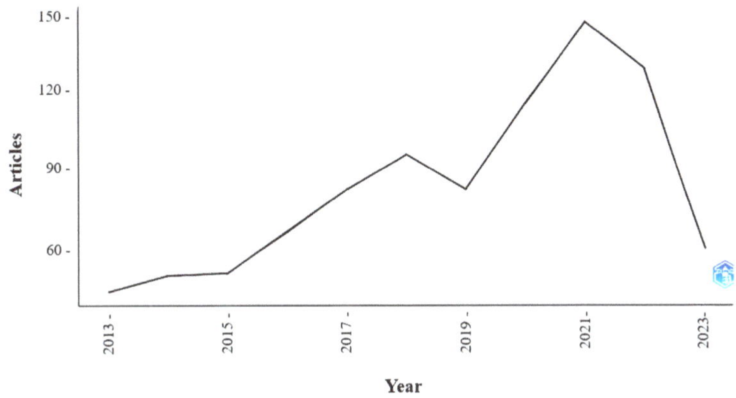

Figure 3. Scientific production on clay bricks with industrial waste from 2013 to 2023.

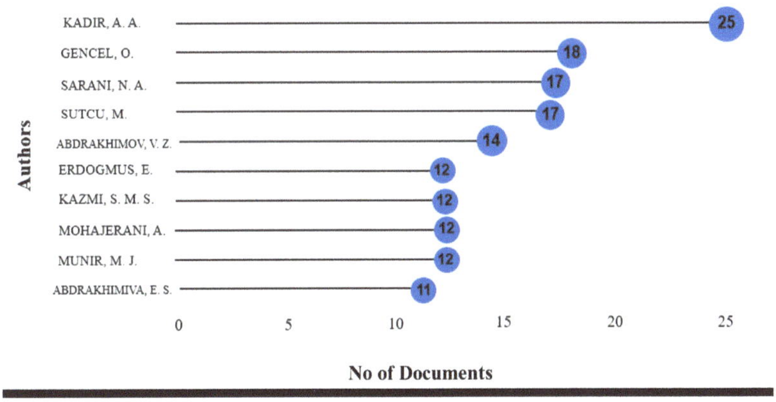

Figure 4. The 10 authors with the highest number of publications.

In total, 41 countries were identified researching clay bricks, and the 10 countries with the highest frequency of publications are presented in Table 3.

Table 3. The 10 countries with the highest frequency of publications.

Region	Frequency
Brazil	62
Spain	52
India	23
Turkey	19
Chile	16
France	11
Australia	10
Italy	10
USA	10
Greece	9

Brazil is the country with the largest number of studies on the destination and recycling of industrial waste into clay products with a frequency of 62 publications, in absolute numbers.

Next is Spain, with a frequency of 52 articles, and India with 23. It is understood that Brazil's leadership can be associated with the incentives of associations regarding the disposal of solid waste. Furthermore, the occurrence of landslide flood disasters in industrial areas of Brazil was considered a reason for implementing waste management policies [17]. Spain and India follow the same perspective and are countries that seek to invest and develop innovation and technology aiming at waste management with a great possibility of reuse [18,19].

The National Solid Waste Policy, enacted in Brazil in 2010, provided a crucial framework for waste management guidelines, facilitating research and technical analyses. Furthermore, Brazil holds a prominent position in research on clay bricks incorporating waste materials, due to its tradition of self-supporting masonry with solid clay bricks, a legacy largely influenced by Portuguese colonization. However, the emergence of new products (different types of blocks and wall systems) has been diminishing the competitiveness of bricks within the sector. The incorporation of waste materials in bricks emerges as a promising solution to address this challenge.

By reducing the extraction of raw materials and waste generated in brick production, companies in the sector can not only mitigate their environmental impacts—a primary effect of implementing more sustainable practices—but also generate substantial economic benefits. Optimizing production processes and seeking innovative solutions can result in secondary impacts such as reduced production costs, new job creation, and the stimulation of technological innovation, configuring a virtuous cycle of sustainable development.

The 10 institutions, from 212, with the most publications and research development in clay bricks and the compatibility of industrial waste are presented in Figure 5. Although Brazil has been identified as the country with the highest frequency of publications (Table 3), the research on this subject is of interest to a large group spread across the globe. Bartin University, Tun Hussein University and RMIT University, located in Turkey, Malaysia and Australia, respectively, are the most relevant in the development of studies on the topic covered in this article.

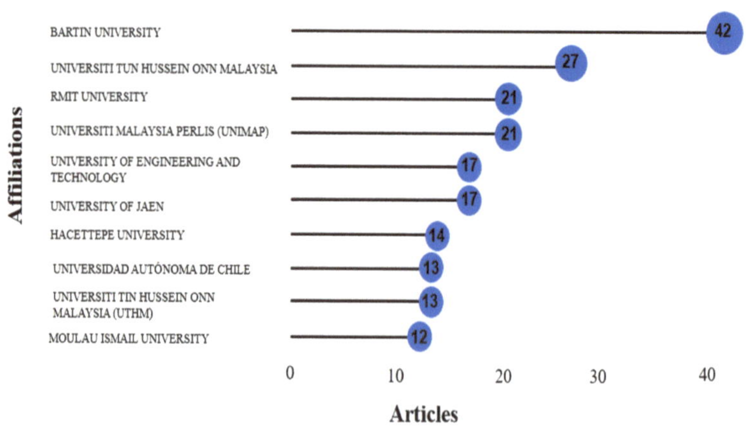

Figure 5. Institutions that publish the most on the topic.

Another important aspect to be analyzed is the choice of journals for publication. Figure 6 lists the 10 (out of 60) most relevant journals regarding the impact factor, which considers the number of publications and citations per year. Construction and Building Materials stand out with the maximum number in the index. Next, there is the Journal of Cleaner Production, with an impact index of 8, and in third place the Journal of

Building Engineering, with an index of 6. These are the main journals on the recycling of industrial waste in construction materials. However, they are seen as multidisciplinary because they publish research on materials engineering, civil construction, the chemistry of materials, clean production, waste management and treatment and energy efficiency in production processes.

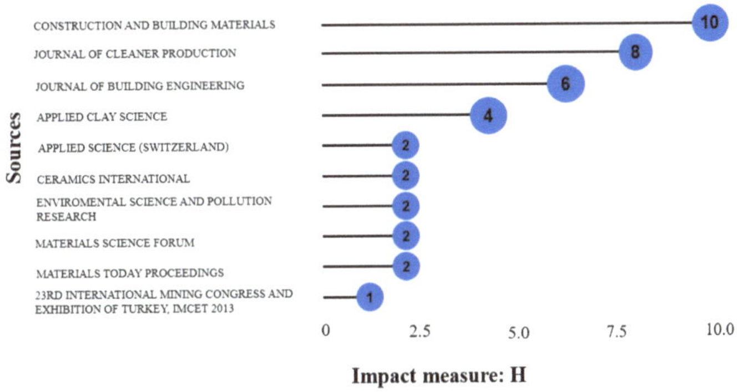

Figure 6. Measurement of the indexing number of the 10 most relevant journals.

The evaluation of keywords was carried out through the visual representation of the word cloud, where the font size determined the relevance of the topic in the searches on the database. Among the 50 words mapped and used in Figure 7, it was observed that the themes with the greatest research relevance are focused on the words bricks, with 504 repetitions, clay, with 206, ceramic materials, with 113, and recycling, with 99 repetitions. The most relevant properties under study were resistance to variation, with 504 repetitions, water absorption, with 235 and thermal conductivity, with 113.

Figure 7. Word cloud referring to the 50 most frequent keywords.

The other keywords had a frequency of repetition lower than 113 times, thus having less relevance and frequency of use in searches. The keywords referring to waste topics, such as industrial waste, had 68 repetitions in the articles found. Furthermore, words such as firing temperature, sintering and particle size were also found, however, with a minimum amount of frequency. In view of this analysis, it is possible to see that the panorama of studies about the recycling of industrial waste is still a subject in development. It also showed an increase in consolidated research areas, like clay bricks, and possibilities to combine industrial waste into ceramic pieces [20,21].

When collecting keywords, it was identified that there was no high frequency of the words aesthetics and texture. Those are important properties to be studied in exposed bricks, since one of the functions of the product is to compose the masonry system and be exposed in facades, forming the composition and aesthetics. Therefore, this topic is a knowledge gap for future research.

A thematic map of niches (Figure 8) was developed, addressing themes that are relevant and under development. In the quadrant referring to "basic" themes, the most frequent themes are about clays, ceramics and energy conservation, as they are underdeveloped due to their consolidated concepts.

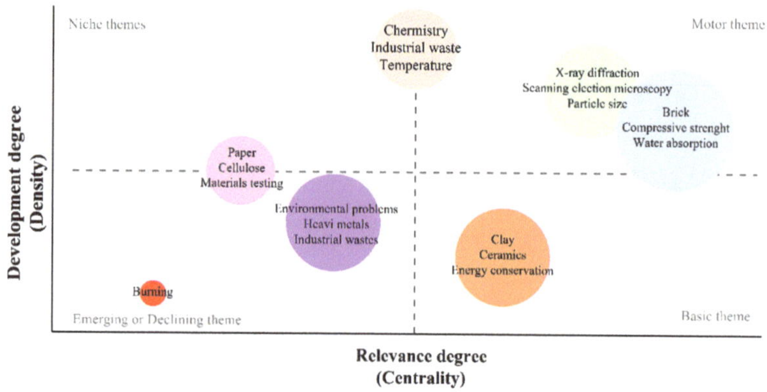

Figure 8. Niche themes chart about the relevance of themes and the development of publications.

In the second quadrant are the "motor" themes, which have continuous relevance and are currently featured in the research and publications found. Considering only clay bricks, compressive strength and water absorption are words with great relevance in the point cloud and, according to the studies, there is a search for improvements and enhancement of these properties. Studies have also been developed on X-ray diffraction (XRD), scanning electron microscopy (SEM) and particle size, which are fundamental for understanding the physical and mineralogical characteristics of types of waste, influencing compatibility with clay matrices.

The topics about chemistry, industrial waste and temperature are seen as "central" and highly relevant in analysis and discussions. These themes are the ones that are most underdeveloped. It shows a need to consolidate scientific studies and methodologies to advance the manufacturing of products, achieving the necessary requirements.

Topics considered "emerging or declining" are presented in the fourth quadrant. It includes paper, cellulose, materials testing, environmental problems, heavy metals and industrial waste. There has been a growing number of publications in the last 10 years about them, which are themes for studies that have emerged (or can still be considered as emerging); however, there is still low production, and there may be advances.

Considering a general view, research has shown that there is a great interest of the construction industry in unraveling studies on the characteristics of the types of industry waste and consolidating methodologies for its application and inclusion in clay matrices.

3.2. Types of Industrial Waste and Their Influence on the Properties of Clay Bricks

Table 4 contains the articles pre-selected by the ProKnow-C method, in which, based on the author's interpretation, we sought those that are most closely aligned with the theme studied and the types of industrial waste most used. This was carried ou by reading the abstracts and titles. These articles (Table 4) were read in full and in detail. The table was organized according to the type of waste evaluated, whether there was a need for treatment for application and what the influence was on the properties of the clay bricks.

Table 4. Industrial waste and the influence on the properties of clay bricks.

n°	Waste	Preparation	Properties	Percentage	References
1	Marble slurry residue	N.T.S (not specified)	Mass loss; Decrease in compressive strength; Increase in water absorption; Reduction in thermal conductivity; Increase in apparent porosity; larger open pore size after firing; Irregular and interconnected pores	5, 10, 15, 20 and 25 wt%	[22]
2	Sulfide Mining Waste	Flotation and comminution	Lower porosity; lower water absorption; greater retraction; higher modulus of elasticity; greater efflorescence; high arsenic leaching	20 and 40 wt%	[16]
3	Fly ash	N.T.S	Decrease in compressive and flexural strength; lower efflorescence; less mass; higher porosity; higher water absorption;	5, 10, 15, 20 and 25 wt%	[23]
4	Fly ash	N.T.S	Decrease in plasticity index; Increase in compressive strength; lower water absorption; higher porosity; higher water absorption	50, 60, 75, 80 wt%	[24]
5	Kraft cellulose pulp residue	Dried to a uniform weight and then dissolved in water	Increase in water requirement to confer plasticity; increase in shrinkage; increase in apparent porosity; decrease in density; increase in water absorption; decrease in compressive strength	0, 2.5, 5 and 10 wt%	[25]
6	Electroplating sludge (10 wt%) + Glass residue	The sludge and pulp were dried at 105 °C for 24 h, then the glass was ground to 74 μm. Everything was mixed in powder form until homogeneity was obtained	Reduction in open porosity; reduction in the surface area of the bricks; increase in compressive strength; higher density; reduction in heavy metal leaching	Content of electroplating sludge of 10 wt% and 5, 10 15, 20, 25, 30 wt% of waste glass powder	[26]
7	Coconut husk	Dried in an oven, ground and classified	Greater mass loss on firing; increase in porosity; increase in water absorption; decrease in compressive strength; decrease in apparent density; decrease in thermal conductivity; decrease in shrinkage, decrease in sintering temperature	0, 10, 20, and 30 wt%	[27]
8	Urban sewage treatment sludge	Anaerobic reactor, anoxic filter, drum filter, secondary and tertiary decanters and calcium hydroxide	Greater mass loss, decrease in compressive strength, decrease in sintering temperature, increase in porosity, increase in water absorption, increase in plasticity	15 wt% sludge	[28]
9	Contaminated glass powder	N.T.S	Decrease in firing temperature, decrease in shrinkage, increase in compressive strength, increase in thermal conductivity, increase in density, decrease in porosity, decrease in water absorption, lower leaching index	5, 10, 15, 25 wt%	[29]

Table 4. *Cont.*

n°	Waste	Preparation	Properties	Percentage	References
10	Water treatment plant sludge	Sieved at 60 mesh, drying at 70 °C for 24 h	Increase in porosity, decrease in flexural strength, increase in water absorption, increase in mass loss, decrease in apparent specific mass	5, 10, 15, 20 and 25 wt%	[30]
11	Silica sand	Drying at 105 °C for 24 h, Comminution up to 1.2 mm, mixed with water for extrusion	Decrease in plasticity, decrease in flexural strength on drying, decrease in shrinkage, decrease in thermal conductivity, decrease in flexural strength, increase in water absorption, decrease in density	11 wt%	[31]
12	Sawdust	Drying at 105 °C for 24 h, Comminution up to 1.2 mm, mixed with water for extrusion	Increase in plasticity, decrease in flexural strength on drying, decrease in shrinkage, decrease in thermal conductivity, decrease in flexural strength, increase in water absorption, decrease in density	3.7 wt%	[31]
13	Bauxite residue	Drying at 105 °C for 24 h, Comminution up to 1.2 mm, mixed with water for extrusion	Decrease in plasticity, increase in flexural strength on drying, decrease in shrinkage, increase in thermal conductivity, increase in flexural strength, increase in water absorption, increase in density	3 wt%	[31]
14	Dolomitic limestone	Drying at 105 °C for 24 h, Comminution up to 1.2 mm, mixed with water for extrusion	Increase in plasticity. Decrease in flexural strength on drying, decrease in shrinkage, increase in thermal conductivity, decrease in flexural strength, decrease in water absorption, increase in density	25 wt%	[31]
15	Paper sludge	Drying at 105 °C for 24 h, Comminution up to 1.2 mm, mixed with water for extrusion	Increase in plasticity, increase in flexural strength on drying, increase in shrinkage, decrease in thermal conductivity, decrease in flexural strength, increase in water absorption, decrease in density	8 wt%	[31]
16	Iron scrap	Drying at 105 °C for 24 h, Comminution up to 1.2 mm, mixed with water for extrusion	Decrease in plasticity, increase in flexural strength on drying, increase in shrinkage, decrease in thermal conductivity, decrease in flexural strength, increase in water absorption, decrease in density	15 wt%	[31]
17	Coal	Drying at 105 °C for 24 h, Comminution up to 1.2 mm, mixed with water for extrusion	Increase in plasticity, decrease in flexural strength on drying, decrease in shrinkage, decrease in thermal conductivity, decrease in flexural strength, increase in water absorption, decrease in density	8 wt%	[31]
18	Olive stone	Drying at 105 °C for 24 h, Comminution up to 1.2 mm, mixed with water for extrusion	Increase in plasticity, increase in flexural strength on drying, decrease in shrinkage, decrease in thermal conductivity, decrease in flexural strength, increase in water absorption, decrease in density	20 wt%	[31]

Table 4. *Cont.*

n°	Waste	Preparation	Properties	Percentage	References
19	Hausmannite	N.T.S	Decrease in water absorption, no change in drying rate, lower mass loss in freeze–thaw test, decrease in mass loss in crystallization test, little change in porosity: more interconnected pores	15 wt%	[32]
20	Industrial ceramic sludge	N.T.S	Greater pore interconnectivity, increase in water absorption index, increase in drying rate, Greater mass loss in freeze–thaw test, Mass gain in salt crystallization test, little change in porosity: greater number of open pores	10 wt%	[32]

According to Table 4, mining, water and sewage treatment waste, glass, fly ash and biomaterials were used in the last research. Mining residues are wastes from the industrial processing of ores and, as they have a composition closer to the raw material used in bricks, they are the first to be investigated and considered ideal additives. Industrial ceramic waste is an example of application, as it has a majority composition of aluminum-silicon and has a chemical composition similar to that of the clay used to manufacture bricks, influencing the reduction in defects [6].

Researchers [22] studied the marble mud waste, rich in calcite, and observed an increase in apparent porosity, a larger pore size and irregular and interconnected pores. The results were influenced by the combustion of calcite that decomposes into CO_2, creating pores in the brick structure and CaO prone to expansion, forming pores. Porosity allows the brick to be lighter and provides better thermal insulation.

Other identified wastes were silica sand, bauxite residue, dolomitic limestone, iron scrap and coal [31]. Bricks using silica sand, bauxite residue and iron scrap showed a decrease in plasticity and a reduction in flexural strength during drying and in the product. Those with bauxite residue had a higher flexural strength. These wastes influenced shrinkage positively, reducing the risk of defects; however, the use of scrap residue had the opposite effect. Bricks with dolomitic limestone and coal achieved plasticity with better processing conditions and an increase in thermal conductivity proportional to the increase in density. All samples, except for the one with dolomitic limestone, had an increase in water absorption.

The incorporation of industrial ceramic sludge waste [28] and waste from the production of iron alloys and manganese oxide batteries (hausmannite) [32] was also considered. The samples with ceramic sludge showed greater pore interconnectivity, resulting in a higher water absorption rate and better drying rate. The samples with hausmannite had a decrease in water absorption, while the drying rate remained almost the same.

Water treatment and sewage industries' sludge is a mixture of organic and mineral waste, presenting pollution risks for rivers and other bodies of water. The biggest concern when it comes to incorporating this waste is its toxicity and risk to human and environmental health. The presence of heavy metals and other metals is considered toxic but during the burning process, the organic matter combusted and released enough energy to cause the inclusion of these metals in the crystalline structure of the brick. This combustion process helped with the energy efficiency of burning, increasing the temperature inside the brick. Properties such as linear retraction were not influenced, however, water absorption increased, and compressive strength reduced. The incorporation of water treatment sludge interfered with the loss of mass during burning [19], influencing the pores and reducing the density of the material, which caused thermal insulation and the reduction in mechanical resistance. However, even with the lowest mechanical resistance, the bricks with up to 15% incorporation of this residue were in accordance with Brazilian safety standards [30].

Glass waste is considered a flow agent and directly influences the reduction in the sintering temperature, with the introduction of a liquid phase filling the pores and densifying the brick during firing. Research shows that this material immobilizes heavy metals through the vitrification process, being able to reduce or eliminate the leaching of these metals in bricks. The glass was able to fuse, acting as an encapsulation of metallic atoms, called spinels. Once the glass creates the liquid phase, it fills the pores, increasing density and mechanical resistance. A study showed that this addition reduced porosity, burning temperature for sintering, and shrinkage. It also caused an increase in compressive strength and density, in addition to reducing water absorption and the leaching rate [26–29].

Fly ash, a waste identified among the 50 keywords in the word cloud presented in Figure 7, comes from the combustion of coal in thermoelectric plants. A study incorporated 50%, 20% and 25% of fly ash in the clay composition and identified different responses regarding compressive strength [23,24]. However, more studies are still needed to establish a pattern.

Biomaterials are classified as organic matter wastes and are considered combustible agents, reducing the energy and temperature required for sintering. Due to the cause of combustion, it is common to identify the formation of pores and a lower density. The addition of Kraft cellulose pulp waste [25], in ratios of 2.5%, 5% and 10%, increased the clay brick porosity. However, the greater the content of the waste, the density, water absorption and compressive strength were more negatively affected. Another modified property was plasticity, increasing the need for water in the extrusion process.

The use of coconut shells, and other biomaterials, in clay matrices was evaluated and there was a greater loss of mass during burning due to the combustion of the material [27]. The result of this is increased porosity, thermal conductivity and water absorption and lower density and compression resistance. This combustion effect caused internal centers of high temperature, reducing the sintering temperature. Sawdust waste, paper sludge and olive pits have also been studied and in all cases, there is an increase in plasticity, porosity water absorption, and a decrease in thermal conductivity and flexural strength [31].

4. Technological Prospection

The research began with the selection of keywords to be inserted into the database Lens.org. Table 5 lists the attempts made. The set of words that provided the greatest number of patents should be the chosen one. However, by reading part of the texts, the Boolean operator "NOT" was included due to the large number of works that use Portland cement as a binder, which is not the focus of this study.

Table 5. Key words used to search the database Lens.org.

Key Words	Results
Faced Brick AND Exposed AND clay AND industrial waste AND aesthetic ANDNOT cement	62
Exposed Brick AND clay AND industrial waste AND aesthetic	79
Faced Brick AND clay AND industrial waste NOT cement	435
Exposed Brick AND clay AND industrial waste NOT cement	712
Exposed Brick AND clay AND industrial waste	2211

The word "faced brick" provided a satisfactory number of documents, however, most patents were related to the production of refractory bricks for the manufacturing industry, and not for the construction of walls. The final selection was exposed brick AND clay AND industrial waste NOT cement, totaling 712 patent texts, as listed in Figure 2.

Figure 9 shows the number of patents linked to the topic over the years, together with the equation obtained from the simple regression analysis.

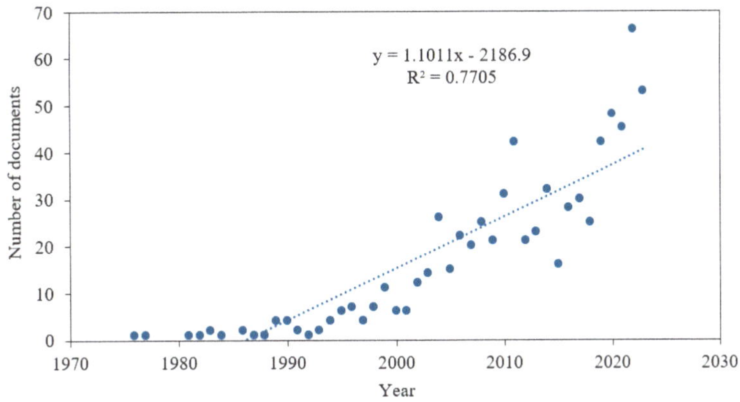

Figure 9. Number of patents linked to the topic, throughout the years.

The analysis yielded an R^2 value of 77.05% with a significance level of 5%. The *p*-value for the "year" variable was lower than the significance level (5.213×10^{-15}), indicating that the variable has a significant effect on the number of patents. Despite this, we concluded that to accurately forecast the number of patents based on the year, a larger dataset should be analyzed. This is because the residuals exhibited in Figure 9 show a certain trend up to the year 2000, with greater dispersion observed after that year. The seemingly parabolic trend suggests that the relationship between the variables may be nonlinear. In other words, the number of patents does not increase at a constant rate over time, but rather with a variable growth rate.

There has been a considerable increase since the early 1990s, coinciding with the growing global interest in the preservation of natural resources and sustainable development. It is important to mention the ECO-92 event (or Earth Summit), held by the United Nations Conference on Environment and Development in the city of Rio de Janeiro, in 1992. This conference is understood as a milestone, helping to consolidate and expand the concept of "sustainable development", making it a central part of the international development and environmental agenda [33]

Comparing the results shown in Figures 3 and 9, it is possible to observe that the growth patterns of scientific research and technological work are similar after 2013. In both cases, the production peaks occurred in 2021, during the pandemic. Thereby, the production of clay bricks incorporating industrial waste has become a topic of interest to the scientific and technical community, also evidenced by the increase in concessions and patent applications related to the topic (Table 6). The COVID-19 pandemic has underscored the urgent need to re-evaluate and modify existing production and consumption models, prompting a heightened focus on sustainable practices and the exploration of alternative materials. In this context, the incorporation of industrial waste in brick manufacturing emerges as a solution with dual benefits: it reduces reliance on natural resources, such as virgin clay, and contributes to effective waste management, thereby minimizing the environmental impact of the construction industry.

Figure 10 presents the jurisdiction of the patents identified in this study. American patents fall under the jurisdiction of the United States Patent and Trademark Office (USPTO), which guarantees intellectual property rights granted by the U.S. government. These patents provide protection concerning sale, use, and manufacturing for the domestic market, as well as importation into the U.S. This protection is time-limited (typically 20 years) in exchange for public disclosure of the invention upon patent grant [34]. The European Patent Office (EPO) has enhanced its collaborations with major foreign corporations such as Bosch, Toyota, Nissan, and Phillips, among others. The EPO's patent network is a more globalized and interconnected framework [35]. The World Intellectual Property Organization (WIPO) serves as the global forum for services, policies, information, and cooperation in the field of intellectual property. The WIPO became a specialized agency of the United Nations in 1974. This entity comprises various sectors responsible for administering numerous international treaties and ensuring protection for inventions across multiple countries [35].

Table 6. Number of patents granted and patents applications per year.

Year	Granted Patent	Patent Application
2016	9	19
2017	12	18
2018	11	14
2019	15	27
2020	12	36
2021	15	30
2022	23	42
2023	25	28

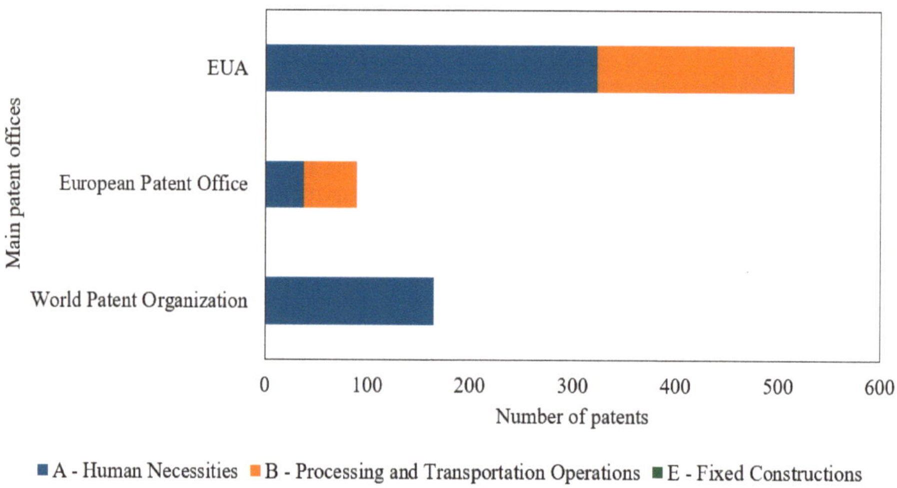

Figure 10. Number of patents according to their jurisdiction and IPC classification.

The results presented in Figure 10 differ from those from Table 3. The countries with the most scientific research on exposed clay bricks with waste are not the ones that develop the most technology with this material.

The nationality of inventors matches the country that has the most patents (Figure 11). However, when comparing the data obtained in the survey of articles, there is a difference in the results. At least part of the top 10 researchers are known to be non-American (Figure 4), even though the USA is the country with the most patent inventors.

Despite the significant number of American inventors, many patent holders are Japanese and Canadian entities. These organizations hold exclusive commercialization rights for these inventions (Figure 12. Number of patents according to the nationality of the patent holders).

The main corporations holding patents are shown in Figure 13. The different areas in which the companies operate can also be seen. The patents associated with the keywords selected are focused on water treatment and the uses of clay materials as thickeners or rheology modifiers [36]. For example, the company Boral Industries INC., a construction company, holds a patent that use clay materials to manufacture polyurethane composites filled with light fillers [37]. However, we also identified patents for different uses than listed as those held by the Fluid Energy Group, a Canadian company. Their patents are related to fine particle technology for different industrial applications [38].

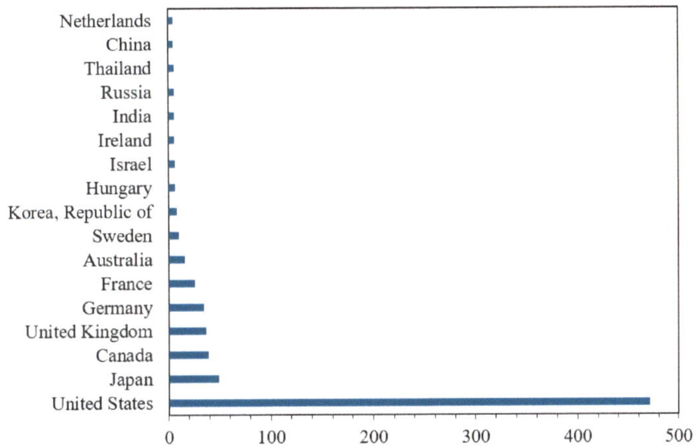

Figure 11. Patent numbers according to the nationality of the inventors.

Figure 12. Numerical distribution according to the nationality of patent holders.

Regarding the relevance of the study, the citation was also considered as a research variable. Figure 14 demonstrates the growing interest in this topic over recent decades, as evidenced by the numerous citations of the documents selected for this study. Patents classified under section A (Human Needs) are more prevalent. However, this category encompasses various purposes. Examples include adsorbent ceramic particles [39]; processes for the chemical binding of heavy metals from sludge into the silicate structure of clays and shales for building manufacturing material [40]; a method for forming bricks, tiles, and similar products by treating clay, shale, or other clay ceramic raw materials containing pyrite [41]; and a formula for sintered brick with red mud [42].

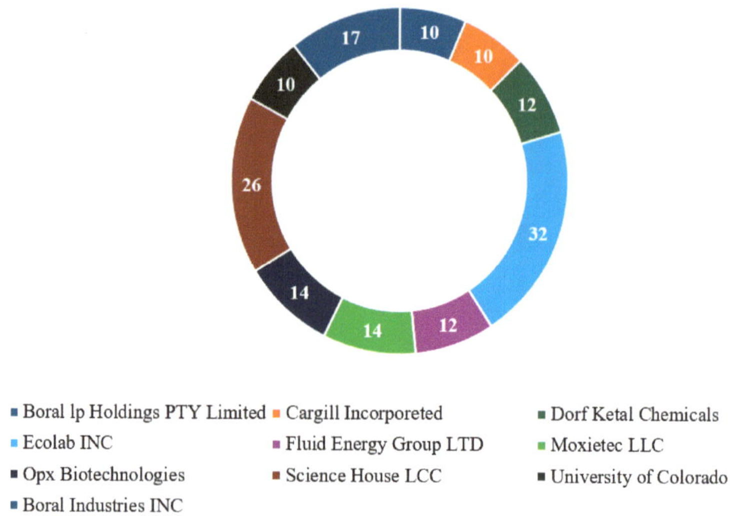

Figure 13. Major patent-holding corporations.

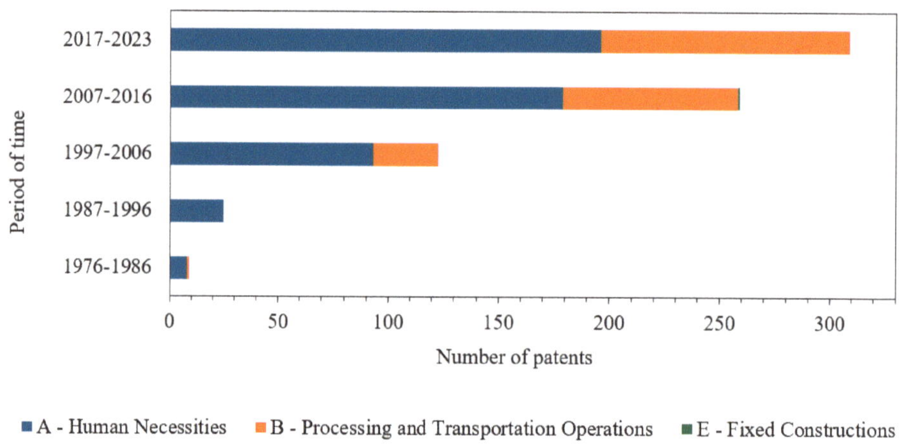

Figure 14. "Cited by Patents" tab as an indicator of the relevance of articles according to the IPC classification.

Among the most relevant patents in class B are those that propose advancements in raw material processing, material recovery in effluent and waste treatment, and the synthesis of new materials. Examples include the composition and manufacturing process of building bricks and tiles [43], clay fiber filtration tubes containing flocculant wound on a mandrel [44], linear hearth kiln system and related methods [45], and method for processing clay ceramic materials [41].

The legal status of the surveyed patents is in Figure 15. It is evident that even patents filed over 20 years ago remain active, reflecting the ongoing interest of the holders. Notably, there is a significant number of patents with pending filing confirmations. Specifically, between 2017 and 2023, there are 121 patents awaiting filing confirmation. Again, a confirmation of the interest of the industry and the market on the use of waste in exposed clay bricks.

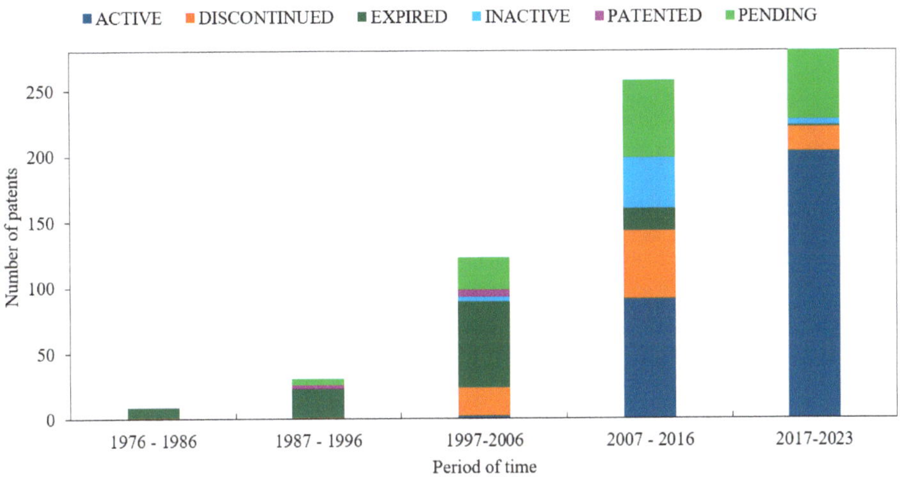

Figure 15. Number of patents over the years according to their legal status.

The recurring keywords found in the patent abstracts are shown in Figure 16. The words "embodiment", "chemical product", "article", "mixture" and "coating" highlight the focus of the patents: the search for optimized processes and compositions. These keywords are strongly connected. From this central core, connections radiate to more specific areas of research. For example, the term "mixture" is linked to "catalyst" and "chamber", suggesting that the patents explore chemical reactions in controlled environments to create innovative mixtures.

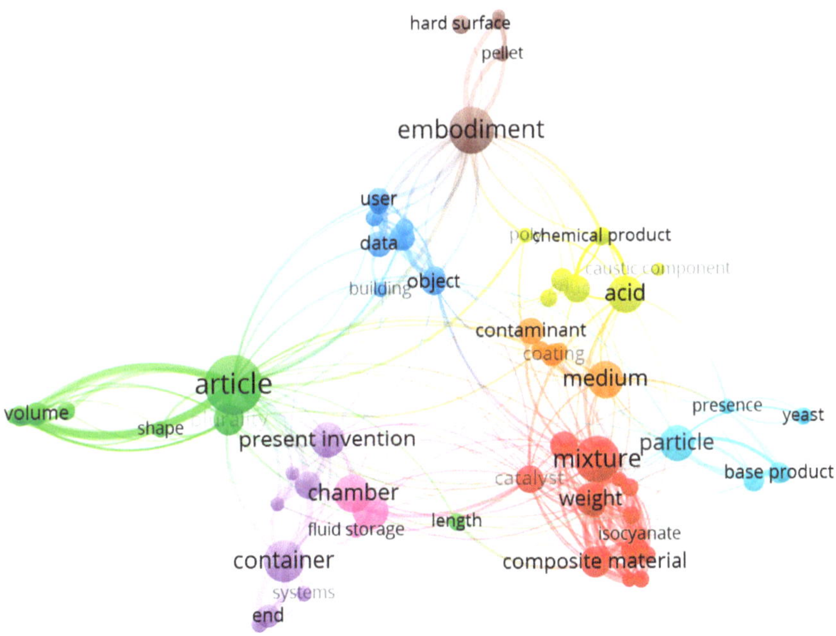

Figure 16. Most frequent words and their connections.

The connection between "mixture" and "particle" and "weight" indicates the investigation of the impact of particle size and proportion on the composition of clays. The presence of the term "contaminant" connected to "mixture" and "coating" highlights the concern with the safety and sustainability of the inventions.

The word "contaminant" is most strongly connected to "amount", "pellets" and "particle", which are believed to represent inventions about the synthesis of particles to the encapsulation of contaminating substances. The pellets produced are then incorporated into the manufacturing of bricks or other building materials.

It can be observed that the scientific community, regarding the production of clay bricks, seeks to understand the impact of the incorporation of different types of industrial waste on the properties of the brick, with emphasis on compressive strength. However, from the analysis of the selected patents, the inventors focus on the processes of transformation of clays as a mineral and on the creation of methods that allow the synthesis of clay particles associated with contaminants [46,47]. For example, the US patent 4882067 [40] discusses the chemical bonding of heavy metals from sludge in the silicate structure of clays and shale for the manufacture of construction materials.

5. Conclusions

This study has revealed a diverse range of research over the last decade on the incorporation of waste materials in clay bricks, highlighting a growing trend in publications and patents. This field demonstrates significant potential for promoting sustainability within the construction industry, driven by increasing interest in waste recycling and reuse.

Analysis of the most cited keywords in scientific articles has shown a focus on the properties of water absorption, strength, and thermal conductivity of bricks incorporating waste materials. Various industrial wastes, such as those from mining, water treatment, glass, and fly ash, have proven promising for this application, significantly impacting brick properties such as plasticity, shrinkage, and water absorption. However, the impact on these properties varied according to the waste studied, as shown in Table 2. This is associated with the variation in the chemical and mineralogical composition of the waste materials.

Patent analysis has revealed trends and opportunities for innovation in the field, particularly highlighting the use of waste contaminated with pyrite, red mud, and ash from various sources.

An interesting point identified was the divergence of interests between academia and industry. While academic research focuses on the influence of waste on brick properties, the industry seeks to optimize production processes. This divergence points to the need for greater alignment between research and development, aiming to accelerate innovation and the application of new technologies.

Finally, the study identified a knowledge gap regarding the aesthetics of bricks incorporating waste materials, both in scientific articles and patents. Future research could explore the impact of sustainability and the use of clean energy on the aesthetics of bricks, in addition to investigating the life cycle of these materials.

Funding: This research received no external funding.

Conflicts of Interest: The authors declare no conflict of interest.

References

1. Al-Fakih, A.; Mohammed, B.S.; Liew, M.S.; Nikbakht, E. Incorporation of waste materials in the manufacture of masonry bricks: An update review. *J. Build. Eng.* **2019**, *21*, 37–54. [CrossRef]
2. Padmalosan, P.; Vanitha, S.; Kumar, S.V.; Anish, M.; Rajesh Tiwari, R.; Dhapekar, N.K.; Yadav, A.S. An investigation on the use of waste materials from industrial processes in clay brick production. *Mater. Today Proc.* 2023; *in press*. [CrossRef]
3. Jiménez-Quero, V.; Maza-Ignacio, O.T.; Guerrero-Paz, J.; Campos-Venegas, K. Industrial wastes as alternative raw materials to produce eco-friendly fired bricks. *J. Phys. Conf. Ser.* **2017**, *792*, 012065. [CrossRef]
4. Raj, A.; Sharma, T. *Durability Characteristics of Unfired Earth Blocks Influenced by the Addition of Industrial Waste and Synthetic Fibre*; AIP Publishing: Melville, NY, USA, 2024; p. 030003. [CrossRef]

5. Alonso-Santurde, R.; Coz, A.; Quijorna, N.; Viguri, J.R.; Andrés, A. Valorization of Foundry Sand in Clay Bricks at Industrial Scale. *J. Ind. Ecol.* **2010**, *14*, 217–230. [CrossRef]
6. Zhang, L. Production of bricks from waste materials—A review. *Constr. Build. Mater.* **2013**, *47*, 643–655. [CrossRef]
7. Lenkiewicz, Z. Global Waste Management Outlook 2024—Beyond an Age of Waste: Turning Rubbish into a Resource. Available online: https://wedocs.unep.org/20.500.11822/44939 (accessed on 17 November 2024).
8. Ministério Do Meio Ambiente (MMA). Relatório Nacional de Gestão de Resíduos Sólidos. 2024. Available online: https://sinir.gov.br/paineis/inventario/ (accessed on 17 November 2024).
9. Brasil. Ministério do Desenvolvimento Regional (MDR). Sistema Nacional de Informações sobre Saneamento: Diagnóstico do Manejo de Resíduos Sólidos Urbanos—2019. Available online: https://www.gov.br/cidades/pt-br/acesso-a-informacao/acoes-e-programas/saneamento/snis/diagnosticos-anteriores-do-snis/residuos-solidos-1/2019/Diagnostico_RS2019.pdf (accessed on 17 November 2024).
10. Bonet-Martínez, E.; Pérez-Villarejo, L.; Eliche-Quesada, D.; Castro, E. Manufacture of Sustainable Clay Bricks Using Waste from Secondary Aluminum Recycling as Raw Material. *Materials* **2018**, *11*, 2439. [CrossRef]
11. Ribeiro, M.J.; Tulyaganov, D.U.; Ferreira, J.M.; Labrincha, J.A. Recycling of Al-rich industrial sludge in refractory ceramic pressed bodies. *Ceram. Int.* **2002**, *28*, 319–326. [CrossRef]
12. Osman, S.; Firnando, M.F.P.; Zakaria, M.N.; Ahmad, M. Physical and Mechanical Properties of Fired Industrial Waste-Clay Brick from Sugarcane Bagasse. *Environ. Behav. Proc. J.* **2024**, *9*, 11–16. [CrossRef]
13. Mesquita, D.F.S.; Brito, J.D.E. Tijolos Face à Vista (TFV), Outra Forma de Encarar a Alvenaria. *Eng. Vida* **2008**, *IV*, 44–49.
14. Martins, A.P.G.; Vasconcelos, G.; Costa, A.C. Caracterização Experimental do Comportamento de Ligadores em Paredes de Tijolo Face à Vista à Tração e à Compressão. 2016. Available online: https://www.researchgate.net/publication/305411935_CARACTERIZACAO_EXPERIMENTAL_DO_COMPORTAMENTO_DE_LIGADORES_EM_PAREDES_DE_TIJOLO_FACE_A_VISTA_A_TRACAO_E_A_COMPRESSAO (accessed on 17 November 2024).
15. Raut, S.P.; Ralegaonkar, R.V.; Mandavgane, S.A. Development of sustainable construction material using industrial and agricultural solid waste: A review of waste-create bricks. *Constr. Build. Mater.* **2011**, *25*, 4037–4042. [CrossRef]
16. Simão, F.V.; Chambart, H.; Vandemeulebroeke, L.; Nielsen, P.; Adrianto, L.R.; Pfister, S.; Cappuyns, V. Mine waste as a sustainable resource for facing bricks. *J. Clean. Prod.* **2022**, *368*, 133118. [CrossRef]
17. da Silva, O.H.; Locastro, J.K.; Umada, M.K.; Polastri, P.; De Angelis Neto, G. Legislação e normatização técnica aplicáveis às etapas do gerenciamento de resíduos sólidos industriais. In *Qualidade e Sustentabilidade na Construção Civil*; Editora Científica Digital: São Paulo, Brazil, 2021; pp. 208–220. [CrossRef]
18. Cortez, A.T.C. Gerenciamento de resíduos sólidos urbanos: A experiência de barcelona (Espanha) como contribuição as cidades brasileiras. *Estud. Geográficos Rev. Eletrônica De Geogr.* **2013**, *11*, 54–65.
19. Machado, H.H.; Sgorlon, J.G.; Altoé, S.P.S.; Meneguetti, K.S.; Oliveira, J.C.D.; Martins, C.H.; Tavares, C.R.G. A gestão dos resíduos sólidos industriais aplicada em países desenvolvidos e em desenvolvimento. In Proceedings of the Congreso Latinoamericano de Ecología Urbana, Buenos Aires, Argentina, 14–15 June 2012.
20. Wang, S.; Gainey, L.; Mackinnon, I.D.R.; Allen, C.; Gu, Y.; Xi, Y. Thermal behaviors of clay minerals as key components and additives for fired brick properties: A review. *J. Build. Eng.* **2023**, *66*, 105802. [CrossRef]
21. Sun, J.; Zhou, H.; Jiang, H.; Zhang, W.; Mao, L. Recycling municipal solid waste incineration fly ash in fired bricks: An evaluation of physical-mechanical and environmental properties. *Constr. Build. Mater.* **2021**, *294*, 123476. [CrossRef]
22. Munir, M.J.; Kazmi, S.M.S.; Wu, Y.-F.; Hanif, A.; Khan, M.U.A. Thermally efficient fired clay bricks incorporating waste marble sludge: An industrial-scale study. *J. Clean. Prod.* **2018**, *174*, 1122–1135. [CrossRef]
23. Abbas, S.; Saleem, M.A.; Kazmi, S.M.S.; Munir, M.J. Production of sustainable clay bricks using waste fly ash: Mechanical and durability properties. *J. Build. Eng.* **2017**, *14*, 7–14. [CrossRef]
24. Lingling, X.; Wei, G.; Tao, W.; Nanru, Y. Study on fired bricks with replacing clay by fly ash in high volume ratio. *Constr. Build. Mater.* **2005**, *19*, 243–247. [CrossRef]
25. Demir, I.; Baspınar, M.S.; Orhan, M. Utilization of kraft pulp production residues in clay brick production. *Build. Environ.* **2005**, *40*, 1533–1537. [CrossRef]
26. Mao, L.; Guo, H.; Zhang, W. Addition of waste glass for improving the immobilization of heavy metals during the use of electroplating sludge in the production of clay bricks. *Constr. Build. Mater.* **2018**, *163*, 875–879. [CrossRef]
27. Moujoud, Z.; Harrati, A.; Manni, A.; Naim, A.; El Bouari, A.; Tanane, O. Study of fired clay bricks with coconut shell waste as a renewable pore-forming agent: Technological, mechanical, and thermal properties. *J. Build. Eng.* **2023**, *68*, 106107. [CrossRef]
28. Areias, I.O.R.; Vieira, C.M.F.; Colorado, H.A.; Delaqua, G.C.G.; Monteiro, S.N.; Azevedo, A.R.G. Could city sewage sludge be directly used into clay bricks for building construction? A comprehensive case study from Brazil. *J. Build. Eng.* **2020**, *31*, 101374. [CrossRef]
29. Xin, Y.; Robert, D.; Mohajerani, A.; Tran, P.; Pramanik, B.K. Transformation of waste-contaminated glass dust in sustainable fired clay bricks. *Case Stud. Constr. Mater.* **2023**, *18*, e01717. [CrossRef]
30. da Silva, E.L.G.; Maciel, A.P. Uso de resíduos sólidos de estação de tratamento de água como carga em blocos cerâmicos. *Cerâmica Ind.* **2019**, *24*, 29–36. [CrossRef]
31. Makrygiannis, I.; Tsetsekou, A. Efficient Recovery of Solid Waste Units as Substitutes for Raw Materials in Clay Bricks. *Recycling* **2022**, *7*, 75. [CrossRef]

32. Coletti, C.; Maritan, L.; Cultrone, G.; Mazzoli, C. Use of industrial ceramic sludge in brick production: Effect on aesthetic quality and physical properties. *Constr. Build. Mater.* **2016**, *124*, 219–227. [CrossRef]
33. de Oliveira, L.D. A Geopolítica do Desenvolvimento Sustentável na CNUMAD—1992 (ECO-92): Entre o Global e o Local, a Tensão e a Celebração. *Rev. De Geopolítica* **2011**, 2, 43–56.
34. Maria, T.; Alex, S.; Panos, A.; Shlomo, H. Globalization emergence in the European Patent Office (EPO) patent network. *Phys. Soc.* 2020; *Pre print version*. [CrossRef]
35. World Intellectual Property Organization. WIPO External Offices. Available online: https://www.wipo.int/about-wipo/en/offices/ (accessed on 18 October 2024).
36. Gupta, A.; Lohokare, H.R.; Bhole, Y.S. Non-Chlorinated Oxidizing Biocide Chemistries, Their Methods of Production, Application and Methods of Feed Thereof. WO Patent 2019/213483A1, 7 November 2019.
37. Kumar, A.; Ai, L.; Hill, R.L. Filled Polyurethane Composites with Lightweight Fillers. U.S. Patent 10030126 B2, 24 July 2018.
38. Clay, P.; Markus, W. Nova Composição Cáustica Sintética. WO Patent 2019/095035 A1, 23 May 2019.
39. Hiroshi, T. Cleaning Agent. EP Patent 1437397 A1, 14 July 2004.
40. Barrett, J.; Charles, B.R. Process for the Chemical Bonding of Heavy Metals from Sludge in the Silicate Structure of Clays and Shales and the Manufacture of Building and Construction Materials Therewith. U.S. Patent 4882067 A, 21 November 1989.
41. Brosnan, D.A.; Frederic, J.C., Jr.; Sanders, J.P., III. Method for Processing Clay Ceramic Materials. U.S. Patent 6548438 B2, 15 April 2003.
42. Zheng, H. Formula of Sintered Brick with Red Mud. CN Patent 101747018 A, 23 June 2010.
43. Theophilus, A.D. Composition and Process for Making Building Bricks and Tiles. U.S. Patent 6440884 B1, 27 August 2002.
44. Theisen, M.S.; Spittle, K.S. Mandrel-Wound Flocculant-Containing Fiber Filtration Tubes. U.S. Patent 7883291 B2, 8 February 2011.
45. Bleifuss Rodney, L.; Englund David, J.; Kiesel Richard, F. Linear Hearth Furnace System and Methods Regarding Same. U.S. Patent 7875236 B2, 25 January 2011.
46. Hagen, D. Clay Composition. U.S. Patent 7323429 B2, 29 January 2008.
47. Byung Ok, P.; Sulk, K.Y. Clay Bricks and Pavers Using Industrial Waste. KR Patent 20100130855 A, 14 December 2010.

sustainability

MDPI

Article

Reliability of Reusing Gypsum Flat Board Grinded Waste as a Conventional Plaster Replacement for Buildings

Julian David Puerto [1,*], Sandra Uribe [2], Luis Ayala [2], Alexander Padilla [2] and Alvaro Rodriguez [2]

1 Departamento de Ingeniería Civil y Agrícola, Universidad Nacional de Colombia, Bogotá 111321, Colombia
2 Departamento de Ingeniería Civil, Universidad de la Salle, Bogotá 111711, Colombia; sluribe@unisalle.edu.co (S.U.); layalar@unisalle.edu.co (L.A.); edpadilla@unisalle.edu.co (A.P.); arodriguezpaez@unisalle.edu.co (A.R.)
* Correspondence: jdpuertos@unal.edu.co

Abstract: This research work focuses on the recycling of gypsum, a key component of drywall panels commonly used in construction. It proposes this practice as a technically and economically viable strategy that also aligns with the circular economy model. In this context, waste loses its status as refuse and is transformed back into raw material. A technical feasibility analysis was conducted based on a sample extracted from a specific project, which was subjected to characterization tests and mechanical behavior assessments in both laboratory and real construction conditions. Additionally, an economic feasibility analysis was performed by comparing the budgets of the same project in two scenarios: the traditional plastering process and the use of recycled gypsum. This analysis highlighted the fiscal and legal benefits that adopting the circular model could offer to stakeholders in the construction industry. This study began with a market analysis to determine the availability of recyclable material and to assess the multiple benefits that its reuse can provide. Based on the characterization of the material obtained from the construction site and the mechanical tests conducted, the economic advantages were evaluated for contractors as well as for the potential establishment of companies focused on gypsum recycling. All of this analysis was framed within the context of sustainability, emphasizing the positive environmental impacts of this practice, as well as the development of a strategy that serves as a valuable proposal for the construction sector. This work concludes that recycling gypsum in construction projects represents a technically, environmentally, and economically sustainable alternative that can positively transform the industry.

Keywords: feasibility analysis; gypsum board recycling; circular economy; construction waste potential; gypsum characterization; construction strategy

Citation: Puerto, J.D.; Uribe, S.; Ayala, L.; Padilla, A.; Rodriguez, A. Reliability of Reusing Gypsum Flat Board Grinded Waste as a Conventional Plaster Replacement for Buildings. *Sustainability* **2024**, *16*, 7889. https://doi.org/10.3390/su16187889

Academic Editors: Anibal C. Maury-Ramirez and Nele De Belie

Received: 6 August 2024
Revised: 4 September 2024
Accepted: 5 September 2024
Published: 10 September 2024

1. Introduction

In recent decades, the concept of a circular economy has gained prominence as a promising approach to promoting sustainability and the efficient use of resources, challenging the traditional linear model of "extract, produce, dispose" [1]. Construction, being one of the sectors with the highest consumption of raw materials and generation of waste, represents a key area for implementing circular economy practices [2]. The extensive use of gypsum boards (also known as drywall) in partition systems, flooring, ceilings, and wall coverings has raised growing concerns about the volumes of waste generated. This waste, resulting from cutting, damage during handling and dismantling, is often discarded as common debris, wasting its potential to be recycled and reintegrated into production cycles. Numerous studies have explored the recycling of gypsum boards, primarily focusing on the recovery of gypsum through wet separation processes, calcination, and thermal treatments. Although these methods have demonstrated the viability of high-quality recycled gypsum, they involve additional costs, complexity, and, in some cases, environmental impacts associated with energy consumption and emissions [3]. Recognizing these limitations, this paper proposes an alternative strategy for the direct recycling of gypsum board waste at construction sites, assessing its technical and economic feasibility as a substitute for conventional gypsum used in the manufacture of stucco and finishes. This approach avoids complex and costly treatments, facilitating the on-site reuse of generated waste and aligning with the principles of a circular economy and sustainable construction. In addition to the potential environmental and economic benefits, the recycling of gypsum boards helps address the challenges associated with their final disposal, as this waste can generate pollution problems if not managed properly. In this context, various international initiatives and regulations have promoted responsible management and recycling of these construction wastes [4]. This study contributes to this trend by providing a practical and cost-effective solution for utilizing gypsum board waste, fostering the transition toward a more circular and sustainable construction industry [5].

2. Materials and Methods

The main objective of the implemented methodological strategy was to transform gypsum board waste, through mechanical grinding and filtering processes, into recycled gypsum that met the minimum characteristics required to be used as a substitute for conventional commercial gypsum in plastering and finishing applications in construction projects. Once a suitable recycled material was obtained, its properties were thoroughly studied and its potential for application as finishing plaster was evaluated through a series of standardized technical tests, comparing it with a reference sample of standard commercial gypsum widely used in Colombian construction projects for plastering and finishing processes.

To ensure the representativeness and practical relevance of the study, samples of gypsum board waste were selected from an ongoing commercial project in the city of Bogotá, Colombia, where this type of material was being intensively used. In the first stage, these samples were homogenized through a grinding and filtering process (Figure 1). The waste material underwent a total of 4 cycles of grinding and sieving in order to achieve an acceptable degree of fineness, meeting the requirement established by the Colombian standard [6], which defines a maximum particle size of 600 μm, equivalent to the mesh opening of the No. 30 sieve in the standard series.

Figure 1. Detail of waste uniformization and homogenization.

This initial preparation of the samples was crucial to ensure that the obtained recycled material met the necessary fineness standards for its use in finishing processes, laying the groundwork for its subsequent characterization and performance evaluation.

Once a suitable recycled material was obtained, with particles smaller than the size required for use as stucco plaster, a thorough characterization process was carried out to evaluate its physical and mechanical properties. This stage was crucial in determining the technical feasibility of using recycled gypsum as a substitute for conventional gypsum.

To carry out this characterization, a battery of standardized tests widely recognized in the construction industry were applied. Table 1 presents a detailed list of the test methods used, which are based on current international and Colombian technical standards.

Table 1. List of applied characterization tests.

Test	Characteristic	International Reference Standards/Colombia
Flow	Determination of water-to-gypsum ratio	ASTM C230/NTC 111 [7,8]
Purity	Chemical analysis	ASTM C471M/NTC 5227 [9,10]
Compressive strength	Compressive strength	ASTM C472/NA [11]
Setting time	Setting time	ASTM C472/NTC 490 [12]
Density	Density	ASTM C188/NTC 221 [13,14]

These tests, conducted strictly following the protocols established in the mentioned standards, allowed for precise and reliable characterization of key aspects of the recycled material's behavior, such as its workability, mechanical strength, setting times, and fundamental physical properties. The results obtained were compared with those of a reference sample of standard commercial gypsum, widely used in Colombian construction projects, in order to evaluate the technical feasibility of recycled gypsum as an effective substitute.

In the first characterization stage, the index properties of both recycled and reference gypsums were established. Subsequently, the fresh state properties of pastes made with both materials were determined (Figures 2 and 3). These tests allowed for the evaluation of crucial aspects such as initial and final setting times, as well as consistency and workability indicators—such as the optimal water/gypsum ratio for mixing—through the determination of paste fluidity. Additionally, a series of mechanical tests were performed on the mixtures in the hardened state to characterize fundamental properties such as compressive and tensile strength (Figure 4). These mechanical tests were complemented by detailed observations made using scanning electron microscopy (SEM), which allowed for an in-depth analysis of the morphology and microstructure of the materials, providing valuable information on the interaction between the components of recycled gypsum. Finally, as a key stage of the methodology, performance tests were carried out in which both recycled and reference gypsums were applied under real-use conditions (Figure 5). These tests aimed to evaluate the behavior and durability of the materials when used for plastering and finishing surfaces, simulating the conditions they would be exposed to in a construction project. This comprehensive methodology, which combines standardized characterization tests with performance tests under real conditions and microstructural analysis, allowed for a deep and reliable understanding of the behavior of recycled gypsum, laying the foundation for evaluating its technical viability as a substitute for conventional gypsum in finishing applications in the construction industry.

Figure 2. Details of the consistency test.

Figure 3. Details of the flow test.

Figure 4. Details of the mechanical test.

Figure 5. Details of performance tests of gypsum plasters.

3. Results, Discussion, and Analysis

3.1. Testing Phyisical and Mechanical Properties

Table 2 presents a comparative summary of the main parameters resulting from the characterization process carried out for recycled gypsum and conventional reference gypsum. One of the fundamental aspects that influenced the behavior of recycled gypsum was the presence of remaining cardboard particles in the mixture, which could not be completely removed during the sieving process. This presence of organic matter was clearly evidenced in the results of the chemical analysis, which reported values close to 4% for recycled gypsum. While this percentage is well below the 7% that corresponds to the typical cardboard content in a standard gypsum board [15], it is significant and reflects that the material used in this study, being homogenized only through grinding and screening processes, retained an appreciable fraction of cardboard particles smaller than the meshes used.

Table 2. Summary of characterization results for reference and recycled gypsum.

Parameter	Standard Gypsum	Standard Recycled
Density (g/cm^3)	2.66	2.56
Organic matter (%)	0	3.8
Optimal ratio A/Y	0.5	0.7
Compressive strength (MPa)	7.6	5.2
Tensile strength (MPa)	1.67	0.5
Initial setting time (Min)	3	150
Final setting time (Min)	10	>1000

It is worth noting that this minimal processing approach, avoiding wet separation methods or thermal treatments commonly applied in recycling plants [16,17], was implemented with the aim of simulating the processing that could be performed on waste directly at the construction site, facilitating its on-site reuse. The remaining cardboard particles, composed mainly of cellulose fibers, had a significant impact on the hydration kinetics of recycled gypsum. This retarding effect of cellulose on gypsum hydration has been widely documented in previous studies [18–20] and was reflected in the setting times obtained. While the conventional reference gypsum set in normal times, recycled gypsum experienced a substantial delay in its initial and final setting times, as observed in the evolution of the setting process presented in (Figure 6). This phenomenon can be explained according to the setting stages described [21], where the presence of cellulose fibers delays the transition between the fluid phase and the moldable phase in which the paste acquires a plastic consistency. Although these prolonged setting times exceed the limits established in some regulations [22], recent research has shown that gypsum mixtures with delayed hydration kinetics can be used in applications requiring greater workability and extended use times, avoiding the use of retarding additives [23].

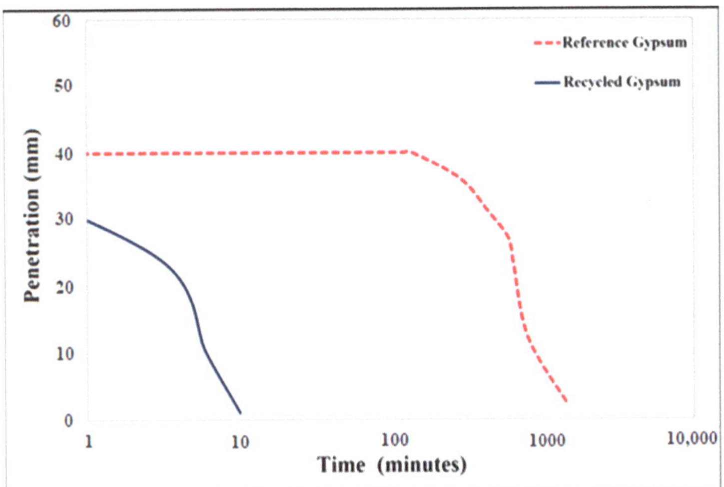

Figure 6. Analysis of setting evolution.

The setting of recycled gypsum with cardboard particles experienced a significant delay, requiring almost a full day to complete its setting. This behavior is explained by the retarding effect caused by cellulose from the cardboard, as has been reported in numerous studies where cellulose pulp has been incorporated to generate composite materials with gypsum [18–20]. Although these prolonged setting times exceed the maximums allowed by the current Colombian Technical Standard [6], which establishes a limit of 50 min, it is important to note that the final state of the stucco with cardboard particles after completing its setting was acceptable and did not present significant surface effects. This finding suggests that the recycled material could be used in complex work areas that require longer utilization times compared to conventional stuccos, providing greater workability without the need to incorporate additional retarding additives. Regarding mechanical properties, the test results revealed that the presence of cardboard particles generated a 31.5% reduction in compressive strength and a 70% reduction in tensile strength, compared to the standard sample of conventional gypsum. These values, although lower than those of the reference sample, are in line with those reported in previous studies that have evaluated the effect of incorporating natural fibers into gypsum mixtures.

For example, Refs. [24,25] have documented similar reductions in compressive strength when incorporating different types of natural fibers into gypsum, attributing this behavior to the lack of adequate integration between the fibers and the gypsum matrix, as evidenced in the scanning electron microscopy observations carried out in the present study. However, it is important to highlight that, despite these decreases in mechanical properties, the compressive strength values obtained for recycled gypsum (5.2 MPa) meet the minimum requirements established by the European standard EN 13279-1 [26] (2 MPa) for construction gypsums. This result supports the technical feasibility of using recycled material in plastering and finishing applications, coinciding with the findings of other studies that have evaluated mixtures of recycled and conventional gypsums.

For example, compressive strength values between 6 and 7 MPa have been reported for mixtures of conventional and recycled gypsums (70/30 ratio) that were subjected to thermal processes between 100 and 140 °C to remove cardboard. In comparison, the variation with respect to the recycled gypsum in the present study would be only 13–25%, which is considered favorable given that the recycled material was only subjected to grinding processes, without additional thermal treatments. It is important to note that, as in the study by [27], the reference sample of conventional gypsum used in this work did not meet the compressive strength requirement established by the Colombian standard (12.4 MPa), adopted from ASTM standards. This fact suggests that national standards might be excessively strict compared to international requirements, such as those of European regulations.

3.2. Performance Test

In order to comprehensively and objectively evaluate the performance of recycled gypsum under real-use conditions, extensive application and on-site tests were carried out. These tests were conducted in both indoor and outdoor environments (Figure 7), preparing the mixtures with the optimal dosage previously determined in the characterization tests.

The main purpose of these tests was to evaluate the final appearance of the stucco made with recycled gypsum and to monitor any changes in appearance that might occur when varying the grinding process and, consequently the fineness of the material. Additionally, the aim was to analyze possible changes in the properties of the stucco over time and its response to wetting and drying processes. To this end, stucco samples were prepared using recycled gypsum subjected to 1, 2, and 3 grinding cycles and were exposed to the outdoor environment for an extended period of 55 days. During this time, periodic and detailed monitoring was carried out, evaluated through visual inspection of any changes in surface texture or alterations in the color of the stucco.

These observations were complemented with an objective analysis of colorimetric variations using advanced digital image analysis techniques. Following the methodology defined by [28], Def-Lab-Geo-Imagen software GEO5 2022 [29] was applied to compare the average colors of the samples in different color spaces (CIELa*b and RGB) before and after the exposure period. Additionally, Euclidean distances between initial and final chromatic values were calculated, using a threshold of 2 units established by [30] as an indication that two materials or bodies present the same color.

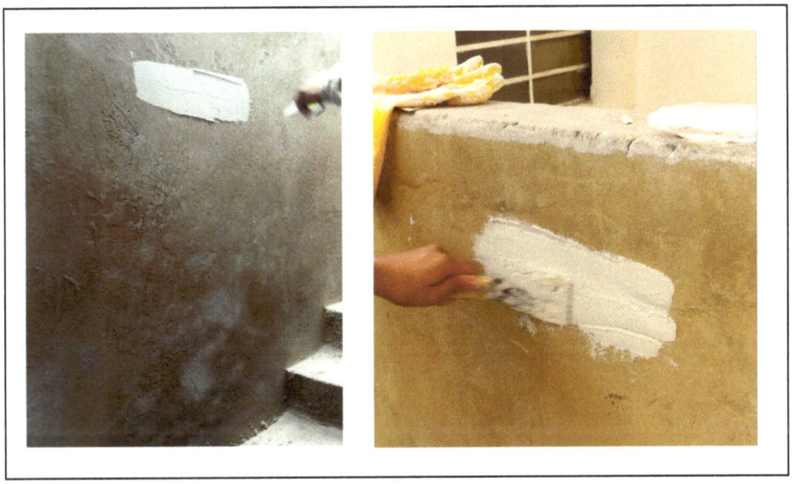

Figure 7. Detail of stucco application during indoor and outdoor performance tests.

The results obtained, both from visual evaluations and image analysis, led to the conclusion that the stucco made with recycled gypsum exhibited excellent stability and durability over time (Figure 8). No significant changes were observed in surface appearance, texture, or color, regardless of the degree of grinding of the recycled material used. Even after prolonged exposure to the elements, the stucco did not show any detachment, alterations, or loss of consistency.

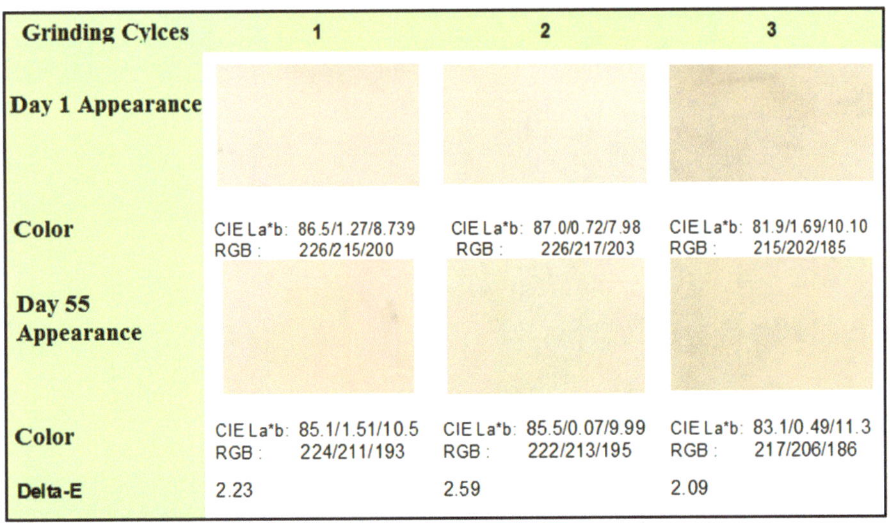

Figure 8. Evolution of color changes in stucco through image analysis.

Additionally, to evaluate the material's resistance to extreme conditions, two stucco samples were subjected to multiple wetting and drying cycles. These tests simulated exposure to sudden changes in humidity, such as those that could occur in real construction conditions due to environmental factors or cleaning processes. Again, the recycled gypsum demonstrated optimal behavior, without presenting significant changes in its appearance or surface properties.

These results support the practical viability of using recycled gypsum in plastering and finishing applications, demonstrating its ability to maintain its integrity and appearance even under adverse conditions. The stability and durability exhibited by the recycled material are consistent with the findings reported by [31] in their technical feasibility analysis of gypsum board recycling, reinforcing the promising potential of this circular economy-based strategy in the construction industry.

3.3. Observation of Morphology and Microstructure

In order to gain a deep understanding of the microstructure and organization of recycled gypsum, as well as the distribution and interaction of remaining cardboard fibers with the present gypsum crystals and their agglomeration levels, detailed observations were carried out using scanning electron microscopy (SEM). These observations were conducted at the facilities of the Geosciences Laboratories at the National University of Colombia. Figure 9 shows a representative micrograph of a recycled gypsum sample obtained through SEM. In it, agglomerated particles with different crystalline morphologies can be observed. Needle-like or prismatic shapes predominate and are characteristic of dehydrated gypsum crystals ($CaSO_4 \cdot 2H_2O$), the main mineralogical component of gypsum. Additionally, crystals with a spherical morphology can be distinguished in smaller proportions, which could indicate the presence of other mineralogical phases or compounds formed during the gypsum hydration process. These observations align with the results reported by other researchers who have studied the microstructure of recycled gypsum mixtures [16,32]. A key aspect evident in the micrographs is the presence of fibers dispersed among the crystalline agglomerations. These fibers correspond to cardboard particles that could not be completely removed during the screening process and were incorporated into the recycled gypsum matrix. Although these cellulose fibers are heterogeneously distributed in the microstructure, no significant integration or embedding of gypsum crystals into them is observed. This phenomenon could explain, at least in part, the observed decrease in the mechanical properties of recycled gypsum compared to conventional gypsum, as discussed earlier.

Figure 9. Detail of recycled gypsum morphology microstructure.

On the other hand, the lack of adequate adhesion and interaction between the cardboard fibers and the gypsum matrix could be due to the hydrophilic nature of cellulose, which hinders the formation of strong bonds with gypsum crystals during the hydration process. This behavior has been previously documented in studies that have analyzed the incorporation of natural fibers into gypsum matrices [24,25]. These microstructural observations provide valuable information about the mechanisms underlying the observed macroscopic behavior in recycled gypsum and will serve as a basis for future research aimed at optimizing the integration of cardboard fibers into the gypsum matrix, in order to improve mechanical properties without compromising workability and setting times. Figure 10 shows a detail of the gypsum crystal agglomerations around the cellulose fibers from the cardboard present in the recycled gypsum board waste. It can be observed that there is no appreciable embedding of gypsum crystals into the cellulose fibers.

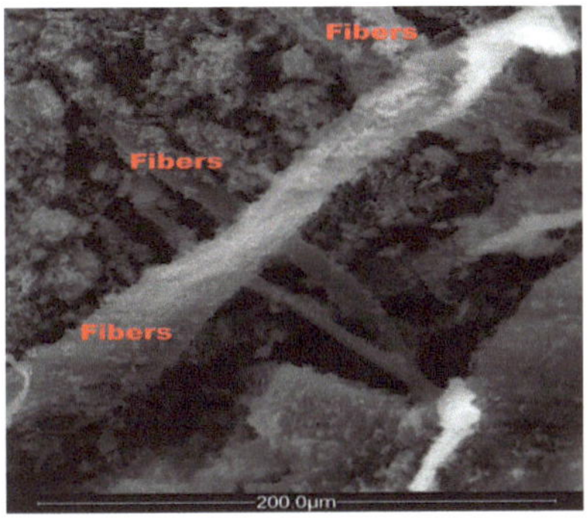

Figure 10. Detail of cardboard fiber presence.

4. Discussion on Circular Economy

In addition to the above, the life cycle analysis of gypsum board supports the technical feasibility of its direct recycling in construction projects, while highlighting the inherent advantages of the circular economy. By avoiding complex and costly treatments for gypsum recovery, this approach significantly reduces the environmental impacts associated with raw material extraction and the production of new materials. Furthermore, by reincorporating gypsum board waste into productive cycles, waste generation is minimized and a more efficient use of resources is promoted, contributing to sustainability throughout the entire value chain of this construction material.

The life cycle analysis of gypsum board, or drywall, spans from raw material extraction to the recycling of waste generated on-site. The production of virgin gypsum, the main component of these boards, leads to high CO_2 emissions due to the use of fossil fuels. However, recycling allows for a 49.8% reduction in greenhouse gas emissions compared to production from primary sources, as well as decreasing energy consumption and depletion of natural resources.

In the same context, among the environmental benefits of recycling gypsum board, a reduction of up to 50% in electricity consumption stands out, as more is used for the production of the original material. Additionally, it promotes the development of ecological awareness, encouraging responsible environmental behaviors from construction companies and contractors. Furthermore, from a slightly broader perspective, using recycled material generates Certified Emission Reductions (CERs), which are units of reduction in greenhouse gas (GHG) emissions resulting from Clean Development Mechanism (CDM) projects. These reductions are expressed in metric tons of carbon dioxide equivalent (tCO_2e), and in the case study featured in this research, we found a reduction of 7.2525277×10^{-5} tCO_2 eq/t (CERs) in the global warming category and 0.001372033271 tCO_2 eq/t (CERs) in the non-renewable energy category. The characterization results were obtained using SimaPro 8.3.0 software [33]. See Table 3.

Now, from an economic standpoint, and considering that this factor is of great importance for builders when making decisions, recycling gypsum board becomes a viable and sustainable alternative to its final disposal in landfills or dumps, while avoiding costly disposal fees. Additionally, there is potential to generate new income by commercializing recovered materials to produce new construction products.

Table 3. Characterization results for natural gypsum.

Impact Category	Unit	Total
Global warming	Kg CO_2. Eq/Kg	0.072525277
Non-renewable energy	M J primaria/Kg	1.372033271

In the same vein, and seeking to incentivize Colombian contracting companies in this activity, in the business context, the adoption of recycling practices not only boosts corporate image by reflecting a commitment to sustainability but also positively influences the loyalty and satisfaction of consumers aware of these causes.

On the other hand, regarding material consumption, it is estimated that approximately 100 million tons of construction materials are used nationally. Of this amount, about 91.5 million tons (equivalent to 91%) are used in the construction of buildings, housing, and civil works in the country. The remaining 2% is destined for export as construction products, while the remaining 7%, equivalent to 7.4 million tons, is considered debris resulting from construction works (Government of the Republic of Colombia, 2019) [34]. Moreover, specific regulations promote the recycling of materials, and complying with these regulations is not only an environmental responsibility but also a smart strategy to avoid fines and penalties, ultimately having a direct and positive effect on the finances of companies in the sector.

In this context, the proposal to recycle gypsum board in the construction industry strategically aligns with the principles of the circular economy promoted by the Colombian government through various initiatives such as the National Circular Economy Strategy 2018–2022, which sought continuous resource valorization, closing material cycles, and above all, the adoption of new, more efficient and sustainable business models; Resolution 1407 of 2018, which sought to regulate environmental waste management, assigning responsibilities to producers; or CONPES 3874 of 2016 [35], which established the National Policy for Integrated Solid Waste Management 2016–2030.

The construction project that served as a case study for this research is located in the city of Bogotá, Colombia. To document the economic benefit, the official budget of the contractor in charge of the activity was worked on, and only the items representing the use of both gypsum board and stucco were impacted, showing that the savings make up only 1% of the total direct cost value of the project when implementing the recycling strategy. However, this work was carried out experimentally without contemplating truly efficient logistics. Furthermore, there are not many studies that focus on the empirical analysis of this type of strategy, and therefore, there was no frame of reference or comparison. Nevertheless, it was also evident that the behavior of several companies in obtaining a typology of implementation of these practices is a gradual process that begins with the implementation of recycling and material reuse measures to continue with minimizing energy consumption and redesigning products (in this case, substituting stucco for gypsum). Only after this process will it be possible to analyze the economic benefits in their full dimension.

5. Conclusions

The on-site application tests conducted over nearly two months with stucco made from recycled gypsum containing around 4% cardboard particles demonstrated that this material is stable over time and does not easily deteriorate. There were no changes in its surface finish, with no detachment or alterations, maintaining its consistency even when exposed to the elements.

The setting of recycled gypsum with cardboard particles is much slower than conventional gypsum, taking almost a full day to complete setting. This is due to the retarding effect caused by the cellulose present in the cardboard. Although the setting times exceed the maximum allowed by Colombian regulations, the final state of the cardboard-containing stucco after setting is acceptable. This slower setting could allow the use of recycled gypsum in work areas that require longer usage times than conventional stuccos, providing greater workability and avoiding the use of retarding additives.

Microstructure observations suggest that the hydrophilic nature of cellulose fibers may affect the adhesion between the fibers and the matrix composed of dehydrated gypsum crystals (gypsita), resulting in poor phase integration. This could explain the reduction in the mechanical properties of recycled gypsum compared to mixtures made with conventional gypsum without cellulose fibers.

Author Contributions: Conceptualization, S.U. and J.D.P.; methodology, S.U.; validation, L.A., A.R. and A.P.; formal analysis, S.U.; investigation, J.D.P.; writing—original draft preparation, L.A.; writing—review and editing, A.R.; visualization, A.P.; supervision, J.D.P. All authors have read and agreed to the published version of the manuscript.

Funding: This Research received no external funding.

Institutional Review Board Statement: The study did not require ethical approval.

Informed Consent Statement: Informed consent was obtained from all subjects involved in the study.

Data Availability Statement: The data that supporting the findings of this research are available from the corresponding author upon reasonable request. However, due to ethical considerations and privacy concerns, these data cannot be made publicly available. Restrictions apply to the availability of these data, which were used under license for the current study, and so are not publicly accessible. Researchers interested in accessing the data should contact the corresponding author, who will consider requests on a case-by-case basis, subject to obtaining appropriate ethical approvals and data sharing agreements.

Acknowledgments: Engineer Cristian Camilo Flores Suesca.

Conflicts of Interest: The authors declare no conflicts of interest.

References

1. Ghisellini, P.; Cialani, C.; Ulgiati, S. A review on circular economy: The expected transition to a balanced interplay of environmental and economic systems. *J. Clean. Prod.* **2016**, *114*, 11–32. [CrossRef]
2. Hossain, M.U.; Ng, S.T.; Antwi-Afari, P.; Amor, B. Circular economy and the construction industry: Existing trends, challenges and prospective framework for sustainable construction. *Renew. Sust. Energ. Rev.* **2020**, *130*, 109948. [CrossRef]
3. Purnell, P.; Renzetti, A. Challenges and opportunities for recycling gypsum waste in construction: A comprehensive review. *J. Clean. Prod.* **2021**, 300.
4. Koçan, S.; Gómez-Camacho, C.E.; Martínez-Rocamora, A. An international review of waste management and recycling strategies in the construction industry. *Waste Manag. Res.* **2021**, *39*, 1012–1028.
5. Yuan, H.; Huang, Z.; Xu, P. A framework for eco-efficiency of construction and demolition waste management. *Waste Manag.* **2022**, *138*, 153–163.
6. *NTC_4914*; Norma Técnica Colombiana NTC 4914: Cementos. Especificaciones para el uso de yeso de estucado para fundido y para modelado. INCONTEC: Bogotá, Colombia, 2001.
7. *ASTM C230*; Standard Specification for Flow Table for Use in Tests of Hydraulic Cement. ASTM International: West Conshohocken, PA, USA, 2023.
8. *NTC 111*; Norma Técnica Colombiana NTC 111: Método para determinar la fluidez de morteros de cemento hidráulico. ICONTEC: Bogotá, Colombia, 2024.
9. *ASTM C471M*; Standard Test Methods for Chemical Analysis of Gypsum and Gypsum Products (Metric). ASTM International: West Conshohocken, PA, USA, 2024.
10. *NTC 5227*; Norma Técnica Colombiana NTC 5227: Yeso y productos de yeso. Métodos de ensayo químicos. ICONTEC: Bogotá, Colombia, 2021.
11. *ASTM C472*; Standard Test Methods for Physical Testing of Gypsum, Gypsum Plasters and Gypsum Concrete. ASTM International: West Conshohocken, PA, USA, 2014.
12. *NTC 490*; Norma Técnica Colombiana NTC 490: Yesos para construcción. Especificaciones físicas y mecánicas. ICONTEC: Bogotá, Colombia, 1971.
13. *ASTM C188-17*; Standard Test Method for Density of Hydraulic Cement. ASTM International: West Conshohocken, PA, USA, 2023.
14. *NTC 221*; Norma Técnica Colombiana NTC 221: Método de ensayo para determinar la densidad del cemento hidráulico. ICONTEC: Bogotá, Colombia, 2019.
15. Kara, S. Gypsum board waste management in construction: A review. *Waste Manag.* **2022**, *138*, 210–224.

16. Arroyo, R.; Horta, R.; Álvarez, G.; Pérez, N.; Aguilera, E. Reciclaje de residuos de placas de yeso laminado: Análisis de métodos y propuesta de proceso optimizado. *Rev. Constr.* **2019**, *18*, 107–122.
17. López, M.; García, A.; Sánchez, J.; Martínez, P. Comparative study of wet and thermal recycling methods for gypsum board waste. *J. Clean. Prod.* 2021.
18. Carvalho, M.A.; Calil Junior, C.; Savastano Junior, H.; Tubino, R.; Carvalho, M.T. Microstructure and mechanical properties of gypsum composites reinforced with recycled cellulose pulp. *Mater. Res.* **2008**, *11*, 391–397. [CrossRef]
19. Cárdenas, H.E.; Arteaga, J.C.; Fernández-Gómez, J.A.; Solarte, N. Efecto de fibras de celulosa en las propiedades físicas y mecánicas de compuestos de yeso. *Rev. Ing. Constr.* **2015**, *30*, 284–294.
20. Muniz-Villarreal, M.S.; Manzano-Ramírez, A.; Sampieri-Bulbarela, S.; Gasca-Tirado, J.R.; Reyes-Araiza, J.L. Influence of cellulose fibers on the hydration kinetics and mechanical properties of gypsum composites. *Constr. Build. Mater.* **2022**, *315*, 125710.
21. Lewry, A.J.; Williamson, J. The setting of gypsum plaster. *J. Mater. Sci.* **1994**, *29*, 6085–6090. [CrossRef]
22. Karni, J. Gypsum in construction: Origin and properties. *Mater. Struct.* **1995**, *28*, 92–100. [CrossRef]
23. Pavlidou, E.; Economidou, A.; Papageorgiou, D.; Efthimiadou, E.; Bakolas, A. Modified gypsum-based plasters with enhanced workability: Effect of organic additives on hydration kinetics and mechanical properties. *Constr. Build. Mater.* **2016**, *121*, 530–539.
24. Ashrapov, A.K.; Abdullaev, I.A.; Abdullaev, A.M.; Umarov, K.S. Effect of natural fibers on mechanical properties of gypsum composites. *Constr. Build. Mater.* **2020**, *252*, 119088.
25. Stães, J.; Van Der Bergh, J.M.; Ascione, E.; Segers, S.; Vandewalle, L.; Van Balen, K. Influence of natural fibers on the mechanical properties of gypsum-based composites: A comprehensive review. *Compos. Part B Eng.* **2021**, *215*, 108790.
26. *EN 13279-1 (2008)*; Gypsum Binders and Gypsum Plasters—Part 1: Definitions and Requirements. European Committee for Standardization (CEN): Brussels, Belgium, 2008.
27. Begliardo, H.F.; Sánchez, M.A.; Panigatti, M.C.; Garrappa, S.E. Recuperación de yeso a partir de placas de yeso laminado de desecho. *Inf. Tecnol.* **2013**, *24*, 53–62.
28. Rincón, J.M.; Ocampo, M.A. Análisis colorimétrico de materiales de construcción mediante técnicas de imagen digital: Metodología y aplicaciones. *Rev. Constr.* **2017**, *16*, 104–114.
29. Def-Lab-Geo-Imagen. *Software de Análisis Digital de Imágenes para Geomateriales*; Universidad Complutense de Madrid: Madrid, Spain, 2022.
30. Rincón, J.M. Evaluación objetiva del color en materiales de construcción: Propuesta de un nuevo método basado en distancias euclidianas. *Mater. Constr.* **2016**, *66*, 30–89.
31. Gómez, A.M.; Arciniegas, C.S. Gestión de residuos de placas de yeso laminado en la construcción: Oportunidades de reciclaje y reincorporación. *Rev. Ing. Construcción* **2017**, *32*, 105–120.
32. López, M.; García, A.; Sánchez, J.; Martínez, P. Estudio comparativo de los métodos de reciclado húmedo y térmico de residuos de placas de yeso laminado. *J. Clean. Prod.* **2021**, *278*, 123948.
33. *SimaPro 8.3.0*, Software de Análisis de Ciclo de Vida; PRé Sustainability: Amersfoort, The Netherlands, 2017.
34. Government of the Republic of Colombia. *Report on Construction Material Consumption and Waste in Colombia*; Ministry of Environment and Sustainable Development: Bogotá, Colombia, 2019.
35. Consejo Nacional de Política Económica y Social (CONPES). *CONPES 3874: Política Nacional para la Gestión Integral de Residuos Sólidos*; Departamento Nacional de Planeación, República de Colombia: Bogotá, Colombia, 2016.

sustainability

Article

Engineering the Tensile Response of Glass Textile Reinforced Concrete for Thin Elements

Sachin Paul * and Ravindra Gettu

Department of Civil Engineering, Indian Institute of Technology Madras, Chennai 600036, India; gettu@iitm.ac.in
* Correspondence: sachinmmp@gmail.com

Abstract: Textile-reinforced concrete (TRC) is a composite made with bi-directional non-metallic fabric embedded in a fine-grained cementitious matrix. When engineered appropriately, these composites can reduce material usage for the desired performance, resulting in slimmer sections and enhanced material efficiency, which in turn lowers the CO_2 footprint. To facilitate the widespread application of TRC in practice, it is crucial to comprehend the material and structural behavior of these composites, which can pave the way toward an optimized design methodology. In this paper, the tensile response of TRC is studied with different textile geometries, volume fractions and matrix strengths. The influence of the coating impregnation on the effectiveness of the textile to enhance the response of the composite is discussed, with complementing evidence from microstructural observations. The results of tests with different textile configurations indicate a transition in the type of stress–strain response from tri-linear to bi-linear, beyond a certain effective volume fraction. The paper also presents a simplified model to predict the bi-linear response from the efficiency factor-based approach. The insights gained can assist in achieving composite designs with optimized sections and limited tensile stress cracking, ensuring the targeted performance in slender elements.

Keywords: textile-reinforced concrete; tensile response; textile coating; fracture; toughening

Citation: Paul, S.; Gettu, R. Engineering the Tensile Response of Glass Textile Reinforced Concrete for Thin Elements. *Sustainability* **2023**, *15*, 14502. https://doi.org/10.3390/su151914502

Academic Editors: Nele De Belie and Anibal C. Maury-Ramirez

Received: 9 August 2023
Revised: 29 September 2023
Accepted: 3 October 2023
Published: 5 October 2023

1. Introduction

The most sustainable options in construction, in terms of reducing the carbon footprint, involve the choice of the right structural system, minimization of the element dimensions (for lower raw material consumption) and increasing the durability (for more efficient exploitation of resources). Textile-reinforced concrete (TRC) presents a unique advantage in this regard, enabling the construction of lightweight, durable and modular structural elements, while eliminating the risk of curtailed service life due to corrosion [1]. Comprising bi-directional non-metallic fabric embedded in a fine-grained cementitious matrix, TRC exhibits high tensile strength, proving to be a viable solution for both the construction of slender elements and the retrofitting of existing structures [1–3]. This innovative combination of the concepts of mesh reinforcement with fiber-reinforced concrete effectively addresses the limitations of durability and crack width control in both of those systems [4].

TRC provides an environmentally sustainable option by substantially reducing concrete usage [4,5], thereby reducing the impacts associated with conventional reinforced concrete, including a reduction in the use of materials like portland cement and steel. The non-corrosive reinforcement (i.e., glass or carbon) in TRC contributes to its durability, which enhances sustainability. Overall, such aspects lead to a significant decrease in embodied emissions and energy, raw material consumption and waste generation [6].

Most applications of TRC have been in thin-walled tensile and flexural members. However, with a better understanding of its response, the rational design of more complex TRC elements is becoming a reality. Substantial work in this area has been reported in many state-of-the-art reports [1–4], and the results are encouraging with respect to the performance of these composites in terms of limiting crack propagation during failure and enhancing tensile strength and strain capacity. The response of TRC is, however, influenced by several factors, such as the material characteristics and geometric parameters of the reinforcement, and its volume fraction. High-moduli, high-strength fabrics, such as those of carbon, glass and aramid, impart high strength and toughness to the composite [7,8], and are consequently more desirable as reinforcement.

The tensile stress–strain response of TRC composites is conventionally represented by three distinct zones [9,10]. The initial phase—Zone I—pertaining to the linear response, is governed by the stiffness of the matrix, until first cracking. Due to the relatively low fiber volume fractions, the contribution of the textile to the overall stiffness of the composite is negligible in this zone. After the first crack, a further increase in load leads to the development of multiple cracks, which manifests as a nonlinear zone of the tensile response (Zone II). A significant drop in stiffness is observed in this zone, with its extent and number of cracks being dependent on the amount of reinforcement, the textile geometry, and the bond between the textile and matrix [11–15]. Often, cracking is observed to occur at almost constant or decreasing stress. The next phase (i.e., Zone III) of the stress–strain response is of the strain-hardening type, with a practically linear response, governed by the properties of the reinforcement. In this stage, all the load is carried by the textile yarns, and the existing cracks widen until the fabric, bridging one of the cracks, ruptures. The final failure could also be accompanied by the pullout of the yarns, depending on the geometry of the composite element [16,17]. In some cases, the response is practically bi-linear, with the absence of a distinct Zone II, yielding higher load-carrying capacity at the same strain, when compared with a composite with a tri-linear response.

The nature of the individual fibers that make up each yarn, the coating material [18–20] and the level of impregnation [21] are crucial for the micro-mechanics of the composite, especially in terms of the stress distribution at the matrix–fabric interface. The interaction between the inner and outer fibers of yarns determines if their response in the composite is monolithic or telescopic [21,22]. A fully impregnated fabric behaves more monolithically, with uniform stress distribution across the yarn, whereas partial impregnation results in higher stresses in the exterior fibers causing a reduction in the ultimate tensile capacity of the composite due to telescopic failure. Stiffer coating materials, such as epoxy, are observed to impart better uniformity of the stress distribution in the yarn, and therefore, yield higher tensile strength, in comparison to softer styrene–butadiene rubber (SBR) or acrylic-based coated systems [19,23,24]. Furthermore, the durability and aging of fibers within the cementitious system can also alter the final behavior of the composite [25–27].

The matrix composition and the specimen geometry can significantly influence the response of TRC composites [11,16,28]. Using a matrix with lower strength can reduce the bond between the textile and the matrix, resulting in potential failure due to fiber pullout or slippage [23,29,30]. The incorporation of short fibers can improve the bond, especially when utilizing low-strength mixtures [31,32].

It should also be noted that the experimental procedure used for characterizing the tensile response of the composite can influence the type of failure. Earlier research predominantly endorsed rotating end conditions during uniaxial tensile testing of TRC [11,33,34]. However, more recent studies apply partially clamped boundary conditions, where the grips mitigate textile slippage, curbing pullout failure [14,21,35]. However, reports indicate that the end conditions are especially important for uncoated textiles, compared to those that are partially or fully impregnated [36,37], or for specimens that are warped or have misaligned reinforcement [38].

Several approaches for the prediction of the overall response of TRC can be seen in the literature. Two early analytical models were the ACK model [39] and the Cuypers model [40], both of which are based on the tri-linear model for the stress–strain behavior, with the initial response based on the method of mixtures and the final strain-hardening zone based only on the textile response. The difference between the two models lies in the second phase, where the ACK model considers multiple cracking at constant stress levels while the Cuypers model contemplates crack formation at progressively increasing stress based on the Weibull distribution. However, several studies indicate that these models may not always represent the true behavior of TRC [41,42]. A more recent analytical approach is based on the global stiffness governed by cracking behavior and stress drops at each crack [43]. Another model accounts for the bond-lag behavior between the core and sleeve filaments to obtain the composite response [21]. Even though both these approaches can predict the TRC behavior with reasonable accuracy, the input parameters for the modeling require complex experimental procedures. The present study attempts to complement the more sophisticated models by developing a simplified methodology for representing the tensile response of TRC based on efficiency factors.

An important aspect of the design of TRC for sustainability is the decision on the number of layers of reinforcement or volume fraction for a given matrix, to provide the desired load-carrying capacity for an allowable strain with the least element thickness. Despite earlier research, there is a deficit in the understanding of the tensile behavior of TRC when considering varying textile configurations and reinforcement ratios, which is critical for an optimized design approach for achieving better sustainability. Accordingly, an objective of this paper is to propose a criterion for choosing the minimum number of layers that result in the highest load-carrying capacity at a given strain. This concept is illustrated here with several types of textiles embedded in two different cementitious matrices, by analyzing the composite tensile behavior.

2. Materials Used

2.1. Textiles

Six types of bi-directional woven glass fiber textiles were used in the study (see Table 1 for the geometry and characteristics). The architecture and weaving patterns of the different textiles (denoted F1 to F6) are shown in Figure 1. Textile F2 is an alkali-resistant (AR) glass textile, with zirconium, and the others are E-glass textiles with alkali-resistant coatings. Fourier transform infrared spectroscopy (FTIR) analysis indicates that the coatings on the F1, F3, F4, F5, and F6 textiles contain styrene–butadiene, and that F2 is coated with an acrylic-based product. The textiles have different mesh opening sizes, yarn densities, and tensile strength. In the current study, testing and analysis focus solely on the weft direction (the widthwise orientation that is strongest for the selected fabrics). The cross-sectional area of the textile yarn was calculated from the TEX value provided by the manufacturer and verified by weighing the fibers after removing the coating by thermal treatment. The tensile strength of the single yarn, in each case, was determined as per ASTM D6637/D6637M-15 (Test Method A) [44].

Table 1. Properties of the textiles used.

Textile	Coating Material [a]	Opening Size [c] [mm × mm]	Mass per Unit Textile Area with the Coating [c] [g/m^2]	Nominal Cross-Section Area of a Single Weft Yarn [b] [mm^2]	Measured Weft Yarn Tensile Strength [MPa]
F1	SBR	8.5 × 6.5	267	0.79	1168 ± 103
F2	Acrylic	25.0 × 25.0	180	0.92	1040 ± 102
F3	SBR	15.7 × 10.1	280	0.92	1144 ± 133
F4	SBR	8.0 × 8.0	178	0.31	1367 ± 86
F5	SBR	9.0 × 10.0	118	0.31	1393 ± 76
F6	SBR	4.0 × 4.5	117	0.13	1483 ± 12

[a] From FTIR analysis; SBR: styrene–butadiene rubber. [b] Calculated based on the TEX value provided by the manufacturer. [c] Details provided by the manufacturers.

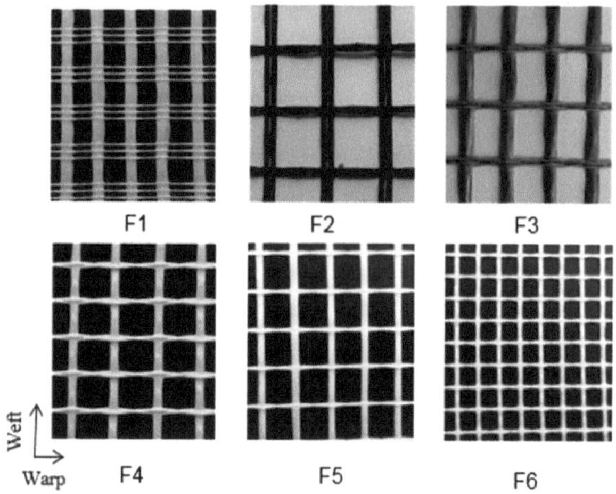

Figure 1. Types of textiles used in the study.

2.2. Cementitious Matrices

The matrix used for TRC is required to possess high flowability for adequate penetration between layers of textile in thin elements. In view of this, two self-consolidating matrices were developed with a relatively high binder content. A polycarboxylate-based superplasticizer was used to obtain the desired rheology. The matrix M1 had a water–binder ratio of 0.4 and a moderate compressive strength of 60.2 ± 2.7 MPa while the higher-strength mix M2 had a water–binder ratio of 0.24 and a compressive strength of 104.1 ± 4.2 MPa. The binder for both mixes was a blend of ordinary portland cement, ASTM Class F fly ash and silica fume. A hydroxypropyl methylcellulose-based viscosity modifying agent (VMA) was used in mix M2 for better stability. The maximum size of aggregates was limited to about 1 mm to facilitate uniform penetration of the matrix between the textile layers. The mix proportions and the mechanical properties of the mixes are given in Table 2. The compression tests were conducted in accordance with IS 4031 on 70.6 mm cube specimens. The flexural test was carried out as per ASTM C348 standards, with specimens of 40 × 40 × 160 mm dimensions. The matrix tensile strength was derived from the uniaxial test following the RILEM TC 232-TDT (2016) guidelines [33]. All samples were tested after 28 days of curing in a mist room, at a temperature of 25 ± 2 °C.

Table 2. Mix proportions and mechanical properties (mean ± standard deviation) of the fine-grained concrete.

Materials/Properties	Mix M1	Mix M2
Cement (kg/m^3)	583	674
Fly ash (kg/m^3)	208	114
Silica fume (kg/m^3)	42	79
Quartz sand, 0.2–1.1 mm (kg/m^3)	595	1037
Quartz powder, 20–160 μm (kg/m^3)	357	207
Water/binder	0.40	0.24
PCE superplasticizer (% solids/binder by weight)	0.15	1.30
VMA (% solids/binder by weight of binder)	-	0.08
28-day cube compressive strength (MPa)	60.2 ± 2.7	104.1 ± 4.2
28-day flexural strength (MPa)	7.2 ± 0.1	11.8 ± 0.2
28-day tensile strength (MPa)	3.9 ± 0.2	5.8 ± 0.3
Modulus of elasticity (GPa)	27.1 ± 1.8	34.6 ± 1.7

3. Experimental Programme

3.1. Specimen Preparation

All the specimens, of 500 mm length and 60 mm width, were molded as per the geometry specified by RILEM TC 232-TDT (2016); see Figure 2. The thickness was kept at 10.0 mm, except for the five-layer configurations of F2 and F3 textiles for which it was increased to 11.2 mm. The specimens were cast in steel molds with screw-down end plates. The textiles were positioned horizontally in layers with steel spacers at the edges, and the matrix mix was poured into the mold. Five specimens were cast for each configuration and maintained at room temperature (about 25 °C) for 24 h, after which they were demolded, and cured in a mist room at 25 ± 2 °C for 28 days. The nomenclature of the specimens presented in the study is T-FM-nL: where T indicates the tensile test, F—the textile type (i.e., F1 to F6), M—matrix used (i.e., M1 or M2) and nL—number of layers (i.e., 1 to 5).

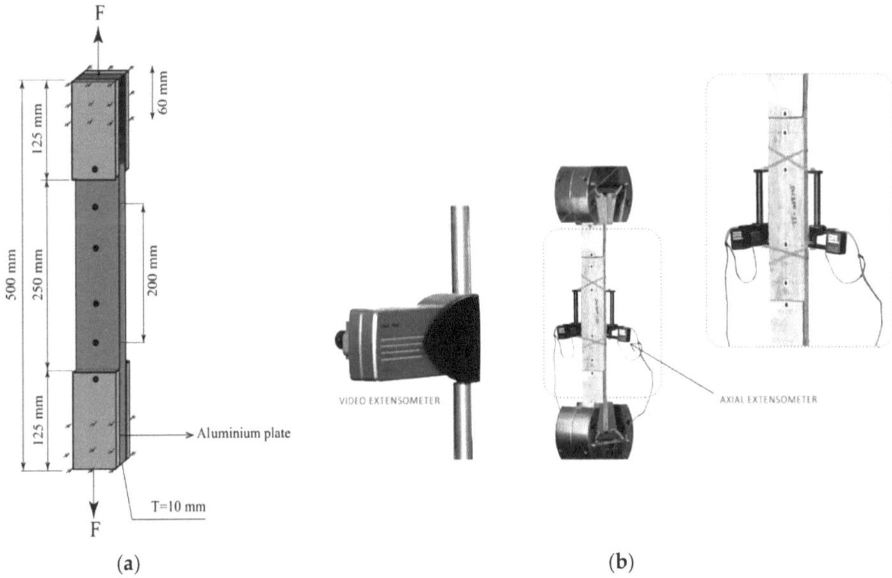

Figure 2. (**a**) Geometry of the specimen and (**b**) uniaxial tensile test setup.

3.2. Experimental Setup

The uniaxial tensile tests (Figure 2) were carried out in a servo-controlled electromechanical system with a 50 kN load cell, as shown in Figure 2b. Aluminum end plates of 2 mm thickness with rectangular geometry were glued to the gripping zone to avoid local crushing at the supports during testing, and screw grips were used to mount the specimens on the machine. The tests were performed at a constant displacement rate of 0.8 mm/min. A combination of signals from video and strain gauge-based axial extensometers was used to obtain the displacement of the specimen. The axial extensometer was used to measure the strain in the composite until the first crack, and the video extensometer, measuring the relative displacement over a 200 mm gauge length, was used for the remaining portion of the curve. This method was adopted to mitigate the influence of the noise in the video extensometer signal at the low displacement range. It is to be noted that though some cracking occurred outside the gauge length, the strains calculated were consistent. Specimens exhibiting failure in the end zone or clamping area were discarded from the analysis.

4. Results and Discussions

In all the cases, the stress values were obtained by dividing the measured load by the gross cross-sectional area of the composite. The stress corresponding to the first crack was identified from the first drop in the initial linear portion of the stress–strain response, after which there was a sudden change in the slope.

4.1. Effect of Textile Geometry on Stress-Response and Crack Formation

The textiles F1 and F2 are both leno woven but have different geometries and reinforcement ratios (see Table 1). The typical stress–strain responses of the composites with one to four layers of textile F1 and one to five layers with textile F2 in the matrix M1 are shown in Figure 3, and the average stress and strain values are reported in Table 3. In these cases, the specimens exhibited distributed cracking, with the number of cracks increasing with the reinforcement ratio and the crack widths consequently reducing. For lower reinforcement ratios, such as in F1M1-1L and -2L, and F2M1-1L and -2L, the ultimate rupture occurred near the mid-length of the specimen. However, for higher reinforcement ratios, the failure occurred in the vicinity of the endplates. Visual examination revealed that the first crack occurs near the fill yarns, possibly due to weakening of the cross-section at these locations, especially for high fill yarn volume fraction or fill yarns with larger cross-sections. However, at higher reinforcement ratios, additional cracks form between the fill yarns.

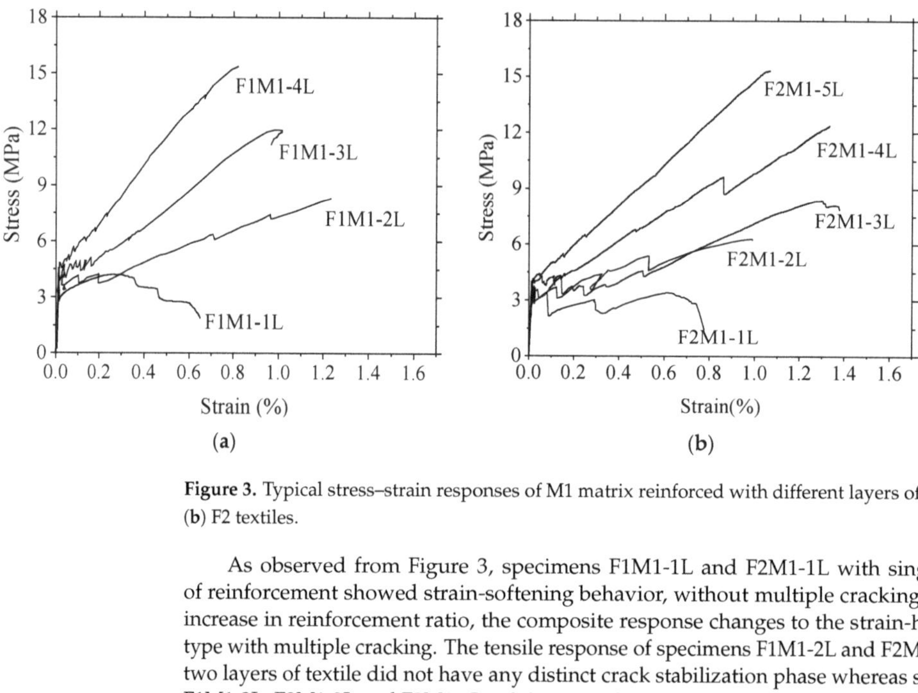

Figure 3. Typical stress–strain responses of M1 matrix reinforced with different layers of (**a**) F1 and (**b**) F2 textiles.

As observed from Figure 3, specimens F1M1-1L and F2M1-1L with single layers of reinforcement showed strain-softening behavior, without multiple cracking. With an increase in reinforcement ratio, the composite response changes to the strain-hardening type with multiple cracking. The tensile response of specimens F1M1-2L and F2M1-2L with two layers of textile did not have any distinct crack stabilization phase whereas specimens F1M1-3L, F2M1-3L and F2M1-4L exhibit typical tri-linear behavior with multiple cracking under almost constant stress, followed by the widening of the cracks under increasing stress. However, a few new cracks were observed to form in some of the specimens during the strain-hardening regime. Further increase in reinforcement (i.e., in specimens F1M1-4L and F2M1-5L) leads to a transition in the nature of the stress–strain response from tri-linear to bi-linear, with multiple cracking occurring as the stress increases. It is evident that composites with bi-linear response exhibit much better load-carrying capacities at the same strains when compared to those with tri-linear behavior.

Considering the stress and strain values given in Table 3, it is seen that the ultimate strains are higher for lower reinforcement ratios and that an increase in the reinforcement ratio leads to higher stiffness in the strain hardening regime, as expected, with a reduction in the ultimate strain of the composite.

Table 3. Parameters of tensile response of the composites.

Specimen	First Peak		Intermediate Stress at Different Strains				Ultimate Peak		Number of Cracks
	Stress	Strain	0.2%	0.4%	0.8%	1.2%	Stress	Strain	
F1M1-1L	3.38 ± 0.51	0.013 ± 0.0014	3.73 ± 0.21	3.47 ± 0.73	1.75 ± 0.78	-	4.08 ± 0.23	0.31 ± 0.11	1
F1M1-2L	3.4 ± 0.65	0.0126 ± 0.0015	3.66 ± 0.37	3.89 ± 0.47	5.49 ± 0.41	6.46 ± 0.44	6.84 ± 0.36	1.33 ± 0.15	7
F1M1-3L	4.03 ± 0.75	0.0164 ± 0.0052	5.092 ± 0.60	6.704 ± 0.45	10.12 ± 0.44	-	11.37 ± 0.44	1.10 ± 0.041	10
F1M1-4L	5.15 ± 0.41	0.0198 ± 0.0026	7.84 ± 0.24	10.51 ± 0.77	15.31 ± 0.25	-	15.80 ± 0.36	0.91 ± 0.054	15
F2M1-1L	3.11 ± 0.62	0.0114 ± 0.0025	2.8 ± 0.30	2.5 ± 0.76	1.37 ± 1.30	-	3.14 ± 0.30	0.63 ± 0.23	1
F2M1-2L	3.86 ± 0.426	0.0142 ± 0.003	3.67 ± 0.34	4.42 ± 0.4	4.88 ± 1.00	3.98 ± 1.6	5.62 ± 0.9	0.99 ± 0.22	7
F2M1-3L	3.51 ± 0.26	0.0126 ± 0.0005	4.17 ± 0.66	4.47 ± 0.53	6.02 ± 0.13	7.96 ± 0.15	8.45 ± 0.10	1.40 ± 0.10	7
F2M1-4L	4.05 ± 0.20	0.0142 ± 0.0014	4.53 ± 0.28	5.78 ± 0.55	9.04 ± 1.02	11.99 ± 0.90	12.26 ± 0.91	1.26 ± 0.04	10
F2M1-5L	4.39 ± 0.28	0.0165 ± 0.002	6.24 ± 0.5	9.56 ± 0.86	14.67 ± 0.47	-	15.13 ± 0.54	1.09 ± 0.03	15
F3M1-4L	3.83 ± 0.37	0.013 ± 0.002	4.02 ± 0.43	6.03 ± 0.36	10.53 ± 0.42	12.21 ± 0.86	12.58 ± 1.03	1.23 ± 0.13	13
F3M1-5L	4.32 ± 0.52	0.016 ± 0.006	6.83 ± 0.38	9.43 ± 0.92	15.32 ± 1.12	-	16.63 ± 0.86	1.06 ± 0.09	17
F4M1-4L	3.32 ± 0.61	0.0118 ± 0.0025	3.52 ± 0.36	3.88 ± 0.73	5.43 ± 1.30	-	7.35 ± 0.35	1.13 ± 0.048	9
F4M1-5L	3.46 ± 0.43	0.0124 ± 0.0018	3.72 ± 0.24	4.13 ± 0.23	6.96 ± 0.83	-	9.34 ± 0.56	1.15 ± 0.032	13
F1M2-3L	5.62 ± 0.45	0.016 ± 0.0012	5.96 ± 0.52	7.68 ± 1.01	10.54 ± 1.31	-	11.13 ± 0.76	0.816 ± 0.016	10
F1M2-4L	5.67 ± 0.38	0.0168 ± 0.0026	7.67 ± 0.24	10.56 ± 0.77	-	-	15.32 ± 0.36	0.789 ± 0.034	14
F2M2-4L	5.23 ± 0.23	0.0152 ± 0.0018	5.82 ± 0.27	6.96 ± 0.32	11.56 ± 52	-	11.63 ± 0.36	0.802 ± 0.05	
F2M2-5L	5.63 ± 0.34	0.0171 ± 0.002	8.93 ± 0.42	11.32 ± 0.86	-	-	15.03 ± 0.54	0.74 ± 0.02	15
F3M1-4L	5.33 ± 0.42	0.016 ± 0.003	5.86 ± 0.46	7.56 ± 0.58	-	-	11.03 ± 1.03	0.89 ± 0.16	11
F3M2-5L	5.16 ± 0.18	0.015 ± 0.003	7.96 ± 0.36	10.86 ± 0.75	15.83 ± 1.36	-	16.13 ± 1.43	0.88 ± 0.12	16

The effect of concrete strength on the stress–strain response of the composites can be studied by comparing the responses of specimens with the F1 and F2 textiles in the M1 (60 MPa) and M2 (104 MPa) matrices. It is seen from the typical curves shown in Figure 4 and the data in Table 3 that the first-crack strength of the composite was 30–55% higher for the M2 mix in comparison to the M1 mix and the ultimate strain was lower, as expected. However, the number of cracks was observed to be in the same range for both matrices, which indicates that the textile characteristics govern the spacing of the cracks. Evidently, the composites with the M2 matrix also exhibit a transition in the response from softening to tri-linear to bi-linear.

In general, the first-crack stress is observed to be marginally less for a TRC composite with a low reinforcement ratio than that of plain mortar. This could be attributed to the reduction of the cross-section near the fill yarns. However, with an increase in the reinforcement ratio, the first-crack stress is found to be in the same range or higher for the composite in comparison to that of the mortar. Further, it is seen that the first-crack stress increases with reinforcement ratio in closely spaced textile configurations. For example, F1M1-4L exhibits 51% higher first-crack stress than the composite with the two-layer configuration of the same textile (F1M1-2L). However, the first-crack response seems to be strongly influenced by the textile geometry, as observed in the cases of textiles F2 and F3 with yarn spacing 25 mm and 16.5 mm, respectively, where there is no significant enhancement in the first-crack stress even with 5 layers of textiles.

To further explore the influence of the textile geometry on the first-crack stress, tests were performed with two textiles F5 and F6 having a similar reinforcement ratio but different yarn spacing. Textile F5 had a relatively larger opening size of 9.0 × 10.0 mm and F6 had an opening size of 4.0 × 4.05 mm. The typical stress–strain responses of F5TM1-4L and F6TM1-4L are shown in Figure 5; it is observed that the first-crack load with 4 layers of F6 textiles was 40% higher than that observed with 4 layers of F5 textiles. The reason for the higher first-crack stress with the textiles of closer yarn spacing can be attributed to more effective arresting of the microcracks, thereby avoiding interconnected cracking and a drop in the strength of the composite.

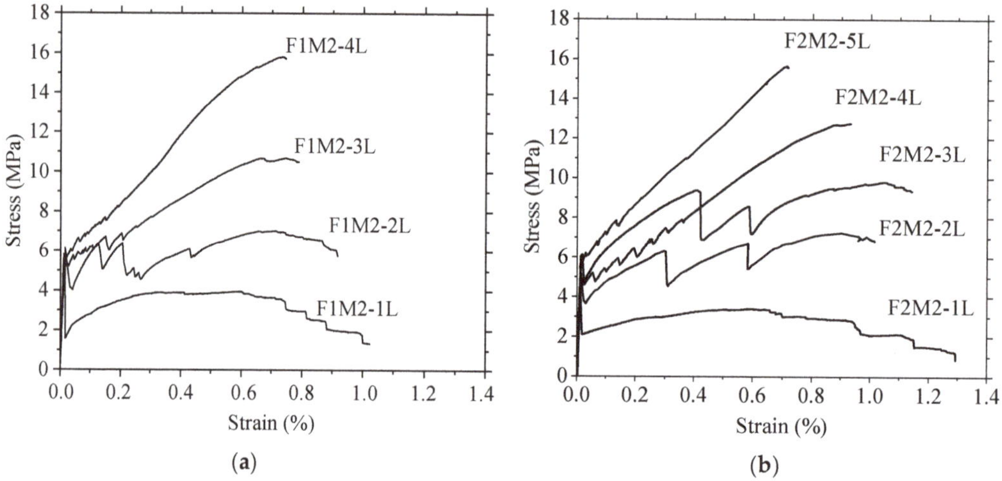

Figure 4. Typical stress–strain responses of specimens with M2 (104 MPa) matrix and different layers of (**a**) F1 and (**b**) F2 textiles.

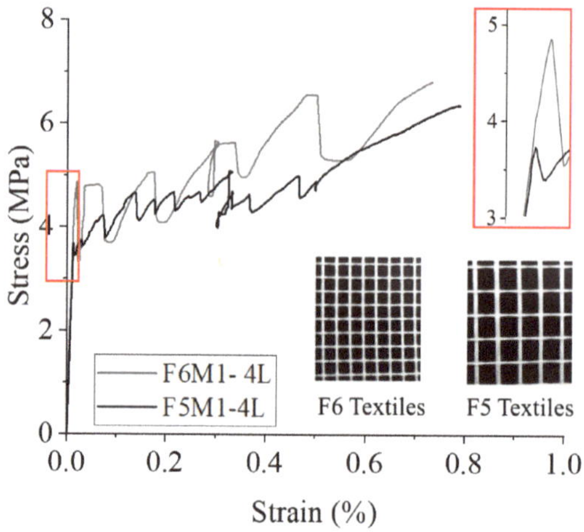

Figure 5. Typical stress–strain response of composite with 4 layers of F5 and F6 textiles.

4.2. Efficiency Factor

The relation between the nominal stress in the textile in the composite at failure and the tensile strength of the yarn is often expressed as the efficiency of the textile in terms of a factor, which can be defined as

$$k = F_{ct}/(V_t \times f_{tt}) \quad (1)$$

where V_t is the volume fraction of the textile in the direction of tensile loading, and F_{ct} is the tensile capacity of the composite and f_{tt} is the tensile strength of the textile used. The efficiency factor depends on the uniformity of the stress distribution between the outer (sleeve) and the inner (core) fibers of a yarn (Figure 6). Since the outer fibers are bonded to the cementitious matrix and the inner fibers are free to slip, the crack bridging generates tensile stresses mostly in the sleeve fibers, and when these rupture, the inner fibers are progressively stressed until the ultimate collapse of the composite. Therefore, the efficiency of reinforcement by the textile yarns in the composite is largely dependent on the relative amounts of sleeve and core fibers in the yarn.

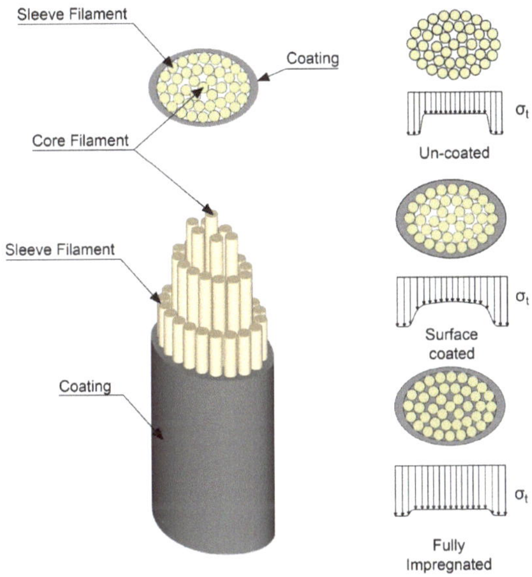

Figure 6. Graphical representation of the core and sleeve fibers in concrete matrix.

The efficiency factors obtained for the different configurations (of textile and matrix) tested here are shown in Figure 7, in terms of average values. It is evident that the values fall in two ranges indicated by the dashed lines. For the SBR-coated textiles, the efficiency factor has an average value of 0.45 ± 0.038, and for the acrylic-coated textiles, it is observed to be 0.62 ± 0.020. This implies that only about 45% and 62% of the strength, respectively, are effectively reached in these textiles when the composite ruptures.

An important reason for the difference in the efficiencies of the two types of textile coatings appears to be the uniformity of the coating material on the yarn. The scanning electron microscope (SEM) images in Figure 8 show the cross-section of the yarns of the F1 (Figure 8a), F2 (in Figure 8b), F3 (in Figure 8c) and F4 (in Figure 8d) textile. It should be noted that F2, F3 and F4 have SBR coating, and F2 has an acrylic-based coating. A closer examination of the cross-section of F1 (Figure 8a) reveals loosely held interior fibers surrounded by sleeve fibers, suggesting the occurrence of slip between the core and sleeve, and within the core itself. This is reflected by the lower efficiency factors, with F1 having the lowest average efficiency factor of 0.42. Textile F2 (Figure 1) has two yarns per roving in the weft direction, and the SEM image of one of the yarns, in Figure 8b, shows an evenly distributed coating with good penetration of the acrylic-based material through the yarn cross-section. Consequently, there is proper adhesion between the interior and exterior fibers and an average efficiency factor of 0.62. Textile F3 (Figure 8c) is seen to have a thicker surface coating with low penetration of coating material into the interior in comparison with F2, though better than F1. The larger cross-section of the textile yarn, the elongated shape and the thick coating result in a hollow central core, with the average efficiency factor for F3 textiles being only 0.47. The efficiency factor for F4 (Figure 8d) textiles is marginally higher than the F1 textiles, which can be attributed to the thinner rovings. Similarly, the efficiency factor for F1 textiles with larger cross-sections is 0.42 whereas composites with the textile F6 with the smallest cross-sectional area have an efficiency factor of 0.47. This is due to the reduction of the sleeve-to-core ratio with an increase in the cross-sectional area. Therefore, it can be generalized that the efficiency factor is influenced by the coating material, the uniformity of the coating and the cross-sectional area of the yarn.

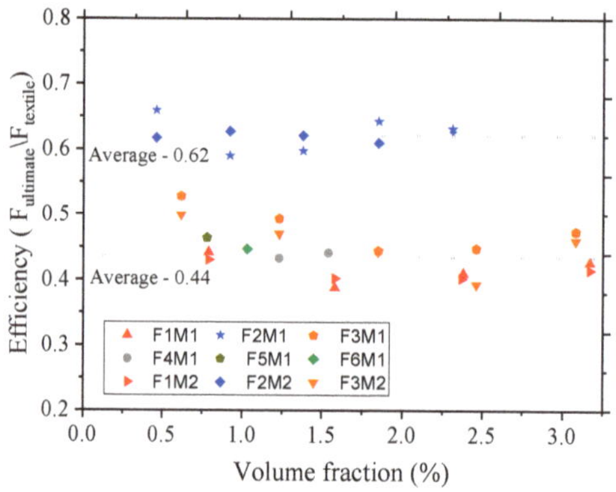

Figure 7. Efficiency factor for the different configurations.

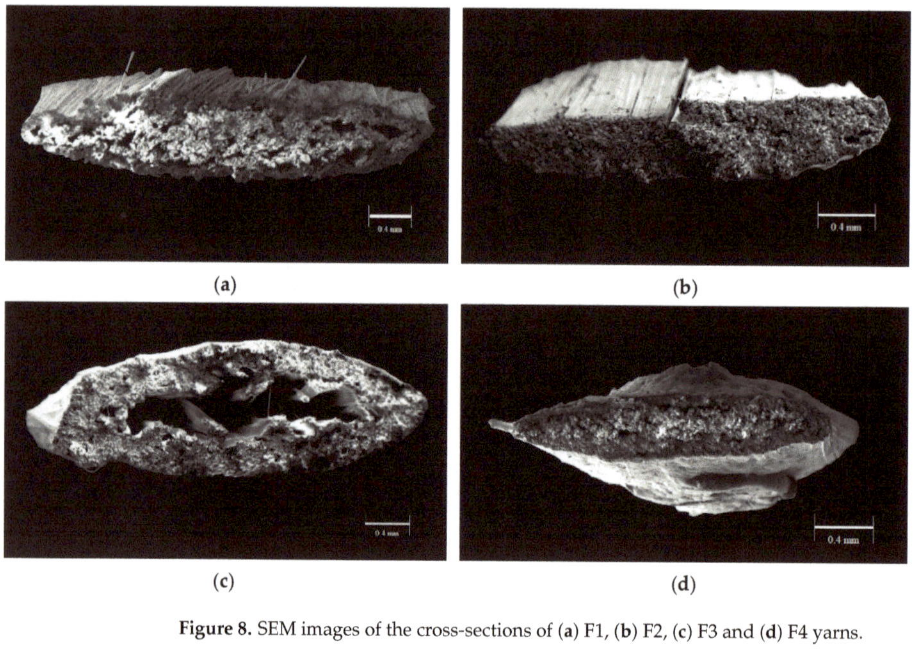

Figure 8. SEM images of the cross-sections of (**a**) F1, (**b**) F2, (**c**) F3 and (**d**) F4 yarns.

5. Transition in Tensile Response from Tri-Linear to Bi-Linear

As seen earlier, the tensile behavior of textile-reinforced concrete can be idealized as being tri-linear or bi-linear, except when low reinforcement causes strain softening. As seen in the tests performed here, the response of the composite is characteristically bi-linear, beyond a certain reinforcement ratio, with multiple cracking occurring as the stress level progressively increases after the first crack, resulting in a strain-hardening type response, without any plateau. For such specimens, crack stabilization (or the absence of new cracks) occurs at a higher strain level than those with a tri-linear response. With a higher strength matrix, the bi-linear response occurs with fewer layers, suggesting that the better bond between the textile and yarn enhances the toughening in the composite. On the other hand, textiles with lower cross-section areas need more layers for the transition to occur (e.g., F4M1-5L versus F1M1-4L).

For any given matrix, the transition in the stress–strain behavior can be related to an effective volume fraction, defined as the product of the efficiency factor and volume fraction (see Table 4), which indicates the proportion of fibers that contribute to the tensile resistance of the composite. It is observed, in the present study, that the transformation from tri-linear to bi-linear occurs when the effective volume fraction crosses a threshold of about 1.3, except for F2M2-4L, which can be attributed to the higher bond strength and better performance of the acrylic coating.

Table 4. Effective volume fraction.

Composite	Volume Fraction (Vt)	Average Efficiency Factor (k)	Average Effective Volume Fraction	Behavior
F1M1-3L	2.38	0.43	1.02	tri-linear
F1M1-4L	3.17	0.41	1.30	bi-linear
F2M1-4L	1.85	0.64	1.18	tri-linear
F2M1-5L	2.31	0.63	1.46	bi-linear
F3M1-4L	2.46	0.45	1.11	tri-linear
F3M1-5L	3.08	0.47	1.45	bi-linear
F4M1-4L	1.23	0.42	0.52	tri-linear
F4M1-5L	1.54	0.43	0.66	tri-linear
F1M2-3L	2.38	0.42	1.00	tri-linear
F1M2-4L	3.17	0.41	1.30	bi-linear
F2M2-4L	1.85	0.63	1.17	bi-linear
F2M2-5L	2.31	0.61	1.41	bi-linear
F3M1-4L	2.46	0.46	1.13	tri-linear
F3M2-5L	3.08	0.44	1.36	bi-linear

The transition of the tensile response from tri-linear to bi-linear can be explained based on the phenomena involved in the cracking and its propagation. As the tensile stress increases, cracking is initiated at some regions that could be statistically weaker than other regions with similar tensile stresses. The crack propagates (with a drop in load-carrying capacity) until it is arrested from progressing further by the textile yarn(s) that bridge(s) the crack tip. For the crack to propagate beyond the yarn(s), a much higher stress would be required. Consequently, cracks initiate at other sections of the composite member as the local tensile strengths at these points are reached. This leads to multiple cracking in TRC with little or no significant increase in load-carrying capacity. Later, mobilization of energy dissipation mechanisms, such as crack bridging and pullout, increases the crack resistance and the load-carrying capacity, resulting in the tri-linear response. However, when the reinforcement level is higher (or denser), the toughening is more effective, and as each crack initiates and is arrested, the stress progressively increases, resulting in a strain-hardening type or bi-linear response.

Considering the bi-linear behavior to be desirable in the composite due to the phenomena discussed earlier, it can be idealized as consisting of two phases for the purposes of structural design [45,46]. The first phase, until the first crack, can be characterized using the law of mixtures based on the moduli of elasticity of the textile yarns (E_t) and matrix (E_m), with the modulus of the elasticity of the composite represented by:

$$E_c = E_t\, V_t + E_m\, (1 - V_t) \quad (2)$$

where V_t is the volume fraction of the textile in the direction of loading. Since V_t is generally small, the initial response is dominated by the properties of the matrix. The first-crack stress can be considered for practical purposes to be the tensile strength of the plain matrix (f_{mt}).

The second (strain-hardening) part of the composite response is taken to be the product of the elastic modulus of the textile, the efficiency factor and the volume fraction of the fabric:

$$E_h = k\ V_t\ E_t. \tag{3}$$

The ultimate tensile stress of the composite or its tensile strength (F_{ct}) can be obtained from the efficiency factor, the volume fraction of the textile and the strength of the textile (f_{tt}), as follows:

$$F_{ct} = k\ V_t\ f_{tt}. \tag{4}$$

Several bi-linear responses modeled with the above equations are compared with the corresponding experimental results in Figure 9.

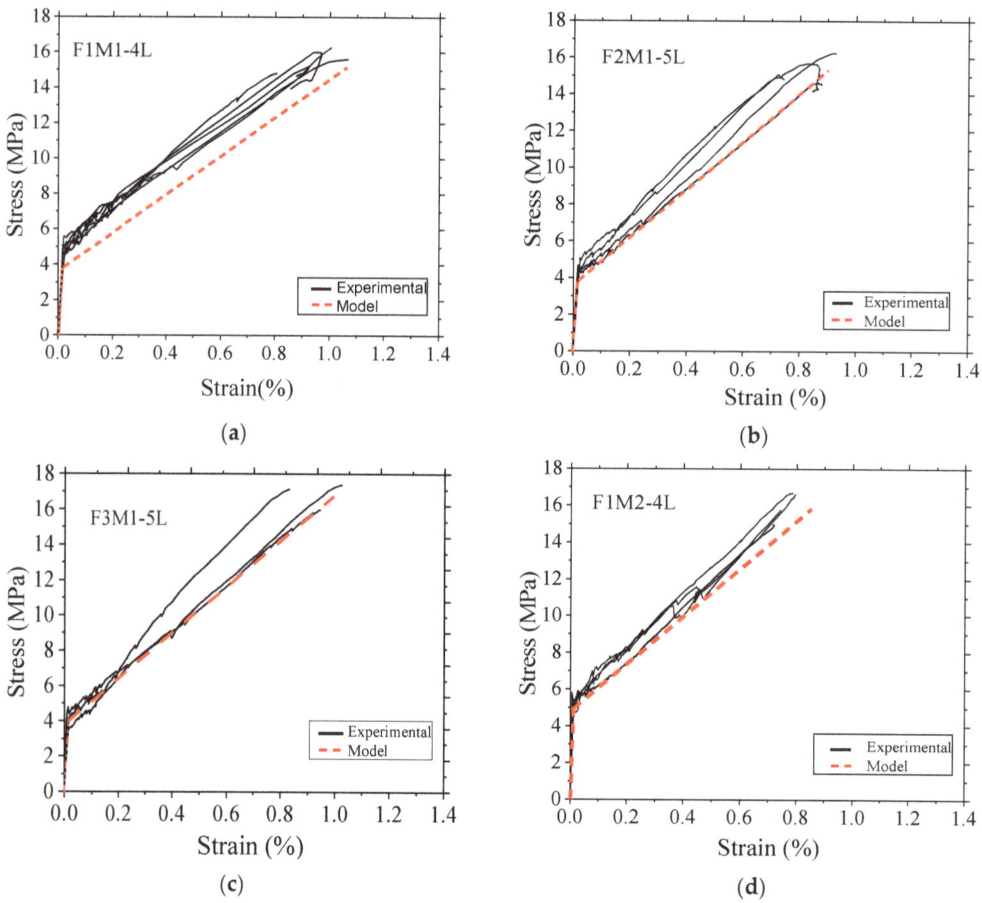

Figure 9. *Cont.*

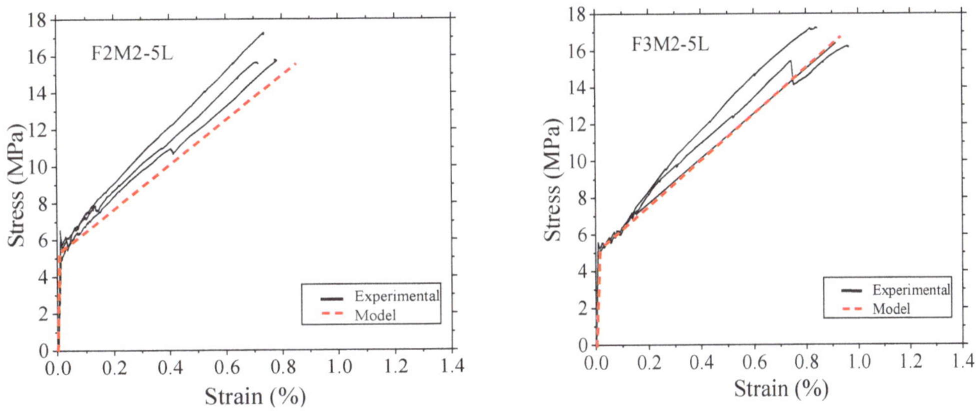

(e) (f)

Figure 9. Comparison of the stress–strain response from the model with experimental data for composites with matrices M1 (**a–c**) and M2 (**d–f**).

From the results, it can be observed that the proposed model can represent the overall tensile behavior of the composite within acceptable limits. It should be noted that the model is conservative in cases where the first-crack strength is enhanced due to the closer grid spacing of yarns at a high reinforcement ratio (as in Figure 9a). More importantly, the model can be used to design the composite system even though test data are not available for the volume fractions or number of layers that are required. This could lead to more efficient material utilization with the appropriate composite thickness and textile layers.

6. Summary and Conclusions

This paper investigates the tensile response of textile-reinforced concrete (TRC) with six types of glass fabrics in two different fine-grained concrete matrices. Composites with multiple layers of textiles were tested to evaluate the influence of the reinforcement ratio on the tensile response. Based on the experimental results, an efficiency factor is obtained for determining the maximum contribution of a given type of textile in a certain matrix. Further, it is seen that a simple model based on this factor can represent the bi-linear response of TRC under tension. Using the model and the understanding gained, the potential to engineer thin elements that are not only optimized for higher performance but also embody principles of sustainability by reducing raw material consumption and cement usage.

Some specific findings from the study are summarized below:

- The first crack in the composite develops mostly in the vicinity of the cross-yarn, especially at lower reinforcement ratios. However, at higher reinforcement ratios, cracks were observed to develop between the cross yarns.
- Though the first-crack strength depends mainly on the matrix properties, it was observed that it could be enhanced by the geometry of the fabric. A closer yarn configuration at a higher reinforcement volume was seen to result in higher first-crack stress than the matrix tensile strength.
- For the textiles used in the study, the tensile response of the composite changes from strain-softening to strain-hardening as the number of layers increases, with a transition from a tri-linear to a bi-linear response.
- The effectiveness of the textiles in the composite is influenced significantly by the type and extent of the penetration of the coating material into the yarns.
- An efficiency factor has been defined as the ratio between the nominal tensile stress in the textile at the failure of the composite and its tensile strength. This seems to be

independent of the matrix strength or the volume fraction of the particular textile used in the TRC composite.

- From the present study, an effective volume fraction or cross-sectional area of textiles was identified based on the efficiency factor for predicting the threshold for the tri-linear to bi-linear transition.
- A simple model for the bi-linear response of TRC was developed for possible use in design methods, based on the efficiency factor, the volume fraction of the textile used, tensile strength and modulus of elasticity of the matrix, and the modulus of elasticity and strength of the textile. The prediction model compares satisfactorily with the experimental results. This approach would aid in the design for appropriate functionality of these elements with low material usage, leading to better sustainability.

Author Contributions: Conceptualization, S.P. and R.G.; methodology, S.P. and R.G.; formal analysis, S.P.; investigation, S.P.; resources, R.G.; data curation, S.P.; writing—original draft, S.P.; writing—review and editing, R.G.; visualization, S.P. and R.G.; supervision, R.G.; project administration, R.G.; funding acquisition, R.G. All authors have read and agreed to the published version of the manuscript.

Funding: Partial funding for this study was provided by the CRG/2019/004267 grant from the Science and Engineering Research Board (SERB), Government of India.

Institutional Review Board Statement: Not applicable.

Informed Consent Statement: Not applicable.

Data Availability Statement: Not applicable.

Acknowledgments: The first author is thankful to the Ministry of Education, Government of India, for providing an assistantship during his doctoral studies, under the QIP scheme. Partial funding for this study was provided by the CRG/2019/004267 grant from the Science and Engineering Research Board (SERB), Government of India. The FIST Grant SR/FST/ETII-054/2012 is gratefully acknowledged for the equipment in the Laboratory for Mechanical Performance of Civil Engineering Materials, IIT Madras. This work was carried out in the Centre of Excellence on Technologies for Low-carbon and Lean Construction at IIT Madras.

Conflicts of Interest: The authors declare no conflict of interest.

References

1. Brameshuber, W. (Ed.) *Textile Reinforced Concrete: State-of-the-Art Report of RILEM Technical Committee 201-TRC*; RILEM Publications: Bagneux, France, 2006; Volume 36.
2. Mobasher, B. *Mechanics of Fiber and Textile Reinforced Cement Composites*; CRC Press: Boca Raton, FL, USA, 2011.
3. Mechtcherine, V.; Slowik, V.; Kabele, P. (Eds.) *Strain-Hardening Cement-Based Composites, SHCC4*; Springer: Berlin/Heidelberg, Germany, 2017; Volume 15.
4. Peled, A.; Bentur, A.; Mobasher, B. *Textile Reinforced Concrete*; CRC Press: Boca Raton, FL, USA, 2017; Volume 19.
5. Tomoscheit, S.; Gries, T.; Horstmann, M.; Hegger, J. Project life INSUSHELL: Reducing the carbon footprint in concrete construction. *Int. J. Sustain. Build. Technol. Urban Dev.* **2011**, *2*, 162–169. [CrossRef]
6. Williams Portal, N.; Lundgren, K.; Wallbaum, H.; Malaga, K. Sustainable potential of textile-reinforced concrete. *J. Mater. Civ. Eng.* **2015**, *27*, 04014207. [CrossRef]
7. Peled, A.; Mobasher, B. Pultruded fabric-cement composites. *ACI Mater. J.* **2005**, *102*, 15–23.
8. Peled, A.; Mobasher, B.; Cohen, Z. Mechanical properties of hybrid fabrics in pultruded cement composites. *Cem. Concr. Compos.* **2009**, *31*, 647–657. [CrossRef]
9. Häußler-Combe, U.; Jesse, F.; Curbach, M. Textile reinforced concrete-overview, experimental and theoretical investigations. In *Fracture Mechanics of Concrete Structures, Proceedings of the Fifth International Conference on Fracture Mechanics of Concrete and Concrete Structures, Ia-FraMCos, Vail, CO, USA, 12–16 April 2004*; AEDIFICATIO Publications: Freiburg, Germany, 2004; Volume 204, pp. 12–16.
10. Jesse, F.; Will, N.; Curbach, M.; Hegger, J. Load-bearing behavior of textile-reinforced concrete. *Spec. Publ.* **2008**, *250*, 59–68.
11. Colombo, I.G.; Magri, A.; Zani, G.; Colombo, M.; Di Prisco, M. Erratum to: Textile Reinforced Concrete: Experimental investigation on design parameters. *Mater. Struct.* **2013**, *46*, 1953–1971. [CrossRef]
12. Rambo, D.A.S.; de Andrade Silva, F.; Toledo Filho, R.D.; Gomes, O.D.F.M. Effect of elevated temperatures on the mechanical behavior of basalt textile reinforced refractory concrete. *Mater. Des. (1980–2015)* **2015**, *65*, 24–33. [CrossRef]

13. Peled, A.; Bentur, A. Fabric structure and its reinforcing efficiency in textile reinforced cement composites. *Compos. Part A Appl. Sci. Manuf.* **2003**, *34*, 107–118. [CrossRef]
14. D'Antino, T.; Papanicolaou, C. Mechanical characterization of textile reinforced inorganic-matrix composites. *Compos. Part B Eng.* **2017**, *127*, 78–91. [CrossRef]
15. Hartig, J.; Häußler-Combe, U.; Schicktanz, K. Influence of bond properties on the tensile behaviour of textile reinforced concrete. *Cem. Concr. Compos.* **2018**, *30*, 898–906. [CrossRef]
16. Häußler-Combe, U.; Hartig, J. Bond and failure mechanisms of textile reinforced concrete (TRC) under uniaxial tensile loading. *Cem. Concr. Compos.* **2007**, *29*, 279–289. [CrossRef]
17. Arboleda, D.; Carozzi, F.G.; Nanni, A.; Poggi, C. Testing procedures for the uniaxial tensile characterization of fabric reinforced cementitious matrix (FRCM) composites. *J. Compos. Constr.* **2016**, *20*, 04015063. [CrossRef]
18. Peled, A.; Zaguri, E.; Marom, G. Bonding characteristics of multifilament polymer yarns and cement matrices. *Compos. Part A Appl. Sci. Manuf.* **2008**, *39*, 930–939. [CrossRef]
19. Messori, M.; Nobili, A.; Signorini, C.; Sola, A. Mechanical performance of epoxy coated AR-glass fabric Textile Reinforced Mortar: Influence of coating thickness and formulation. *Compos. Part B Eng.* **2018**, *149*, 135–143. [CrossRef]
20. Bentur, A.; Tirosh, R.; Yardimci, M.; Puterman, M.; Peled, A. Controlling bond characteristics by impregnation. In Proceedings of the International RILEM Conference on Material Science, Aachen, Germany, 6–9 September 2010; RILEM Publications SARL: Paris, France, 2010; pp. 23–33.
21. Valeri, P.; Ruiz, M.F.; Muttoni, A. Tensile response of textile reinforced concrete. *Constr. Build. Mater.* **2020**, *258*, 119517. [CrossRef]
22. Donnini, J.; Corinaldesi, V.; Nanni, A. Mechanical properties of FRCM using carbon fabrics with different coating treatments. *Compos. Part B Eng.* **2016**, *88*, 220–228. [CrossRef]
23. Hegger, J.; Will, N.; Bruckermann, O.; Voss, S. Load–bearing behaviour and simulation of textile reinforced concrete. *Mater. Struct.* **2006**, *39*, 765–776. [CrossRef]
24. Raupach, M.; Orlowsky, J.; Büttner, T.; Dilthey, U.; Schleser, M.; Hegger, J. Epoxy-impregnated textiles in concrete-load bearing capacity and durability. In Proceedings of the 1st International RILEM Conference, Aachen, Germany, 6 September 2006; RILEM Publications SARL: Paris, France, 2006; pp. 77–88.
25. Paul, S.; Gettu, R.; Arnepalli, D.N.; Samanthula, R. Experimental evaluation of the durability of glass Textile-Reinforced Concrete. *Constr. Build. Mater.* **2023**, *406*, 133390. [CrossRef]
26. Butler, M.; Mechtcherine, V.; Hempel, S. Experimental investigations on the durability of fibre–matrix interfaces in textile-reinforced concrete. *Cem. Concr. Compos.* **2009**, *31*, 221–231. [CrossRef]
27. Kong, K.; Mesticou, Z.; Michel, M.; Larbi, A.S.; Junes, A. Comparative characterization of the durability behaviour of textile-reinforced concrete (TRC) under tension and bending. *Compos. Struct.* **2017**, *179*, 107–123. [CrossRef]
28. Caggegi, C.; Carozzi, F.G.; De Santis, S.; Fabbrocino, F.; Focacci, F.; Hojdys, Ł.; Lanoye, E.; Zuccarino, L. Experimental analysis on tensile and bond properties of PBO and aramid fabric reinforced cementitious matrix for strengthening masonry structures. *Compos. Part B Eng.* **2017**, *127*, 175–195. [CrossRef]
29. Mobasher, B.; Peled, A.; Pahilajani, J. Distributed cracking and stiffness degradation in fabric-cement composites. *Mater. Struct.* **2006**, *39*, 317–331. [CrossRef]
30. Bentur, A.; Yardımcı, M.Y.; Tirosh, R. Preservation of telescopic bonding upon aging of bundled glass filaments by treatments with nano-particles. *Cem. Concr. Res.* **2013**, *47*, 69–77. [CrossRef]
31. Barhum, R.; Mechtcherine, V. Effect of short fibers on fracture behaviour of textile reinforced concrete. *Carbon* **2010**, *7*, 3950.
32. Barhum, R.; Mechtcherine, V. Influence of short dispersed and short integral glass fibers on the mechanical behaviour of textile-reinforced concrete. *Mater. Struct.* **2013**, *46*, 557–572. [CrossRef]
33. RILEM Technical Committee 232-TDT (Chair: W. Brameshuber). Recommendation of RILEM TC 232-TDT: Test methods and design of textile reinforced concrete: Uniaxial tensile test: Test method to determine the load bearing behavior of tensile specimens made of textile reinforced concrete. *Mater. Struct.* **2016**, *49*, 4923–4927. [CrossRef]
34. Hegger, J.; Voss, S. Investigations on the bearing behaviour and application potential of textile reinforced concrete. *Eng. Struct.* **2008**, *30*, 2050–2056. [CrossRef]
35. De Santis, S.; Carozzi, F.G.; de Felice, G.; Poggi, C. Test methods for Textile Reinforced Mortar systems. *Compos. Part B Eng.* **2017**, *127*, 121–132. [CrossRef]
36. Hartig, J.; Jesse, F.; Schicktanz, K.; Häußler-Combe, U. Influence of experimental setups on the apparent uniaxial tensile load-bearing capacity of Textile Reinforced Concrete specimens. *Mater. Struct.* **2012**, *45*, 433–446. [CrossRef]
37. Leone, M.; Aiello, M.A.; Balsamo, A.; Carozzi, F.G.; Ceroni, F.; Corradi, M.; Gams, M.; Garbin, E.; Gattesco, N.; Krajewski, P.; et al. Glass fabric reinforced cementitious matrix: Tensile properties and bond performance on masonry substrate. *Compos. Part B Eng.* **2017**, *127*, 196–214. [CrossRef]
38. Contamine, R.; Larbi, A.S.; Hamelin, P. Contribution to direct tensile testing of textile reinforced concrete (TRC) composites. *Mater. Sci. Eng. A* **2011**, *528*, 8589–8598. [CrossRef]
39. Aveston, J.; Kelly, A. Theory of multiple fracture of fibrous composites. *J. Mater. Sci.* **1973**, *8*, 352–362. [CrossRef]
40. Cuypers, H.; Wastiels, J. A stochastic cracking theory for the introduction of matrix multiple cracking in textile reinforced concrete under tensile loading. In *Finds and Results from the Swedish Cyprus Expedition: A Gender Perspective at the Medelhavsmuseet*; RILEM: Stockholm, Sweden, 2006; pp. 193–202.

41. Li, B.; Xiong, H.; Jiang, J.; Dou, X. Tensile behavior of basalt textile grid reinforced Engineering Cementitious Composite. *Compos. Part B Eng.* **2019**, *156*, 185–200. [CrossRef]
42. Deng, M.; Dong, Z.; Zhang, C. Experimental investigation on tensile behavior of carbon textile reinforced mortar (TRM) added with short polyvinyl alcohol (PVA) fibers. *Constr. Build. Mater.* **2020**, *235*, 117801. [CrossRef]
43. Saidi, M.; Gabor, A. Iterative analytical modelling of the global behaviour of textile-reinforced cementitious matrix composites subjected to tensile loading. *Constr. Build. Mater.* **2020**, *263*, 120130. [CrossRef]
44. *ASTM D6637*; Standard Test Method for Determining Tensile Properties of Geogrids by the Single or Multi-Rib Tensile Method. American Society for Testing and Materials: West Conshohocken, PA, USA, 2015.
45. Soranakom, C.; Mobasher, B. Correlation of tensile and flexural responses of strain softening and strain hardening cement composites. *Cem. Concr. Compos.* **2008**, *30*, 465–477. [CrossRef]
46. Mobasher, B.; Dey, V.; Cohen, Z.; Peled, A. Correlation of constitutive response of hybrid textile reinforced concrete from tensile and flexural tests. *Cem. Concr. Compos.* **2014**, *53*, 148–161. [CrossRef]

sustainability

MDPI

Article

Optimising the Circular Economy for Construction and Demolition Waste Management in Europe: Best Practices, Innovations and Regulatory Avenues

Rachida Idir [1], Assia Djerbi [2] and Nacef Tazi [1,*]

[1] Cerema, Université Gustave Eiffel, UMR MCD, 77171 Sourdun, France; rachida.idir@cerema.fr
[2] Cerema, Université Gustave Eiffel, UMR MCD, 5 Boulevard Descartes Champs-Sur-Marne, 77454 Marne-La-Vallée, Cedex, France; assia.djerbi@univ-eiffel.fr
* Correspondence: nacef.tazi.2019@utt.fr

Abstract: This article explores the sustainable management of construction and demolition waste (CDW) as part of a circular economy in Europe. It provides an overview of current European practices, identifies the main challenges associated with collecting, sorting and recycling waste, and highlights the need to design buildings and construction that encourage the reuse of materials. The study also draws on best practice from internationally recognised examples such as Japan, Singapore, California, the Netherlands and China, which highlights the effectiveness of a combination of strict regulations, economic incentives and advanced recycling technologies. These international cases provide valuable lessons that can be adapted to the European context to improve the situation and fill gaps in policy, innovation and education. This article recommends targeted measures to strengthen circular practices, such as harmonising European standards, promoting eco-design principles in public procurement, investing in research and development (R&D) and establishing green administrative practices. By adopting these strategies, Europe can significantly improve the management of CDW, fostering a more resilient, sustainable and integrated circular economy.

Keywords: circular economy; construction and demolition waste (CDW); sustainable waste management; recycling building materials; environmental regulation; waste recovery; waste management policy; sorting and recycling technology

Academic Editors: Nele De Belie and Anibal C. Maury-Ramirez

Received: 6 March 2025
Revised: 10 April 2025
Accepted: 11 April 2025
Published: 16 April 2025

Citation: Idir, R.; Djerbi, A.; Tazi, N. Optimising the Circular Economy for Construction and Demolition Waste Management in Europe: Best Practices, Innovations and Regulatory Avenues. *Sustainability* **2025**, *17*, 3586. https://doi.org/10.3390/su17083586

1. Introduction

The circular economy is an innovative economic model that aims to minimise the use of resources and reduce the production of waste, while promoting the reuse, recycling and recovery of materials. In contrast to the traditional linear economy, which follows the "produce, consume and dispose" pattern, this model promises a more sustainable and environmentally friendly approach [1]. The transition to a circular economy is particularly relevant in the construction and demolition sector in Europe, where construction and demolition waste (CDW) accounts for around 37% of all waste produced, with an upward trend projected in the coming years [2]. This significant proportion demonstrates the substantial environmental impact of this sector, which is often considered to be one of the most polluting in Europe. The generation rates of CDW also vary considerably across Europe, ranging from just 1% to almost 90% of all waste generated depending on the country, reflecting major disparities in construction practices and national regulations [3].

In response to this problem, the European Union has stepped up its efforts by introducing new regulations, directives and targets to encourage Member States to further adopt the principles of the circular economy. Recent measures include, for instance, improving the recycling and reuse rate of CDW to 70%, reducing landfill waste and increasing the quality of recycled materials—initiatives that are part of the new European Green Deal Action Plan for the Circular Economy [4–6]. Other examples include the new taxonomy regulation aimed at directing investments towards sustainable activities and the newly adopted construction products regulation for clean and smart products [7–9]. Other recent initiatives by the European Commission Joint Research Centre on modular and reusable panels for buildings are also under assessment [10].

However, despite these advances, considerable challenges remain, particularly in terms of collection, sorting and design for circularity. CDW recovery rates vary widely between countries, from around 60% in Finland to almost 99% in the Netherlands, reflecting differences in national policies, the effectiveness of waste management systems and the commitment of local actors to the circular economy [2]. Recent studies indicate that many aspects of the circular economy value chain still require improvement to achieve optimal efficiency [11–15].

This study examines the current state of CDW management in Europe, highlights examples of best practices from around the world, including the Japanese and Singaporean models, and proposes concrete solutions for improving circularity in this sector. The aim is twofold: to assess the effectiveness of current practices in Europe and to propose specific measures to address existing gaps, thus to optimise the management of CDW. Ultimately, this article aims to demonstrate how, through combined regulations, technological innovation and economic incentives, Europe can become a world leader in the sustainable management of construction and demolition waste.

The methodology adopted to structure the path followed in this article is based on an analytical and comparative approach (developed in Figure 1), making it possible to link local European issues to proven global solutions. First, a contextual data analysis was carried out to identify the key challenges facing the management of CDW in Europe, such as disparities in recycling rates, gaps in standards and the fragmentation of practices in flow management. This was followed by an in-depth literature review, which provided an overview of existing policies and initiatives in Europe. Next, an international comparative analysis was conducted, based on case studies identified as best practices from regions that are leaders in the management of CDW (Japan, Singapore, California, the Netherlands and China), in order to extract lessons that can be applied to the European context. Finally, the solutions proposed in this article have been developed through a critical synthesis of the strengths and weaknesses of European practices, coupled with an adaptation of international best practices, thus guaranteeing, in principle, operational and strategic relevance for strengthening the circular economy in Europe.

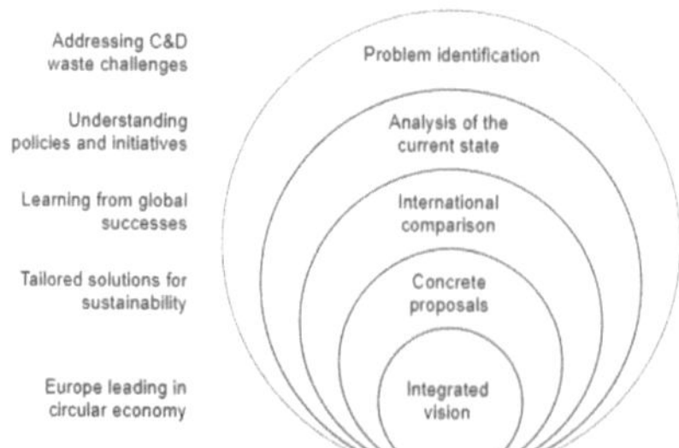

Figure 1. Methodology and analytical framework for this study.

2. Review of Policies, Initiatives and Best Practices for Managing Construction and Demolition Waste in Europe

The management of construction and demolition waste (CDW) has become a strategic priority for Europe, where the construction sector is both an economic mainstay and a major source of waste. In response to growing environmental pressures and sustainability requirements, the European Union has developed an ambitious regulatory framework to transform the way CDW is managed. EU policies aim not only to minimise the amount of waste sent to landfill but also to encourage innovation throughout the product lifecycle, notably in the sorting, recycling and reuse of construction materials. Different Member States, depending on their national contexts and capacities, have implemented these directives with varying results. As a result, Europe is becoming a hub of initiatives and good practices, offering valuable lessons on effective strategies for the sustainable management of CDW. However, CDW generation continues to increase over time, with a significant proportion still being diverted to landfill. According to Eurostat, the EU-wide recycling rate of CDW reached around 88% in 2020; yet, this figure is misleading, as a large share is downcycled into low-value uses (e.g., backfilling) rather than high-quality reuse.

The country cases selected in this section are not intended to be exhaustive. They have been chosen to illustrate a diversity of countries regulatory frameworks, policy maturity levels and geographical contexts across Europe. This approach aims to highlight representative best practices without claiming full coverage of all Member States.

One of the European Union's current targets is to improve the recycling and reuse rate of construction and demolition waste (CDW) to 70% by weight, with a focus on reducing landfill waste and increasing the quality of recycled materials, which is in line with the new Circular Economy Action Plan (CEAP) introduced as part of the European Green Deal [4]. This initiative is part of the Waste Framework Directive 2008/98/EC [16], which aims to promote waste prevention and improve resource efficiency in Europe. Member countries are required to put in place waste management plans that include measures for the sorting and recycling of construction materials [6]. However, the implementation of these measures is uneven, and many Member States struggle to meet the 70% target with high-quality reuse.

CDW is also addressed in policies relating to critical raw materials, as it represents an important urban source for recovering critical, strategic and precious materials, such as copper, aluminium or niobium [17–21]. Construction and demolition waste also contains significant quantities of raw materials, such as steel, iron, aggregates and glass, which are essential for various industries. Recovering them from CDW not only helps to reduce the dependence on imports but also strengthens EU competitiveness and secures the supply of materials needed for European strategic sectors such as electronics, automotive and renewable energy. It also contributes to the decarbonisation roadmap highlighted within the European Clean Industrial Deal [22].

The implementation of CDW management policies varies considerably across Europe. For example, the Netherlands has developed a robust waste management infrastructure, with selective collection systems and advanced recycling facilities. The country has also introduced financial incentives to encourage businesses to adopt sustainable practices [12]. In Belgium, CDW management is also well regulated, with initiatives such as the certification of recycled materials and the creation of secondary market platforms for reused construction materials. On the other hand, countries such as Italy and Poland still face significant challenges in implementing these advanced policies due to fragmented regulations and a lack of appropriate infrastructure [13].

Across Europe, some countries are leading the way with innovative approaches and effective strategies to maximise recycling, reuse and waste reduction.

In the Netherlands, known for its advanced sorting systems and financial incentives, CDW management policy is one of the most advanced in Europe. The country has set up specialised sorting centres that enable recyclable materials to be effectively separated at building sites. In addition, financial and tax incentives are offered to companies that adopt sustainable construction practices. For example, the Dutch government subsidises projects using recycled materials and imposes strict recycling quotas [12].

In Belgium, recognised for its certification systems and reuse platforms, CDW management policy is well regulated and among the most structured in Europe. The country has developed several programmes and strategies to promote a circular economy [23,24], including a certification system for recycled building materials. This certification guarantees that the materials meet strict quality standards, enhancing their acceptance and use in new construction projects. In addition, the country has created online platforms where companies can buy and sell recycled materials, thereby facilitating their reuse [13].

In France, where regulations and training initiatives support the circular economy, CDW management policy is reinforced by strict measures and innovative practices. The country reinforces the management of CDW through strict regulations and initiatives supported by EPR schemes (Extended Producer Responsibility) [25]. This initiative requires producers to finance waste collection and recycling, encouraging companies to adopt sustainable design practices and minimise waste during the planning phase of construction projects. The country also encourages the creation of sorting centres at building sites and offers subsidies for construction projects that use recycled materials [25]. In addition, France has developed training programmes for construction workers to promote sustainable practices and efficient waste management [25]. Furthermore, the RE2020 regulation, introduced to improve the energy efficiency of buildings, also supports the use of greener materials by prioritising low-carbon construction products [26]. In principle, RE2020 could contribute to better end-of-life management by encouraging design practices that facilitate deconstruction, reuse and recycling.

3. Discussion: Challenges, Opportunities and Strategies for CDW Management

Despite the significant progress made in Europe in integrating the circular economy into the management of construction and demolition waste (CDW), a number of challenges persist throughout the value chain. These challenges are not only technical but also regulatory, economic and behavioural. They constitute major obstacles to the wider adoption of circular practices, but they also offer opportunities for future innovation and improvement.

3.1. Locks and Challenges in the Circular Economy Value Chain

To achieve a true circular economy in CDW management, it is crucial to overcome existing shortcomings at every stage of the value chain. Key challenges include inefficiencies in waste collection and sorting, lack of design for circularity, lack of harmonised standards, insufficient economic incentives as well as the lack of awareness and training.

Inefficient collection and sorting of waste on construction sites remains a fundamental problem. Much recyclable material is often contaminated with other types of waste, reducing its quality and limiting its potential for reuse [12,27]. This leads to the degradation of recycled materials, affecting their acceptability for new construction projects. Improving selective collection, supported by stricter regulations, could significantly increase the purity of recovered materials. Practices such as sorting at the site and the use of specialised sorting centres are examples of approaches that can maximise the quality of recycled materials.

The lack of design for circularity is another major challenge. Buildings are often designed without consideration for the end-of-life of the materials used, which limits the possibility of reusing or recycling these materials. To maximise the recyclability of materials, it is essential to incorporate eco-design principles, such as the use of modular and demountable materials [20,28]. In addition, designing for circularity should include strategies to minimise the contamination of materials during use, making them easier to recycle.

The lack of harmonised Europe-wide standards for the recycling of construction materials remains a significant obstacle. Differences in standards and quality output requirements limit the market acceptance of recycled materials and their integration into new projects. Updated, clear and uniform standards would help to guarantee the quality of recycled materials and stimulate demand for them. This could also facilitate cross-border trade in recycled materials, contributing to a more integrated circular economy in Europe [29].

Economic incentives play a crucial role in promoting circular practices, but they are often insufficient or poorly targeted. Subsidies for projects using recycled materials, taxes on virgin materials and tax credits for companies adopting sustainable practices are some examples of policies that could encourage a faster transition to a circular economy [25]. A better design of these incentives, tailored to the economic realities of different countries, could stimulate the wider adoption of recycling and reuse practices.

Finally, the lack of awareness and training among those involved in the construction sector represents a major barrier. Waste management operators are usually not aware of all of the materials incorporated in products, which reduces the likelihood of their recovery at the end of life. Knowledge sharing should thus be encouraged; it is also a pillar of the European Construction Sector Observatory [30]. Many professionals are not sufficiently informed about the benefits of the circular economy and modern techniques for recycling and reusing materials. Ongoing training programmes and awareness-raising campaigns are needed to fill this gap and encourage the wider adoption of circular practices [25]. These efforts should aim to integrate practical knowledge on the economic and environmental benefits of the sustainable management of CDW.

3.2. *International Practices for Managing CDW: Relevant Examples from the Rest of the World*

As Europe continues to make progress in implementing the circular economy for the management of construction and demolition waste (CDW), it is instructive to consider best practices from other parts of the world that could be adapted or adopted in Europe. Countries such as Japan, Singapore, California, the Netherlands and China have developed particularly effective systems that overcome some of the challenges identified in Europe, including strict regulation, economic incentives and advanced technologies.

Japan is renowned for its recycling and recovery model in waste management, achieving remarkable results through a combination of rigorous standards, cutting-edge technologies and strategic planning. The country recycles more than 95% of its construction waste thanks to supporting policies and innovative technologies [29]. This success relies on a combination of stringent regulation and continuous innovation. Japan's Construction Materials Recycling Law, implemented in 2000, requires companies to recycle specific materials such as concrete, wood and asphalt. This legislation aims to ensure the separation of recoverable materials from the demolition phase, thereby improving the purity and quality of recycled materials [29].

To reinforce this policy, selective demolition methods, such as the "Kajima Cut and Take Down" technique [31], have been developed to allow buildings to be dismantled in stages while sorting the recoverable materials directly on-site (Figure 2). This approach reduces cross-contamination and maximises the recovery of high-quality materials, which can then be reused in new construction projects. However, despite its technical relevance, the method remains costly, time-consuming and difficult to apply on a large scale. As a result, its deployment in Japan has been limited to a small number of demonstration projects.

In addition, facilities such as the Taisei Ecological Reproduction System (Tecorep) [32] efficiently transform construction waste into high-quality materials, demonstrating the importance of an integrated approach combining recycling and eco-design. The Japanese model also illustrates the importance of long-term planning and support for innovation to optimise CDW management. The government encourages the use of low-carbon-footprint materials and the development of advanced recycling technologies, creating an environment conducive to continuous innovation in the building sector.

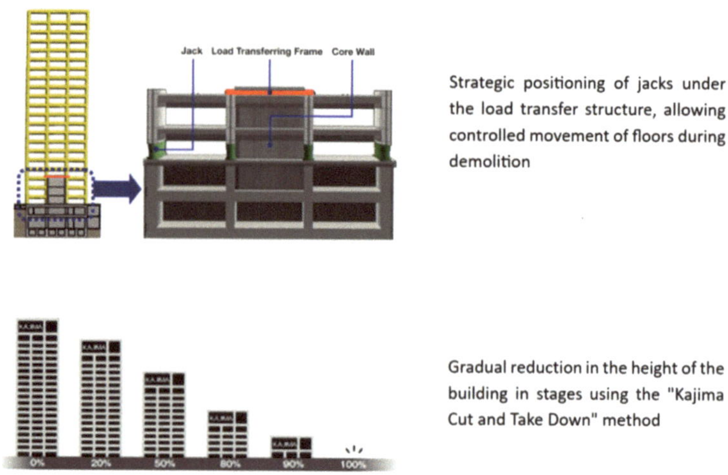

Figure 2. Illustration of the "Kajima Cut and Take Down" method for deconstructing buildings; figure extracted and adapted from [33].

Singapore, renowned for its strategic integration of regulations, economic incentives and advanced technologies, is a successful example of CDW management. This approach effectively combines strict regulations, targeted economic incentives and cutting-edge technologies to address waste management challenges [12]. The government has introduced stringent policies, such as the Green Mark Scheme, which rewards buildings that achieve high standards of sustainability and recycling [34–36]. Similar international frameworks, such as LEED (Leadership in Energy and Environmental Design, USA) and GSAS-CM (Global Sustainability Assessment System for Construction Management, Middle East), also include criteria relevant to CDW management. These frameworks promote the use of recycled materials, construction waste minimisation and design for disassembly, encouraging buildings to be more easily deconstructed and for materials to be reused or recovered at end-of-life.

For example, Table 1 provides an overview of the criteria and point allocation system under the Green Mark Scheme in Singapore, which rewards sustainable construction practices. Projects can earn up to three points if the materials used contain at least 30% recycled content by weight or volume, as highlighted in the table (see point c.(ii)) [37,38]. This system encourages the adoption of environmentally friendly materials and aligns with Singapore's broader goals of reducing waste and promoting a circular economy.

Table 1. Sustainable construction criteria and allocation of points in the Green Mark Scheme in Singapore, adapted from [36–38]. Each score on the right column corresponds to the mentioned practice on the left column. The use of recycled content in products is highlighted by light blue in the table.

Part 3–Environmental protection	**Green Mark Points**
RB 3-1 Sustainable Construction: The aim is to encourage the adoption of building designs, construction practices and materials that are environmentally friendly and sustainable. This can be as following:	
(a) More efficient concrete usage for building components.	0.1 point for every percentage reduction in the prescribed Concrete Usage Index (CUI) limit for residential buildings. (up to 4 points)
(b) Conservation of existing building structure. Applicable to existing structural elements or building envelope.	Extent of coverage: Conserve at least 50% of the existing structural elements or building envelope (by area) 2 points
(c) Use of sustainable materials and products in building construction such as:	
(i) Environmentally friendly products that are certified under the Singapore Green Labelling Scheme (SGLS).	1 point for high Impact item 0.5 point for low impact item (cap at 3 points)
(ii) Products with at least 30% recycled content by weight or volume (applicable only to non-structural elements).	1 point for high Impact item 0.5 point for low impact item (cap at 3 points)
Note (2): for products that are certified under SGLS and with at least 30% recycled contents, points can only be scored either from RB 3-1 (c)(i) or (c)(ii)	(up to 6 points)

Construction companies must comply with waste management standards, and severe penalties are applied for non-compliance. In addition to these regulations, Singapore uses state-of-the-art technology to sort and recover waste. Facilities such as the Tuas South Incineration Plant effectively separate recyclable materials from waste, reducing the amount of waste sent to landfill [12]. Singapore is also investing in energy recovery technologies, converting non-recyclable waste into energy, thereby reducing reliance on landfill.

The Netherlands stands out as a pioneer in the circular economy and eco-design, adopting an integrated approach to CDW management with ambitious policies to achieve a circular economy by 2050 [39,40]. The country actively promotes the use of advanced sorting centres, and financial incentives have been introduced to promote source separation and the use of recycled materials [39]. In addition, the concept of 'circular hubs', where companies collaborate to exchange materials and resources, is being actively promoted [39]. These hubs enhance stakeholder collaboration and facilitate the circular economy by encouraging material reuse.

The Netherlands also supports circular design and procurement strategies, which are guided by measurement tools such as the Environmental Cost Indicator (ECI) and the Dutch National Environmental Database. These tools help to assess environmental impacts and promote circular products [41]. Furthermore, the Building Circularity Index (BCI) provides an innovative method to measure the circularity of buildings, encouraging eco-design and waste reduction at the source [40]. Table 2 provides a comparative overview of these tools, highlighting their objectives, applications and data requirements.

Table 2. Comparison of measurement tools for the circular economy in the Netherlands.

Measurement Tool	Objective	Application	Types of Data/Indicators Used
Environmental Cost Indicator (ECI)	Assessing the environmental impact of construction materials and products.	Used by companies and local authorities to compare the environmental impact of materials and to choose more sustainable solutions.	CO_2 emissions, energy consumption, the use of resources, the impact on biodiversity and external environmental costs.
Dutch National Environmental Database	Providing a standardised database of the environmental impacts of construction materials.	Used for sustainable building design and calculating the environmental impact of construction projects.	Lifecycle inventory (LCI) data, product profiles, greenhouse gas emissions and raw materials consumption.
Building Circularity Index (BCI)	Measuring the circularity of buildings and promoting eco-design and waste reduction.	Used by architects, engineers and urban planners to assess the circularity of a building and plan more sustainable constructions.	Proportion of materials reused/recycled, the lifespan of materials, the adaptability of structures and the potential for dismantling.

To further drive sustainable practices, the Netherlands has implemented taxes on virgin materials, incentivising the use of recycled alternatives and fostering innovation in construction [40]. Innovations such as material passports make it easier to track materials throughout their lifecycle, supporting reuse and recycling [41]. Pilot projects like the Green House in Utrecht, a demountable circular catering pavilion, exemplify the practical application of circular construction. Additionally, technologies such as SmartCrusher, which recovers sand, gravel and cement from concrete, highlight the potential for advanced recycling techniques in the building sector [41].

California illustrates how a mix of federal regulations and local incentives can create a robust framework for CDW management. California legislation requires construction projects to meet certain recycling standards, notably under the California Green Building Standards Code [42], which mandates that 65% of construction waste be recycled or diverted from landfill since 2010.

In addition to CALGreen's statewide mandate, local governments have introduced complementary initiatives to further encourage recycling and material reuse. Cities such as San Francisco and Los Angeles have established municipal programmes that promote the reuse of materials recovered from construction sites and offer economic incentives to contractors who comply with recycling guidelines.

The Californian government's green procurement policies, which impose sustainable building standards for municipal projects, are also driving change in the private sector. As shown in Figure 3, municipal recycling programmes in San Francisco and Los Angeles set ambitious targets for material reuse and incentivise contractors who meet these guidelines. These policies encourage private developers to adopt similar standards, such as LEED (Leadership in Energy and Environmental Design) certification, thereby expanding the market for sustainable building materials and practices [43].

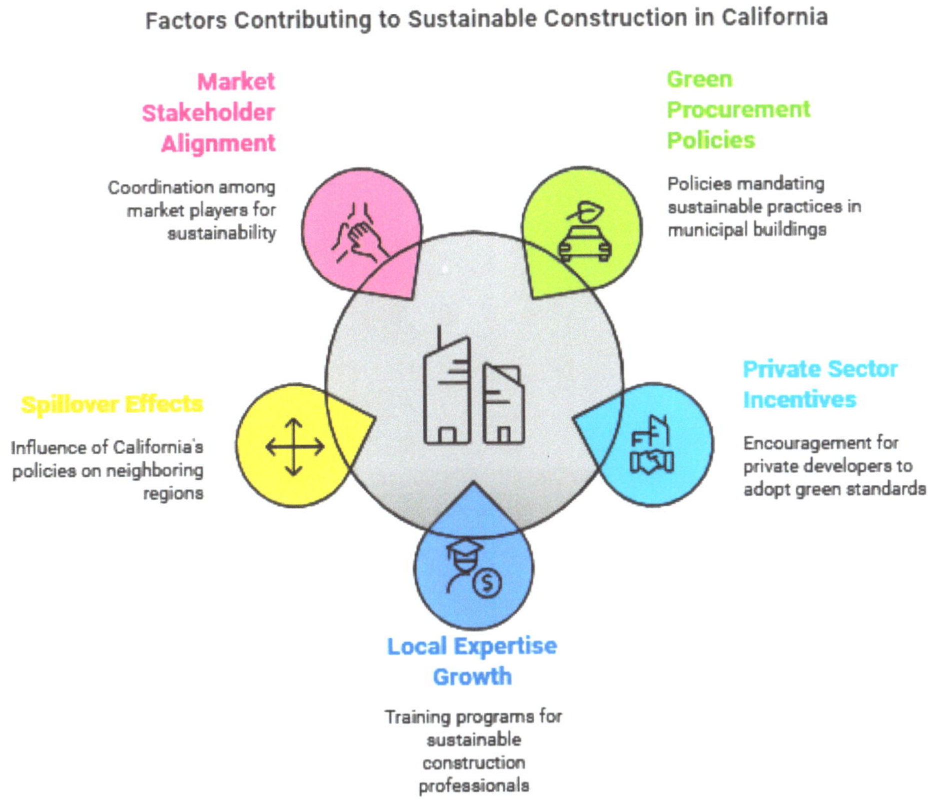

Figure 3. Impact of green building policies in California on the reuse of materials and economic incentives.

Furthermore, these initiatives are fostering the development of local expertise in sustainable construction. For instance, the number of Leadership in Energy and Environmental Design Accredited Professionals (LEED APs) has increased significantly in local markets, as developers align with green building standards [43]. California's policies also have a regional impact, influencing private sector practices and LEED AP certifications in neighbouring cities. This ripple effect demonstrates how well-designed local policies can influence the broader adoption of sustainable practices, thereby extending their benefits beyond state boundaries.

China demonstrates the challenges and opportunities of scaling up CDW management in the context of rapid urbanisation. As one of the largest producers of construction and demolition waste (CDW) in the world, it has historically faced significant challenges in its management, including economic feasibility and the implementation of effective recycling practices [44]. Since 2015, China has intensified its efforts by introducing a national policy encouraging CDW recycling, with a particular focus on Zero-Waste City Demonstration Zones [45–47]. These initiatives aim to integrate waste management practices into urban development plans by promoting source separation, the recycling of construction materials and energy recovery from waste.

As part of its 14th Five-Year Plan (2021–2025), China has expanded its "zero-waste cities" pilot programme to include 113 cities and 8 special areas. This programme focuses on reducing solid waste at the source, thus achieving full recycling, and ensuring safe waste management to reduce pollution and carbon emissions while promoting green and sustainable urban development [46]. The government has introduced a 'zero-waste city indicator system' to assess progress and provide guidance, enabling cities to adapt their strategies to local conditions [46].

Figure 4 illustrates the chronology of the key steps and initiatives taken by China to strengthen CDW management and achieve zero-waste goals. These steps include the launch of initial research, the development of policies and the implementation of pilot and demonstration projects.

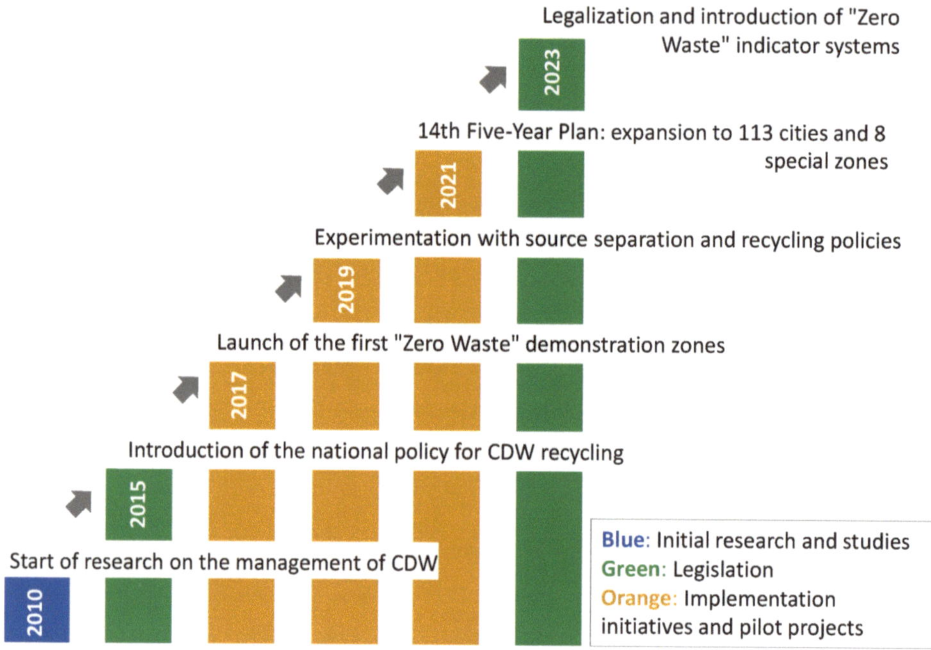

Figure 4. Chronology of China's initiatives and policies for the management of construction and demolition waste (CDW) and the "zero-waste" objective.

Technologies such as concrete recycling and steel waste treatment are being actively encouraged by the government, with subsidies and incentives for companies adopting these practices. In addition, the programme initiated a legalisation process in 2021, introducing laws and regulations that require local governments to integrate solid waste management into economic development and environmental protection plans [46]. However, scaling up remains a significant challenge due to heterogeneous local capacities, a lack of awareness and the need for better coordination to synergistically "reduce pollution and carbon emissions" [46].

3.3. Potential Inputs for European CDW Management

In order to further clarify the relevance of international benchmarks for Europe, a comparative synthesis of the case studies is proposed in Table 3. This table highlights the main strengths and weaknesses observed in each country's approach—Japan, Singapore, California, the Netherlands and China—with a specific focus on regulatory frameworks, economic incentives, technological tools and governance models.

Table 3. Comparative synthesis of international case studies: strengths, weaknesses and transferable elements for EU CE implementation.

Country	Strengths	Weaknesses	Transferable Elements for the CE in EU
Japan	Rigorous national regulation, high recycling rates, selective demolition, integration of innovation	High costs of selective demolition, not easily scalable	Selective deconstruction, legal obligations for material separation
Singapore	Certification schemes (Green Mark), advanced technologies, penalties for non-compliance	Dependence on incineration, limited space for sorting	Incentive systems, integration of sustainability in building codes
California	Mix of federal and local policies, green procurement spillovers, widespread LEED certification	Disparities between municipalities, moderate recycling targets	Local empowerment, green procurement, spillover strategies
Netherlands	Circular hubs, material passports, advanced tools (ECI and BCI), fiscal incentives	High upfront investment required, complex tracking	Digital tools, inter-company collaboration, national CE roadmap
China	Ambitious top-down policy (zero-waste cities), scaling strategies, subsidies	Regional disparities, lack of enforcement in some areas	National indicator frameworks adaptable at the local level

The comparative perspective allows for the identification of key transferable elements that could inspire the design of a more coherent and effective strategy for implementing circular economy principles in the European construction and demolition sector. These include, for example, selective deconstruction protocols, digital tools for material tracking, incentive-based certification systems or integrated governance frameworks that combine national coordination with local adaptability.

This structured comparison supports the formulation of shared and scalable practices, which can help to bridge current disparities between EU Member States and contribute to the harmonisation of CDW management approaches across Europe.

3.4. Opportunities for Improving the Management of CDW in Europe

The challenges and lessons learned from international practices offer a number of concrete opportunities for improving the management of construction and demolition waste (CDW) in Europe. This section summarises some key areas for improvement.

- Harmonisation of standards and regulations: Europe could work to harmonise CDW management and use standards across its Member States while allowing for local adaptations. The introduction of Europe-wide guidelines, similar to the EU omnibus or the forthcoming Circular Economy Act, or similar to those in California and Japan, combined with local adaptations to meet the specific needs of each region, could enhance the effectiveness of CDW management policies.
- Strengthening economic incentives: Policies, such as subsidies for projects using recycled materials, taxes on virgin materials or tax credits for businesses adopting sustainable practices, are essential to encourage a faster transition to a circular economy. Europe could benefit from Singapore's approach by developing certification and reward schemes for sustainable buildings, which would encourage businesses to adopt circular practices.
- Supporting innovation and developing advanced sorting and recycling technologies: Europe could invest more in advanced sorting and recycling technologies, such as

those used in Japan and Singapore. Targeted investments in selective demolition technologies, such as the "Kajima Cut and Take Down" method, and automated sorting centres could improve the purity of recycled materials and increase the recycling rates.

- Promoting eco-design and design for deconstruction: Incorporating eco-design principles, such as the use of modular and demountable materials, into building regulations could maximise the recyclability of materials. In Europe, the Building Circularity Index (BCI) concept used in the Netherlands could be adopted to assess the circularity of construction projects and encourage design for deconstruction.
- Raising awareness, knowledge sharing and providing training: Europe could step up its efforts to raise awareness and provide training for professionals in the construction sector. Inspired by the educational programmes set up in California and the Netherlands, Europe could develop continuous training programmes and awareness campaigns to promote sustainable CDW management practices. This could also be supported by enhanced knowledge sharing between all actors in the value chain to promote better management of materials throughout their lifecycles.
- Developing public–private partnerships: Partnerships between the public and private sectors, such as the circular hubs in the Netherlands, could be encouraged to foster innovation and efficiency in the management of CDW. These collaborations can lead to the better use of resources, the sharing of best practices and the optimisation of recycling and reuse processes.

The diagram elaborated in Figure 5 summarises these opportunities for improvement by highlighting the potential inter-relationships between the different strategies. For example, the harmonisation of standards and regulations can strengthen economic incentives by facilitating the implementation of common policies at the European level. Similarly, the development of advanced sorting and recycling technologies can support eco-design efforts and design for deconstruction, enabling the better management of materials throughout the lifecycle of buildings.

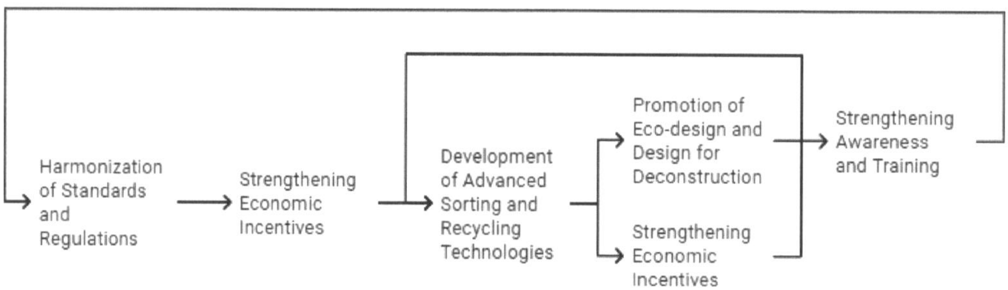

Figure 5. Sequential and interconnected strategies to improve the management of construction and demolition waste (CDW) in Europe.

The process begins with regulatory harmonisation, which plays a fundamental and transversal role across all stages. It then progresses through the activation of economic levers, technological investment and design innovation, culminating in knowledge dissemination and capacity-building. The feedback loop between training, awareness-raising and the earlier stages highlights the iterative nature of these strategies, as well as the importance of stakeholder engagement and policy adjustments throughout the implementation process.

3.5. Practical Measures to Strengthen the Circular Economy of CDW in Europe

To complement the opportunities identified for construction and demolition waste (CDW) management in Europe, a number of practical measures and additional initiatives can be implemented to address the specific challenges. Building on proven approaches and innovations tailored to European realities, these measures aim to optimise waste collection and sorting, promote sustainable design strategies, encourage technological innovation and strengthen administrative and regulatory coordination. By integrating these strategies, Europe can maximise material reuse, reduce the ecological footprint of construction and support the transition to a more circular and resilient economy.

For an overview of these measures, Figure 6 illustrates an integrated process for strengthening the circular economy of CDW in Europe. Although this figure is presented in a circular format, the measures illustrated are not intended to follow a strict chronological sequence. Depending on the national or regional context, they can be implemented in parallel, iteratively or in different orders.

Figure 6. Integrated and non-sequential processes for strengthening the circular economy of CDW in Europe. The circular layout is used for visual clarity and does not imply a fixed implementation order.

3.5.1. Improving Waste Collection and Sorting

Efficient collection and the rigorous sorting of construction and demolition waste (CDW) are essential to guarantee the purity of recycled materials and maximise their reuse. However, the contamination of recyclable materials remains a major obstacle to effective recycling.

- Obligation to sort at source: Construction companies should be required to sort waste directly on-site, separating materials such as concrete, wood and metal [12]. This measure should be supported by frequent inspections and penalties for companies that fail to comply with these obligations.
- Specialised sorting centres: Develop dedicated infrastructure for sorting centres to efficiently process collected waste and ensure the accurate separation of recyclable materials. These centres could be equipped with advanced sorting technologies, such as the automated systems used in Japan and Singapore.

3.5.2. Encouraging Circular Design Through Public Procurement

Incorporating eco-design principles into public procurement is crucial to achieving a truly circular economy.

- Incorporating circular design standards: Public tenders should include eco-design requirements, such as the use of recyclable materials, building modularity and ease of dismantling [28]. This approach could draw inspiration from practices in Singapore

and the Netherlands, which integrate eco-design into their regulations to encourage sustainable construction.
- Supporting innovation in sustainable design: Companies that develop innovative solutions for sustainable design and material reuse should be rewarded through grants, subsidies and tax credits.

3.5.3. Investing in Research and Development (R&D)

Technological innovation and research are essential to improve the recycling and recovery processes for construction and demolition waste (CDW).

- Increased funding for research into new technologies: Greater investment is needed in the development of sorting, recycling and recovery technologies for construction materials, including critical and strategic raw materials, in partnership with universities and research centres [29]. Selective demolition technologies, such as the "Kajima Cut and Take Down" method used in Japan, should be prioritised to optimise material recovery and reduce waste.
- Development of new materials: Encourage the creation of innovative construction materials designed to be more easily recyclable or reusable, thereby reducing the dependence on virgin materials. Substitutes for traditional materials, such as recycled concretes or bio-sourced composites, should be actively supported through economic incentives, such as subsidies or tax credits [28].
- European innovation projects and scientific contributions: Several European research and innovation initiatives aim to accelerate progress in the recycling of construction and demolition waste (CDW). For instance, the ICEBERG Project [48,49], funded by Horizon 2020, contributes to this effort by developing technological solutions to enhance sorting and material recovery. Other projects also support a similar aim, such as [50–52]. In parallel, scientific research is increasingly focusing on advanced processing methods, the design for reuse, material traceability and quality control tools. These ongoing efforts, both experimental and theoretical, help to bridge the gap between technological innovation, regulatory frameworks and industry practices [53–56].

3.5.4. Strengthening Regulation and Harmonising Standards

To achieve a true circular economy in the management of construction and demolition waste (CDW) in Europe, it is crucial to establish effective regulations at every level of the value chain. The following highlights some regulatory frameworks that need to be strengthened.

(a) Legislation level: harmonisation of regulations

Towards a circular economy omnibus: To be potentially explored under the forthcoming EU Circular Economy Act under the EU Competitiveness Compass [57], which would provide a clear long-term circularity roadmap aligned with the omnibus on sustainability [58].

Harmonised European policy files: Implement European regulations requiring the recycling of certain construction materials, inspired by Japan's Construction Materials Recycling law [29]. Such legislation should include recycling quotas for materials such as concrete, wood and asphalt, while allowing for adaptation to local contexts.

Waste prevention strategy: Develop a European strategy for waste prevention, with incentives to reduce waste at the source, such as promoting design for deconstruction [28].

(b) Product level: quality standards and certification

Quality standards for recycled materials: Establish an EU-wide certification scheme for recycled building materials, ensuring they meet strict quality standards [13]. This system could also include eco-labels to help guide consumers and businesses in their purchasing decisions [25].

Environmental Product Declarations (EPDs): In line with the revision of the Construction Products Regulation (CPR), the adoption of Environmental Product Declarations (EPDs) with mandatory carbon footprint indicators is expected to become a standard requirement across the European Union. Integrating these obligations into public procurement and market access policies would serve as a concrete lever to promote the use of low-carbon construction materials and help align national practices with the EU's decarbonisation goals.

Several European countries have already developed national certification systems that integrate environmental performance and resource efficiency. Notable examples include BREEAM (Building Research Establishment Environmental Assessment Method, UK), DGNB (Deutsche Gesellschaft für Nachhaltiges Bauen, Germany) and HQE (Haute Qualité Environnementale, France). These schemes promote sustainable construction through criteria such as energy efficiency, the use of low-impact materials and lifecycle performance. However, the inclusion of specific criteria on construction and demolition waste (CDW) management remains uneven. Strengthening the integration of CDW-related indicators—such as the reuse potential, material traceability or design for deconstruction—within these frameworks could support a more coherent alignment between voluntary certification schemes and European circular economy objectives.

(c) Administrative level: green administrative practices

Mandatory green clauses in calls for tenders and public procurement: Introduce mandatory circularity clauses in public tenders for construction projects, prioritising companies using recycled materials and sustainable practices [25].

Training and awareness-raising: Establish mandatory or standardised continuing education modules on circular economy principles and sustainable construction practices for public project managers, architects and contractors. These modules should be integrated into national professional certification schemes and updated regularly in line with EU regulatory developments and technological innovations. In line with initiatives already launched in some Member States [12], awareness campaigns targeting private stakeholders and SMEs can further accelerate the voluntary adoption of circular practices across the sector.

To complete this analysis, Table 4 provides a detailed overview of practical measures to improve CDW management in Europe. The table identifies concrete actions to be implemented, along with the main stakeholders involved, in order to clarify responsibilities and highlight key levers for action.

Table 4. Practical measures and complementary initiatives for a circular economy for CDW.

Category	Practical Measures	Examples of Actions	Stakeholders Involved
Waste Collection and Sorting	Improving waste collection and sorting on construction sites	Obligation to sort at source and the development of specialised sorting centres	Construction companies, waste operators and local authorities
Circular Design	Incorporating eco-design criteria into public procurement contracts	Circular design standards in calls for tender and support for innovation in sustainable design	Public procurers, architects and design offices

Table 4. *Cont.*

Category	Practical Measures	Examples of Actions	Stakeholders Involved
Research and Development (R&D)	Investing in recycling technologies and developing new materials	R&D funding for recycling and the development of recyclable and bio-sourced materials	Research institutions, private firms and innovation agencies
Regulations	Establishing regulatory frameworks at the European level	Harmonised guidelines for the recycling of construction materials and waste prevention strategies	European Commission, national ministries and regulators
Green Administrative Practices	Introduce environmental clauses in public tenders and ongoing training programmes	Green clauses in calls for tender and training programmes for public project management	Public authorities, procurement officers and training bodies

4. Conclusions

The transition towards a circular economy in the management of construction and demolition waste (CDW) in Europe is essential to meeting current and future environmental challenges. Despite significant progress, much remains to be done to achieve full circularity. European initiatives reflect a growing commitment to circular principles, but persistent disparities in waste collection and sorting, circular building design and recycled material quality underscore the need for more consistent and effective approaches.

Global leaders such as Japan, Singapore, California, the Netherlands and China have demonstrated that a combination of strict regulations, advanced technologies and economic incentives can significantly reduce construction waste and improve the quality of recycled materials. Drawing from these examples, Europe must enhance its efforts to optimise waste management, integrate circular design principles, establish harmonised standards and implement effective economic incentives. Awareness-raising and training for sector professionals remain critical to fostering wider adoption of circular practices.

Achieving these goals will require robust legislation at the political level, the establishment of product standards, the development of alternative materials and the promotion of green practices in public procurement. An integrated and harmonised European approach will be pivotal to the success of these initiatives.

To achieve a true circular economy in the CDW sector, Europe must continue to draw inspiration from global best practices while adapting its policies and regulations to local contexts. Investments in research and development, combined with proactive legislation and targeted economic incentives, can accelerate this transition. Collaboration between governments, businesses and citizens is essential to achieving these ambitious goals.

Although challenges remain, the opportunities for the sustainable and efficient management of CDW in Europe are considerable. By adopting an integrated approach and leveraging technological innovations and global best practices, Europe has the potential to become a world leader in the circular economy for buildings and construction.

Author Contributions: All authors contributed equally to the conceptualisation, methodology, validation, formal analysis, investigation, writing—original draft preparation, writing—review and editing and visualisation. All authors have read and agreed to the published version of the manuscript.

Funding: This research received no external funding.

Institutional Review Board Statement: Not applicable.

Informed Consent Statement: Not applicable.

Data Availability Statement: The original contributions presented in the study are included in the article; further inquiries can be directed to the corresponding author.

Acknowledgments: The authors acknowledge the initial data acquisition and analyses conducted by the first-year students (Class of 2023–2024) from École nationale des ponts et chaussées, under the supervision of the authors Rachida Idir, Nacef Tazi as well as Aphrodite Michali, Academic Director of the First-Year Program, École Nationale des Ponts et chaussées – Institut Polytechnique de Paris.

Conflicts of Interest: The authors declare no conflicts of interest.

Abbreviations

The following abbreviations are used in this manuscript:

BCI	Building Circularity Index
BREEAM	Building Research Establishment Environmental Assessment Method
CALGreen	California Green Building Standards Code
CDW	Construction and Demolition Waste
CEAP	Circular Economy Action Plan
CPR	Construction Products Regulation
CUI	Concrete Usage Index
ECI	Environmental Cost Indicator
EPD	Environmental Product Declaration
EPR	Extended Producer Responsibility
EU	European Union
GSAS-CM	Global Sustainability Assessment System for Construction Management
HQE label	Haute Qualité Environnementale (High Environmental Quality) Label
LCA	LCA
LCI	Lifecycle Inventory
LEED (AP)	Leadership in Energy and Environmental Design (Accredited Professionals)
R&D	Research and Development
SGLS	Singapore Green Labelling Scheme

References

1. Knight, C. What Is the Linear Economy? European Investment Bank. Available online: https://www.eib.org/en/stories/linear-economy-recycling (accessed on 6 March 2025).
2. Statista. Construction Waste as Share of Waste in Europe, by Country. Statista. Available online: https://www.statista.com/statistics/1399131/construction-waste-as-a-share-of-all-waste-generated-in-european-countries/ (accessed on 6 March 2025).
3. Statista. Construction Waste as Share of All Waste EU. Statista. Available online: https://www.statista.com/statistics/1399099/construction-waste-as-a-share-of-all-waste-generated-in-the-eu/ (accessed on 6 March 2025).
4. European Commission. A New Circular Economy Action Plan: For a Cleaner and More Competitive Europe. Available online: https://eur-lex.europa.eu/legal-content/EN/TXT/?uri=CELEX:52020DC0098 (accessed on 6 March 2025).
5. European Commission. Construction and Demolition Waste—European Commission. Available online: https://environment.ec.europa.eu/topics/waste-and-recycling/construction-and-demolition-waste_en (accessed on 6 March 2025).
6. Interreg Europe. Collection and Recycling of Construction and Demolition Waste: Key Learnings. Available online: https://www.interregeurope.eu/find-policy-solutions/webinar/collection-and-recycling-of-construction-and-demolition-waste-key-learnings (accessed on 6 March 2025).
7. European Commission. Regulation (EU) 2020/852 of the European Parliament and of the Council of 18 June 2020 on the Establishment of a Framework to Facilitate Sustainable Investment, and Amending Regulation (EU) 2019/2088 (Text with EEA Relevance). Available online: https://eur-lex.europa.eu/legal-content/EN/TXT/?uri=celex:32020R0852 (accessed on 6 March 2025).
8. European Commission. Review of the Construction Products Regulation. Available online: https://single-market-economy.ec.europa.eu/sectors/construction/construction-products-regulation-cpr/review_en (accessed on 6 March 2025).

9. European Council. Building Materials: Council Adopts Law for Clean and Smart Construction Products—Consilium. Available online: https://www.consilium.europa.eu/en/press/press-releases/2024/11/05/building-materials-council-adopts-law-for-clean-and-smart-construction-products/ (accessed on 6 March 2025).
10. European Commission. Reusable Timber Panels for Safe and Sustainable Buildings. Available online: https://joint-research-centre.ec.europa.eu/scientific-activities-z/iresist-home/ongoing-projects/reusable-timber-panels-safe-and-sustainable-buildings_en (accessed on 6 March 2025).
11. Dziedzic, R.; Pondicherry, P.; Dziedzic, M. Review of national policy instruments motivating circular construction. *Resour. Conserv. Recycl.* **2025**, *215*, 108053. [CrossRef]
12. Bilsen, V.; Kretz, D.; Padilla, P.; Van Acoleyen, M.; Van Ostaeyen, J.; Izdebska, O.; Eggert Hansen, M.; Bergmans, J.; Szuppinger, P. *Development and Implementation of Initiatives Fostering Investment and Innovation in Construction and Demolition Waste Recycling Infrastructure*; Publications Office of the European Union: Luxembourg, 2018. [CrossRef]
13. Mongeard, L. From Demolition to the Production of Recycled Aggregates: Analysis of the Logic of Proximity in a Production Chain in the Urban Area of Lyon. *Flux*, 2017. Available online: https://documentation.insp.gouv.fr/insp/doc/CAIRN/_b64_b2FpLWNhaXJuLmluZm8tRV9GTFVYMV8xMDhfMDA2NA==/from-demolition-to-the-production-of-recycled-aggregates-analysis-of-the-logic-of-proximity-in-a-pro (accessed on 6 March 2025).
14. Baldassarre, B.; Saveyn, H. *A Systematic Analysis of EU Publications on the Circular Economy*; Publications Office of the European Union: Luxembourg, 2023. [CrossRef]
15. Marelli, L.; Trane, M.; Barbero Vignola, G.; Gastaldi, C.; Guerreiro, M.M.; Delgado Callico, L.; Borchardt, S.; Mancini, L.; Sanye Mengual, E.; Gourdon, T.; et al. Delivering the EU Green Deal—Publications Office of the EU. 2025. Available online: https://op.europa.eu/en/publication-detail/-/publication/33a4bbe1-e2b0-11ef-be2a-01aa75ed71a1/language-en (accessed on 6 March 2025).
16. European Commission. Directive—2008/98—EN—Waste Framework Directive. Available online: https://eur-lex.europa.eu/eli/dir/2008/98/oj/eng (accessed on 6 March 2025).
17. Kleemann, F.; Lederer, J.; Rechberger, H.; Fellner, J. GIS-based Analysis of Vienna's Material Stock in Buildings. *J. Ind. Ecol.* **2017**, *21*, 368–380. [CrossRef]
18. Lederer, J.; Gassner, A.; Keringer, F.; Mollay, U.; Schremmer, C.; Fellner, J. Material Flows and Stocks in the Urban Building Sector: A Case Study from Vienna for the Years 1990–2015. *Sustainability* **2020**, *12*, 300. [CrossRef]
19. Lederer, J.; Gassner, A.; Fellner, J.; Mollay, U.; Schremmer, C. Raw materials consumption and demolition waste generation of the urban building sector 2016–2050: A scenario-based material flow analysis of Vienna. *J. Clean. Prod.* **2021**, *288*, 125566. [CrossRef]
20. Tazi, N.; Idir, R.; Ben Fraj, A. Towards achieving circularity in residential building materials: Potential stock, locks and opportunities. *J. Clean. Prod.* **2021**, *281*, 124489. [CrossRef]
21. Callegher, C.Z.; Grazieschi, G.; Wilczynski, E.; Oberegger, U.F.; Pezzutto, S. Assessment of Building Materials in the European Residential Building Stock: An Analysis at EU27 Level. *Sustainability* **2023**, *15*, 8840. [CrossRef]
22. European Commission. Clean Industrial Deal. Available online: https://commission.europa.eu/topics/eu-competitiveness/clean-industrial-deal_en (accessed on 6 March 2025).
23. Bruxelles Environnement. Plan de Gestion des Ressources et des Déchets Pour une Consommation Durable, Sobre, Locale et circulaire Pour une société zéro déchet. Bruxelles Environnement. Bruxelles. Belgique. 2018. Available online: https://document.environnement.brussels/opac_css/elecfile/PLAN_Gestion_DechetHulpbronnen_FR.pdf (accessed on 6 March 2025).
24. Bruxelles Environnement. Le Plan de Gestion des Ressources et Déchets (PGRD). Available online: https://environnement.brussels/citoyen/nos-actions/plans-et-politiques-regionales/le-plan-de-gestion-des-ressources-et-dechets-pgrd (accessed on 6 March 2025).
25. Ademe. (translated from French) Identification of the Constraints and Levers for the Reuse of Construction Products and Materials. 2016. Available online: https://www.actu-environnement.com/media/pdf/news-27317-freins-leviers-reemploi-btp.pdf (accessed on 15 April 2025).
26. Ministry of Ecological Transition. Réglementation Environnementale RE2020 | Ministères Aménagement du Territoire Transition Ecologique. Available online: https://www.ecologie.gouv.fr/politiques-publiques/reglementation-environnementale-re2020 (accessed on 6 March 2025).
27. Condotta, M.; Zatta, E. Reuse of building elements in the architectural practice and the European regulatory context: Inconsistencies and possible improvements. *J. Clean. Prod.* **2021**, *318*, 128413. [CrossRef]
28. Suárez-Eiroa, B.; Fernández, E.; Méndez-Martínez, G.; Soto-Oñate, D. Operational principles of circular economy for sustainable development: Linking theory and practice. *J. Clean. Prod.* **2019**, *214*, 952–961. [CrossRef]
29. Huang, B.; Wang, X.; Kua, H.; Geng, Y.; Bleischwitz, R.; Ren, J. Construction and demolition waste management in China through the 3R principle. *Resour. Conserv. Recycl.* **2018**, *129*, 36–44. [CrossRef]
30. European Commission. European Construction Sector Observatory (ECSO). Available online: https://single-market-economy.ec.europa.eu/sectors/construction/observatory_en (accessed on 6 March 2025).

31. Mizutani, R. Jack Down Building Demolition Method: KC & D (The Kajima Cut and Take Down Method). *J. Soc. Mech. Eng.* **2011**, *114*, 424–425. [CrossRef]
32. Trends in Japan. High-Tech Demolition Systems for High-Rises. Available online: https://web-japan.org/trends/11_tech-life/tec130325.html (accessed on 6 March 2025).
33. Kajima Corporation. The Kajima Cut and Take Down Method. Available online: https://www.kajima.co.jp/english/tech/kcd/index.html (accessed on 6 March 2025).
34. BCA. Built Environment Transformation Gross Floor Area Incentive Scheme. Available online: https://www1.bca.gov.sg/buildsg/sustainability/green-mark-incentive-schemes/built-environment-transformation-gross-floor-area-incentive-scheme (accessed on 6 March 2025).
35. BCA. Green Mark Incentive Scheme for Existing Buildings 2.0 (GMIS-EB 2.0). Available online: https://www1.bca.gov.sg/buildsg/sustainability/green-mark-incentive-schemes/green-mark-incentive-scheme-for-existing-buildings-2.0 (accessed on 6 March 2025).
36. BCA. Green Mark Incentive Schemes. Available online: https://www1.bca.gov.sg/buildsg/sustainability/green-mark-incentive-schemes (accessed on 6 March 2025).
37. BCA. *Certification Standard for New Buildings GM Version 3.0*; Building and Construction Authority (BCA): Singapore, 2010.
38. BCA. *GM NRB 2015 Green Mark for Non-Residential Buildings NRB 2015*; Building and Construction Authority (BCA): Singapore, 2015.
39. Potting, J.; Hekkert, M.; Worrell, E.; Hanemaaijer, A. *Circular Economy: Measuring Innovation in the Product Chain*; PBL Netherlands Environmental Assessment Agency: The Hague, The Netherlands, 2017.
40. van Oorschot, J.; Sprecher, B.; Rijken, B.; Witteveen, P.; Blok, M.; Schouten, N.; van der Voet, E. Toward a low-carbon and circular building sector: Building strategies and urbanization pathways for the Netherlands. *J. Ind. Ecol.* **2023**, *27*, 535–547. [CrossRef]
41. de Graaf, D.; Schuitemaker, S.; Hamada, K.; Gruis, V. *Circular Buildings. Constructing a Sustainable Future*; Netherlands Circular Construction Report; Holland Circular Hotspot: Rotterdam, The Netherlands, 2022.
42. DGS. CALGreen California Green Building Standards Code. Available online: https://www.dgs.ca.gov/BSC/CALGreen (accessed on 6 March 2025).
43. Simcoe, T.; Toffel, M.W. Government green procurement spillovers: Evidence from municipal building policies in California. *J. Environ. Econ. Manag.* **2014**, *68*, 411–434. [CrossRef]
44. Zhao, W.; Leeftink, R.B.; Rotter, V.S. Evaluation of the economic feasibility for the recycling of construction and demolition waste in China—The case of Chongqing. *Resour. Conserv. Recycl.* **2010**, *54*, 377–389. [CrossRef]
45. Meng, M.; Wen, Z.; Luo, W.; Wang, S. Approaches and Policies to Promote Zero-Waste City Construction: China's Practices and Lessons. *Sustainability* **2021**, *13*, 13537. [CrossRef]
46. Qi, S.; Chen, Y.; Wang, X.; Yang, Y.; Teng, J.; Wang, Y. Exploration and practice of 'zero-waste city' in China. *Circ. Econ.* **2024**, *3*, 100079. [CrossRef]
47. General Office of the State Council. *Work Plan on 'Zero-Waste City' Pilot Program in China*; General Office of the State Council: Beijing, China, 2018.
48. Zhang, C.; Hu, M.; van der Meide, M.; Di Maio, F.; Yang, X.; Gao, X.; Li, K.; Zhao, H.; Li, C. Life cycle assessment of material footprint in recycling: A case of concrete recycling. *Waste Manag.* **2023**, *155*, 311–319. [CrossRef] [PubMed]
49. Iceberg. Iceberg—Innovative Circular Economy Based Solutions Demonstrating the Efficient Recovery of Valuable Material Resources from the Generation of Representative End-of-Life Building Materials. Available online: https://cordis.europa.eu/project/id/869336 (accessed on 31 March 2025).
50. SeRaMCo. SeRaMCo: Secondary Raw Materials for Concrete Precast Products. Interreg NWE. Available online: https://vb.nweurope.eu/projects/project-search/seramco-secondary-raw-materials-for-concrete-precast-products/ (accessed on 31 March 2025).
51. SCB. Smart Circular Bridge (SCB) for Pedestrians and Cyclists in a Circular Built Environment. Interreg NWE. Available online: https://vb.nweurope.eu/projects/project-search/smart-circular-bridge-scb-for-pedestrians-and-cyclists-in-a-circular-built-environment/ (accessed on 31 March 2025).
52. Cirmap. CIrcular Economy via Customisable Furniture with Recycled MAterials for Public Places. Materials and Structures. Available online: https://vb.nweurope.eu/projects/project-search/cirmap-circular-economy-via-customisable-furniture-with-recycled-materials-for-public-places/ (accessed on 31 March 2025).
53. Lara, J.C.F.; El Fadel, M.; Khalfan, M.M.A. Integrating Industry 4.0 and circular economy in the UAE construction sector: A policy-aligned framework. *Built Environ. Proj. Asset Manag.* **2025**. [CrossRef]
54. Abdolmaleki, H.; Ahmadi, Z.; Hashemi, E.; Talebi, S. A review of the circular economy approach to the construction and demolition wood waste: A 4 R principle perspective. *Clean. Waste Syst.* **2025**, *11*, 100248. [CrossRef]

55. Chang, C.; Di Maio, F.; Bheemireddy, R.; Posthoorn, P.; Gebremariam, A.T.; Rem, P. Rapid quality control for recycled coarse aggregates (RCA) streams: Multi-sensor integration for advanced contaminant detection. *Comput. Ind.* **2025**, *164*, 104196. [CrossRef]
56. Sáez, P.V.; Barbero-Álvarez, M.A.; Porras-Amores, C.; Alonso, M.Á.; García Torres, Á. Design and Validation of a Mobile Application for Construction and Demolition Waste Traceability. *Buildings* **2023**, *13*, 1908. [CrossRef]
57. European Commission. COM(2025) 30 A Competitiveness Compass for the EU. Available online: https://commission.europa.eu/document/download/10017eb1-4722-4333-add2-e0ed18105a34_en (accessed on 6 April 2025).
58. European Commission. Commission Simplifies Rules on Sustainability and EU Investments, Delivering over €6 Billion in Administrative Relief. Available online: https://finance.ec.europa.eu/publications/commission-simplifies-rules-sustainability-and-eu-investments-delivering-over-eu6-billion_en (accessed on 6 March 2025).

sustainability

MDPI

Article

Climate Responsive Green Building Strategies in Circular Cities: A Comparative Study for Two Regions

Mirjana Miletić [1,*], Dragan Komatina [2] and Marija Mosurović Ružičić [3]

1 Department of Architecture, Faculty of Technical Sciences, University of Priština in Kosovska Mitrovica, Knjaza Miloša 7, 38220 Kosovska Mitrovica, Serbia
2 Department of Architecture, Faculty of Architecture, University of Montenegro, Karađorđeva 5, 81000 Podgorica, Montenegro; komatinadragan@gmail.com
3 Institute of Economic Sciences, Zmaj Jovina 12, 11000 Belgrade, Serbia; marija.mosurovic@ien.bg.ac.rs
* Correspondence: mirjana.miletic@pr.ac.rs; Tel.: +381-638584697

Abstract: This paper provides a comparative analysis of green building strategies in circular cities from an architectural perspective. It focuses on Belgrade, Serbia, which has a temperate continental climate, and Podgorica, Montenegro, with a mild subtropical climate. The data were gathered through an online questionnaire disseminated among 140 architects in both cities. A five-point Likert scale was applied, and the data were analyzed using the Statistical Package for the Social Sciences (SPSS, version 23). Descriptive statistics, factor analysis, reliability testing, and group comparison methods were employed to ensure a valid, reliable, and transparent framework for processing and interpreting the research of data. By analyzing locally available materials, technologies, and climate factors, the research found that the adoption of circular economy principles does not significantly differ between the cities. This suggests that economic and policy-related factors may have a greater influence than initially expected. Additionally, there was no significant difference in the integration of greening strategies integration ($p = 0.08$), challenging the assumption that climate-responsive design would lead to distinct variations in urban form. However, locally available materials and technologies had a stronger impact on green building practices in Serbia ($p = 0.01$). The study highlights that sustainable architecture is shaped by a combination of local resources, regulatory frameworks, and socio-economic conditions rather than climate factors alone. These insights contribute to the theoretical advancement of climate-smart green building strategies in circular cities. They provide valuable guidance for practitioners and policymakers. Future research should further explore the interplay of socio-economic and regulatory influences to refine strategies for climate-responsive and circular architecture.

Keywords: green building strategies; circular cities; sustainable architecture; Belgrade; Podgorica; locally available materials

Academic Editors: Anibal C. Maury-Ramirez and Nele De Belie

Received: 6 March 2025
Revised: 8 April 2025
Accepted: 10 April 2025
Published: 13 April 2025

Citation: Miletić, M.; Komatina, D.; Mosurović Ružičić, M. Climate Responsive Green Building Strategies in Circular Cities: A Comparative Study for Two Regions. *Sustainability* **2025**, *17*, 3469. https://doi.org/10.3390/su17083469

1. Introduction

A circular city operates on the principles of a circular economy, striving to minimize waste and maximize resource efficiency through recycling, repurposing, and sustainable systems [1]. Climate-responsive design, on the other hand, adapts buildings and urban spaces to their specific climate by employing passive strategies such as natural ventilation, thermal mass, and renewable energy to reduce environmental impact [2]. These concepts intersect in their common pursuit of sustainability—by incorporating climate-responsive design, circular cities become more energy-efficient, resilient, and environmentally balanced [3].

1.1. Green Building Practices and Circular Economy

As urbanization accelerates and environmental concerns grow, the integration of sustainable development principles within urban planning and architecture has become critical [4,5]. Circular cities, which aim to eliminate waste and make the most of resources through reuse, recycling, and regeneration, present a forward-thinking model for sustainable urban development [6,7]. In the context of these cities, green building strategies play a key role in achieving environmental goals, particularly by reducing energy consumption, minimizing waste, and enhancing resource efficiency [8–11].

When it comes to green building practices in the context of the circular economy, the literature highlights the increasing implementation of sustainable building solutions. These include the integration of bio-based materials [12,13], digital and Artificial Intelligence (AI) solutions [14,15] and policy-driven sustainability strategies [16] to improve climate resilience and energy efficiency.

However, the literature suggests that sustainable building practices reduce energy consumption, minimize environmental impacts, and improve residents' health and well-being through the use of innovative technologies and materials [17]. Green buildings have become a focal point in recent years because of their ability to reduce environmental impact, improve energy efficiency, and enhance the well-being of their occupants [18]. The increasing body of research highlights different strategies in green building design, such as passive methods to reduce energy use, the incorporation of renewable energy sources, and the use of bio-based and recycled materials to foster sustainability [19].

1.2. Green Building Approaches and Innovative Solutions

Green building approaches play a crucial role in achieving sustainability, especially within the construction industry. Green building is designed to reduce environmental consequences and mitigate effects on occupants. To attain the objectives of green buildings, changes on the management level are needed [20]. Organizational culture and leadership are critical to fostering green innovation. This underscores the need to link organizational principles with sustainability goals to foster a culture that supports innovative practices [21], which foster business performance [22].

Innovative green building solutions integrate advanced technologies to enhance energy efficiency, reduce environmental impact, and improve occupant well-being. Smart building systems, using IoT-enabled devices and sensors, optimize energy consumption by adjusting lighting, heating, and ventilation based on real-time conditions. Building-integrated photovoltaics (BIPV) allow buildings to generate their own energy through solar panels integrated into roofs and facades, contributing to energy independence. Green walls and living facades improve insulation, reduce the urban heat island effect, and enhance air quality by incorporating plants into building exteriors. High-performance glazing systems, such as low-emissivity (Low-E) glass, reduce heat transfer and optimize daylight, boosting energy efficiency and comfort. Rainwater harvesting systems collect and store water for non-potable uses, decreasing reliance on municipal supply. Additionally, zero-carbon buildings, which produce as much energy as they consume, are becoming a key feature of sustainable architecture.

Green building strategies integrate comfort conditions while integrating principles of the circular economy, aiming to reduce waste, promote material reuse, and optimize the life cycle of buildings [23]. Key approaches include passive design techniques such as natural ventilation, shading, and thermal mass to reduce reliance on mechanical heating and cooling systems [24]. Energy efficiency is enhanced through high-performance insulation, smart building systems, and the integration of renewable energy sources like solar panels and geothermal heating [25]. Water conservation measures, such as rainwater harvesting and gray water recycling, contribute to urban sustainability by reducing strain on municipal water supplies [26]. Additionally, green roofs and urban vegetation improve air quality, mitigate the urban heat island effect, and enhance biodiversity [27]. In the broader urban context, green building strategies support resilient cities by promoting adaptable architecture, reducing carbon footprints, and creating healthier environments for residents [28]. They are essential in addressing climate change and fostering long-term urban sustainability. However, the effectiveness of green building strategies is highly influenced by local climatic conditions [29]. In different climate zones, the architectural approaches to sustainability vary significantly due to the need to adapt to specific environmental factors [28].

1.3. Research Questions

This study focuses on two cities, one located in a mild subtropical climate and the other in a temperate continental climate, to examine how these distinct environments shape green building practices within the framework of circular cities

While numerous studies explore green building practices, there is a paucity of comparative analyses focusing on how architects in diverse climatic regions implement climate-responsive strategies. The existing literature often centers on single-case studies or generalized approaches, lacking insights into region-specific challenges and adaptations.

Research questions tailored for the two cities in different climates are:

RQ1: How do architects in Belgrade (Serbia) and Podgorica (Montenegro) with different climates integrate climate-responsive strategies into green building designs, and what challenges do they face in implementing these strategies?

RQ2: What are the key differences in the application of circular economy principles in green building projects between cities with different climates considering their distinct climate zones?

RQ3: How do locally available materials and technologies in cities with different climates influence the adoption of climate-responsive green building practices in line with circular city objectives?

The objective of this research is to compare climate-responsive architectural strategies in these two climates, focusing on energy efficiency, material usage, and resource management. By analyzing the ways in which each city, Belgrade (Serbia) and Podgorica (Montenegro), addresses its unique environmental challenges through green building practices, the study seeks to provide insights into how circular cities can optimize sustainability across different climate zones. The findings will underscore the importance of climate-adaptive design within circular economy frameworks, highlighting how architecture can contribute to the resilience and sustainability of future urban environments. The successful implementation of climate-responsive design strategies can guide architects and urban planners in the optimal use of architectural and urban design techniques to enhance indoor and outdoor thermal comfort while mitigating risks to human health and energy security. These insights will furnish evidence-based and practical design solutions for architects and urban planners throughout the initial planning phase to alleviate the potential effects of increasingly frequent high heat occurrences, energy, water use and resource efficiency.

This research utilizes a mixed-methods approach, combining a structured questionnaire survey to explore how architects, with experience in designing residential and public buildings for a minimum of 5 years, in Belgrade (Serbia) and Podgorica (Montenegro), integrate climate-responsive strategies and circular economy principles in green building designs. The questionnaire focuses on the adoption of climate-responsive strategies, challenges in implementation, and the use of locally available materials and technologies. Data were collected using Likert scale questions, enabling a quantitative analysis of responses, while qualitative interviews will provide deeper insights into specific challenges and strategies employed by architects.

The outcomes include identifying key differences in the application of circular economy principles across cities with different climates, understanding how locally available materials influence sustainable practices, and revealing any variations in urban form and architectural appearance based on climate-responsive strategies. The findings will contribute to the development of more effective sustainable building practices and circular city objectives, offering practical recommendations for architects, urban planners, and policymakers.

2. Literature Review

The key climate challenges consequently incorporated into the questionnaire, which was distributed to architects, serve as the foundation for this research.

2.1. *Climate Adaptive Architecture and Urbanism*

Climate change and urbanization have intensified environmental challenges, requiring climate-responsive and sustainable design strategies [30,31]. Rising global temperatures increase cooling demands, making passive cooling strategies like shading, thermal mass, and natural ventilation essential [32]. High humidity affects both comfort and material durability, necessitating vapor-permeable materials and adaptive ventilation [33]. Increasingly unpredictable precipitation demands stormwater management solutions such as green roofs and permeable surfaces to prevent urban flooding [34]. Urban heat islands intensify urban temperatures due to heat-retaining surfaces, but reflective materials and green infrastructure can mitigate their effects [35]. Energy efficiency remains a key challenge, requiring passive design, smart materials, and renewable energy integration to reduce environmental impact [36]. Addressing these challenges through climate-responsive architecture is essential for resilient and sustainable urban development.

The integration of green building strategies into circular cities is based on the principles of the circular economy, including resource efficiency, waste minimization, and regenerative design [37]. Sustainable architecture in this context involves the use of recycled and bio-based materials, energy-efficient and modular construction, as well as passive design strategies such as natural ventilation and daylight optimization [38–40].

Climate-adaptive architecture is essential for sustainable urban development, as research emphasizes the need for climate-responsive design strategies [41] to enhance building performance and minimize environmental impact [42,43]. In subtropical regions, passive cooling techniques—such as shading, cross-ventilation, and the strategic use of materials—help reduce cooling demands [44]. Studies on vernacular architecture offer valuable knowledge on traditional methods that naturally align with modern sustainable design principles [45].

The application of circular economy principles to architecture and urban planning has gained momentum in recent years [46,47]. Studies emphasize the role of modular design, material recovery, and adaptive reuse in reducing construction waste and extending building [37]. The concept of circular construction, which integrates closed-loop material flows and design-for-disassembly strategies, has been proposed as a key solution for sustainable urbanization [48].

Several comparative studies have analyzed the impact of climatic conditions on sustainable building practices [49]. Findings from comparative studies suggest that tailored strategies based on regional climatic contexts significantly improve energy efficiency and occupant comfort, that subtropical cities prioritize ventilation and solar shading to mitigate overheating, and that temperate continental cities focus on insulation and heat retention to address seasonal temperature variation [50–57]. The integration of circular design strategies into these climate-adaptive approaches provides a comprehensive framework for achieving self-sustaining urban ecosystems. Research supports the argument that a combination of passive design, renewable energy integration, and material circularity fosters long-term resilience and resource efficiency in urban environments [58,59].

The reviewed literature underscores the significance of climate-adaptive and circular design strategies in promoting sustainable urban development [58,60–63]. The intersection of circular economy principles with climate-responsive architecture presents a promising pathway for enhancing energy efficiency, reducing waste, and improving building resilience across diverse climatic conditions [64].

The urban form and integration of greenery in cities that implement circular economy principles are strongly influenced by their respective climate zones. In warmer climates, cities tend to feature compact urban forms with shaded public spaces, green roofs, and vertical gardens to mitigate heat, while in colder climates, urban layouts prioritize solar exposure, wind protection, and the use of seasonal green infrastructure [65,66]. The availability of local materials and sustainable construction technologies further shapes the spatial organization, façade treatments, and landscape strategies, resulting in distinct visual identities and functional characteristics that reflect both ecological and cultural adaptations [67,68].

In addition to field research on the same topic and a comparison of practice from the view of professionals in two regions, the paper provides an overview of indicators based on literature research that are recognized by architects, relating to the circular city, climate, and urban environment.

2.2. Circular City Indicators

Circular city indicators are metrics used to assess the sustainability and resource efficiency of urban environments [69]. These indicators typically include factors such as waste reduction, material recycling, energy efficiency, renewable energy usage, and water conservation. Additionally, they measure the adoption of circular economy principles in areas like transportation, construction, and food systems to promote a closed-loop urban economy [70]. From an architect's perspective, circular city indicators focus on evaluating how urban designs incorporate sustainability and resource efficiency. Key indicators include the use of renewable energy sources, the integration of sustainable building materials, and the implementation of waste reduction strategies such as material recycling and reusing. Additionally, these indicators assess the incorporation of circular economy principles into urban planning, including the design of energy-efficient buildings, green infrastructure, and systems that promote water and resource conservation [71,72].

Urban greenery plays a vital role in addressing the interconnected challenges of circular urban planning, climate resilience, and sustainable development. By integrating greenery into the urban environment, cities can enhance resource efficiency, improve environmental quality, and support social well-being [73,74].

The following table provides an overview of the implementation of key indicators in Belgrade and Podgorica that illustrate the connections between greenery, the circular city concept, climate adaptation, and the urban environment (Table 1). These environmental indicators serve as a framework for understanding how green infrastructure and practices contribute to achieving sustainability goals in diverse urban and climatic contexts as it was presented in the empirical research [75].

Table 1. Implementation of Circular City Environmental Indicators (CCEIs) for Belgrade and Podgorica cities.

<table>
<tr><th colspan="3">CCEIs</th><th>Belgrade</th><th></th><th>Podgorica</th></tr>
<tr><td rowspan="7">Environmental Indicators</td><td>Green Infrastructure</td><td>Horizontal and vertical greenery
Urban green areas</td><td>Yes</td><td rowspan="7">A more intensive application is necessary for both cities for all indicators</td><td>Yes</td></tr>
<tr><td>Microclimate regulation</td><td>Temperature regulation
Humidity control
Different urban climate zones</td><td>Partially</td><td>Partially</td></tr>
<tr><td>Material Circularity in Green structures</td><td>Reuse
Recycling
Adaptive reuse
Landscaping</td><td>No</td><td>No</td></tr>
<tr><td>Energy Efficiency through Bioclimatic design</td><td>Application of trees and green facades to reduce cooling or heating</td><td>Yes</td><td>Yes</td></tr>
<tr><td>Urban Density and Green Spaces</td><td>Urban greenery and its accees</td><td>No</td><td>No</td></tr>
<tr><td>Climate responsive Landscaping</td><td>Selection of plants based on climate conditions</td><td>Yes</td><td>Yes</td></tr>
<tr><td>Water management</td><td>Water recycling
Use of green spaces</td><td>No</td><td>No</td></tr>
</table>

Source: Author's research based on [75].

Green infrastructure indicators in cities refer to the integration of natural elements like parks, trees, and green roofs to manage environmental challenges. Microclimate regulation, through these green areas, helps moderate temperatures, reducing heat island effects, especially in areas without trees or vegetation [76]. Material circularity in green structures focuses on using sustainable, recyclable materials in construction. Urban areas without green spaces may lack this benefit, leading to increased waste and environmental strain [77]. Energy efficiency through bioclimatic design promotes building strategies that adapt to local climates, yet streets without green areas are more likely to have higher energy demands due to poor heat regulation. Urban density and green spaces are closely linked; cities with limited green areas suffer from lower quality of life and reduced environmental resilience, whereas climate-responsive landscaping in these spaces can mitigate heat and improve air quality, which is absent in built-up areas devoid of vegetation [78].

The figure offers a section of the circular city indicators related to recently developed green infrastructure, urban density, and green spaces (Table 1), in both cities, that actually highlights their deficiency in both cities (Figure 1).

Due to the lack of space and construction within the urban core of both cities, after the year 2000, the focus has been placed on residential areas, with less attention given to the urban environment (Figure 1). There is a scarcity of greenery, minimal distance between buildings, and a disregard for the application of bioclimatic parameters [79].

Figure 1. Residential buildings in Belgrade, Serbia (the above images), and Podgorica, Montenegro (the below images), built after 2000. Source: Authors.

3. Methodology

The methodology consisted of multiple elements, including explanations, the development of a questionnaire, data collection, and statistical analysis.

The online survey was specifically designed for practitioners in Serbia and Montenegro and was structured into three logical sections. The study ensured an equal number of participants from both countries, with one hundred forty (140) total responses received, guaranteeing a balanced representation in the dataset. Architectural engineers pursuing doctoral studies in sustainability and resilience of the built environment, as well as professionals with a minimum of five years of work experience, were included. The first section collected general information about the architects, such as their geographical location and years of professional experience. It also explored how architects in Belgrade and Podgorica incorporate climate-responsive strategies into green building designs and the challenges they encounter during implementation, explaining which passive greening strategies they most commonly use in their projects.

The second part examines the key differences in how circular economy principles are applied in green building projects in Belgrade and Podgorica, taking into account the unique characteristics of their respective climate zones, providing an explanation of what the circular city represents for them and how familiar they are with its concepts and indicators.

The third part of the questionnaire provides insights into how locally available materials and technologies in Belgrade and Podgorica impact the adoption of climate-responsive green building practices aligned with circular city objectives.

Architects evaluated five statements related to circular city indicators in architecture, using a five-point Likert scale ranging from 1 (Not at all) to 5 (Extremely). The collected data were analyzed using Statistical Package for the Social Sciences (SPSS) software, with the analysis including descriptive statistics, reliability testing, correlation analysis, and non-parametric tests to assess differences between groups from different countries.

3.1. Research and Questionnaire Design

This research is based on the concepts of sustainable architecture, the circular economy, and climate-responsive strategies in green building design. Sustainable architecture focuses on designing buildings that minimize environmental impact while optimizing resource and energy efficiency [80]. The research questions are built on these theoretical foundations to explore how architects in different climate zones integrate climate-responsive strategies into green building design. Particular focus is on implementation challenges, differences in the application of circular economy principles, and the influence of locally available materials and technologies on adopting sustainable practices in line with the circular city concept [81,82].

The data were collected via a questionnaire in which the respondents assessed various statements on a five-point Likert scale (1—Not at all; 5—Extremely). The statistical analysis was carried out using Statistical Package for the Social Sciences (SPSS) software and involved descriptive statistics, exploratory factor analysis, reliability analysis, parametric and non-parametric tests for comparing differences between groups [83].

A Principal Component Analysis (PCA) was independently conducted on three distinct sets of items to identify key factors within each domain. The first set examined five items measuring the degree of integration of greening strategies. The second set analyzed five items evaluating the degree of application of circular economy principles in projects. The third set assessed five items measuring the impact of the availability of local materials and technologies on adopting green building practices specific to different climate zones.

3.2. Hypothesis Development

This research is based on the concepts of sustainable architecture, the circular economy, and climate-responsive strategies in green building design. Sustainable architecture focuses on designing buildings that minimize environmental impact while optimizing resource and energy efficiency [84]. The research questions are built on these theoretical foundations to explore how architects in different climate zones integrate climate-responsive strategies into green building design. Particular focus is on implementation challenges, differences in the application of circular economy principles, and the influence of locally available materials and technologies on adopting sustainable practices in line with the circular city concept [85–87].

Building upon the insights gained from the literature review and empirical research, this section delves into the core research questions that guide the investigation of climate-responsive strategies and circular economy principles in green building practices in two countries, Serbia and Montenegro.

Defined research questions provide a comprehensive framework for understanding the intersection of climate, design, and sustainability in urban contexts.

The hypotheses are developed based on the idea that climate conditions, local materials, and technology shape circular city strategies [88]. **H1** assumes that differences in material availability and technology influence how circular economy principles are applied in green buildings across climate zones. **H2** suggests that climate-responsive strategies impact urban form and architectural identity, leading to variations in building materials, spatial organization, and esthetics between cities with different climates.

To explore the topics discussed further, the following hypotheses were developed:

H1: *The application of circular economy principles in green building projects differs across climate zones due to variations in locally available materials and technologies, influencing the extent to which climate-responsive and circular city objectives are achieved.*

H2: *The urban form and architectural appearance of cities implementing circular economy principles in different climate zones vary significantly, as climate-responsive strategies shape building materials, design typologies, and spatial organization, leading to distinct esthetic and functional characteristics in the built environment.*

3.3. Sample Description

Architects from Serbia (Belgrade) and Montenegro (Podgorica) who participated in the survey identified key climate challenges specific to their respective cities. Figure 2 illustrates the number of respondents from each country who mentioned each climate challenge.

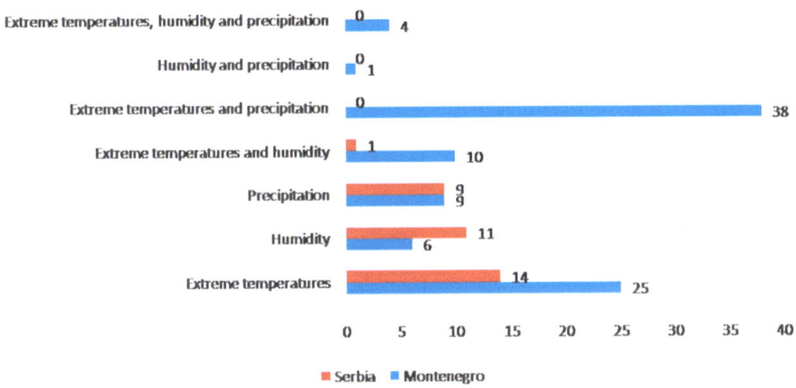

Figure 2. Key climate challenges, results from 140 responses received. Source: Author's research.

Figure 3 illustrates architects' perspectives on what a circular city represents in Montenegro (Podgorica) and Serbia (Belgrade). Among the respondents, 58% from Montenegro and 49% from Serbia indicated that a circular city represents circularity (e.g., circular functioning, reuse, from materials to energy). Additionally, 18% of respondents from Montenegro and 27% from Serbia associated a circular city with sustainability (e.g., sustainable architecture).

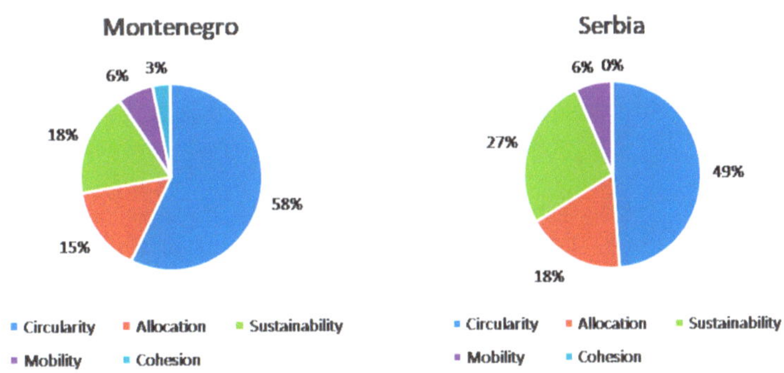

Figure 3. What does a circular city represent for architects? Results from 140 responses received. Source: Author's research.

A smaller proportion—15% from Montenegro and 18% from Serbia—viewed a circular city as representing allocation (effective spatial and functional allocation). Fewer respondents from both countries identified a circular city with mobility (sustainable mobility and traffic) or cohesion (sustainable neighborhoods).

3.4. Data Analysis

The Kaiser–Meyer–Olkin measure confirmed good sampling adequacy for the first set of items (KMO = 0.680), the second set of items (KMO = 0.636), and the third set of items (KMO = 0.793), and Bartlett's test for sphericity ($\chi 2(10) = 103.27$, $p < 0.001$; $\chi 2(10) = 132.89$, $p < 0.001$; $\chi 2(10) = 231.21$, $p < 0.001$) showed that the correlations between the variables were sufficiently high for PCA. Table 2 summarizes exploratory factor analysis results for the degree of integration of greening strategies and presents descriptive statistics for each item.

Table 2. Descriptive statistics for each item and summary of exploratory factor analysis results for the degree of integration of greening strategies.

Item	Factor Loadings	Descriptive Statistics	
		Mean	Std. Deviation
1. How often do you consider local climate conditions in the early stages of design?	0.66	1.93	1.08
2. Are you familiar with the term "passive greening strategies" in building design and construction?	0.68	2.36	1.19
3. Do you use local materials that are adapted to the climate of your region?	0.70	2.05	1.05
4. Do budget constraints affect the implementation of these climate-adapted strategies?	0.70	2.20	1.00
5. Are there specific challenges related to laws and regulations in your city that impact climate-adapted design?	0.55	2.59	1.11
Eigenvalue	2.17		
% of variance	43.39		
Cronbach's Alpha	0.67		

Extraction method: Principal Component Analysis. Source: Author's research.

Eigenvalues were calculated for each component, revealing one component with an eigenvalue exceeding Kaiser's criterion of one (2.17). As a result, a single factor representing the degree of integration of greening strategies was extracted, explaining 43.39% of the variance in the data. The sufficiently high factor loadings indicate strong correlations between the items and the extracted factor, supporting its validity.

Cronbach's Alpha was calculated as 0.67. While values above 0.7 are generally considered acceptable, lower values can be expected in studies measuring a wide range of influences [83]. Additionally, the coefficient is influenced by the number of items within a construct, often increasing as the number of items grows. Given that this construct includes only five items, the obtained value of 0.67 suggests that the measurement instrument demonstrates good reliability.

Table 3 presents descriptive statistics for each item and summarizes exploratory factor analysis results for the degree of application of circular economy principles in projects. Eigenvalues were calculated for each component, revealing one component with an eigenvalue exceeding Kaiser's criterion of one (2.31). As a result, a single factor representing the degree of application of circular economy principles in projects was extracted, explaining 46.16% of the variance in the data. The sufficiently high factor loadings indicate strong correlations between the items and the extracted factor, supporting its validity. Cronbach's Alpha is above the threshold of 0.7, indicating that measuring items have high reliability.

Table 3. Descriptive statistics for each item and summary of exploratory factor analysis results for the degree of application of circular economy principles in projects.

Item	Factor Loadings	Descriptive Statistics	
		Mean	Std. Deviation
1. How familiar are you with the concept of circular cities?	0.56	2.79	1.42
2. Do you think that the climate of your city influences the choice of materials?	0.65	1.99	1.10
3. Are there any local policies or incentives that encourage circular economy practices?	0.70	2.93	1.15
4. Do you face challenges in reducing construction waste during design and construction?	0.75	2.64	1.16
5. Do you ensure the potential for material reuse in your projects in any way?	0.71	2.85	1.17
Eigenvalue	2.31		
% of variance	46.16		
Cronbach's Alpha	0.70		

Extraction method: Principal Component Analysis. Source: Author's research.

Table 4 shows the results of the exploratory factor analysis for the degree of impact of the availability of local materials and technologies on adopting green building practices specific to different climate zones. Descriptive statistics for each item are also presented. One factor was extracted as only one component had an eigenvalue above the Kaiser's criterion of one (2.89). This factor explains 57.70% of the variance in the data. All items have sufficiently high factor loadings, and Cronbach's Alpha is above the threshold of 0.7, indicating a high reliability of the measurement items.

Table 4. Descriptive statistics for each item and summary of exploratory factor analysis results for the degree of impact of the availability of local materials and technologies on adopting green building practices specific to different climate zones.

Item	Factor Loadings	Descriptive Statistics	
		Mean	Std. Deviation
1. How often do you consider passive strategies (e.g., solar shading) when selecting local materials for your projects?	0.66	2.44	1.00
2. How often do you use locally developed technologies in your projects to enhance the building's performance?	0.76	2.59	1.20
3. How often do you collaborate with local manufacturers or suppliers to integrate passive strategies through climate-adapted materials?	0.74	2.71	1.13
4. To what extent does the use of local materials and technologies support your commitment to passive design principles?	0.80	2.58	1.03
5. To what extent does the use of local materials and technologies support your commitment to environmental preservation principles?	0.83	2.57	1.05
Eigenvalue	2.89		
% of variance	57.70		
Cronbach's Alpha	0.81		

Extraction method: Principal Component Analysis. Source: Author's research.

Table 5 contains descriptive statistics for the factor scores. The factor scores are standardized with a mean of zero and a standard deviation of one. However, they show relative differences between respondents in the perceived degree of integration of greening strategies, the perceived degree of application of circular economy principles in projects, and the perceived degree of impact of the availability of local materials and technologies on the adoption of green building practices specific to different climate zones.

Table 5. Descriptive statistics for factor scores with the test of normality.

Factors	Country	Descriptive Statistics			Kolmogorov–Smirnov		Shapiro–Wilk	
		N	Mean	Std. Deviation	Statistic	Sig.	Statistic	Sig.
Degree of integration of greening strategies	Montenegro	93	−0.14	0.86	0.09	0.05	0.94	0.00
	Serbia	44	0.27	1.21	0.13	0.08	0.93	0.01
	Total	137	0.00	1.00	0.10	0.00	0.92	0.00
Degree of application of circular economy principles in projects	Montenegro	94	−0.10	0.91	0.06	0.20	0.99	0.53
	Serbia	45	0.21	1.17	0.08	0.20	0.97	0.44
	Total	139	0.00	1.00	0.07	0.20	0.99	0.18
Degree of impact of the availability of local materials and technologies on adopting green building practices specific to different climate zones	Montenegro	94	−0.17	0.89	0.09	0.09	0.98	0.13
	Serbia	44	0.35	1.13	0.12	0.17	0.93	0.01
	Total	138	0.00	1.00	0.08	0.04	0.97	0.01

The results are significant at the 0.05 level. Source: Author's research.

To address the three research questions, respondents were divided into two groups based on the country (and city) where they work: one hundred and forty architects working in Montenegro (Podgorica) and Serbia (Belgrade). Descriptive statistics for factor scores for each subgroup (Montenegro and Serbia) are presented in Table 5, alongside the results of the Kolmogorov–Smirnov and Shapiro–Wilk tests of normality.

A significant result from these tests ($p < 0.05$) indicates that the data are not normally distributed. The findings suggest that, for both subgroups, the data are normally or approximately normally distributed for the following factors: Degree of application of circular economy principles in projects and Degree of impact of the availability of local materials and technologies on adopting green building practices specific to different climate zones. However, for the factor Degree of integration of greening strategies, the Shapiro–Wilk test indicates that the data are not normally distributed. Similarly, the Kolmogorov–Smirnov test shows that the data for the Montenegro subgroup are not normally distributed, while the data for the Serbia subgroup are normally distributed ($p = 0.05$).

The assumption of normality is required for parametric tests, such as the independent sample t-test, while non-parametric tests, such as the Mann–Whitney test, do not rely on this assumption. Based on the results of the normality tests, both the independent samples t-test (parametric) and the Mann–Whitney test (non-parametric) were employed to examine differences in the integration of greening strategies between architects from different climate zones (Podgorica and Belgrade).

To investigate whether there are differences in the application of circular economy principles in projects and whether locally available materials and technologies influence the adoption of green building practices specific to different climate zones (Podgorica and Belgrade), the independent sample t-test was used. The results of this analysis are presented in Table 6.

Since the t-test has two variations depending on whether the variances of the two groups (Podgorica and Belgrade) are assumed to be equal, Levene's test for equality of variances was conducted. The results of Levene's test were significant for the factors Degree of integration of greening strategies and Degree of impact of the availability of local materials and technologies on adopting green building practices specific to different climate zones. This indicates that the t-test assuming unequal variances should be used for these two factors. Conversely, for the factor Degree of application of circular economy principles in projects, the result of Levene's test was not significant, suggesting that the t-test assuming equal variances is appropriate for this factor.

Table 6. Independent sample *t*-test.

Factors		Levene's Test for Equality of Variances		*t*-Test for Equality of Means	
		F	Sig.	t	Sig. (2-Tailed)
Degree of integration of greening strategies	Equal variances not assumed	6.97	0.01	−2.01	0.05
Degree of application of circular economy principles in projects	Equal variances assumed	3.33	0.07	−1.72	0.09
Degree of impact of the availability of local materials and technologies on adopting green building practices specific to different climate zones	Equal variances not assumed	4.17	0.04	−2.70	0.01

The results are significant at the 0.05 level. Source: Author's research.

The independent sample *t*-test for the factor Degree of integration of greening strategies is marginally significant ($p = 0.05$). Additionally, the Mann–Whitney test yielded a non-significant result ($Z = -1.76$, $p = 0.08$). These findings suggest that there are no statistically significant differences in the integration of greening strategies between architects from different climate zones (Podgorica and Belgrade).

For the factor Degree of application of circular economy principles in projects, the independent sample *t*-test result is not significant ($p = 0.09$), indicating no statistically significant differences in the application of circular economy principles between Podgorica and Belgrade, despite their differing climatic characteristics.

In contrast, the results reveal significant differences ($p = 0.01$) in the extent to which locally available materials and technologies influence the adoption of green building practices specific to these climate zones. In Serbia ($M = 0.35$), locally available materials and technologies have a greater influence on the adoption of green building practices compared to Podgorica ($M = -0.17$).

4. Results and Discussion

Belgrade, as Serbia's capital, has a diverse economy driven by industry, services, and foreign investment, with significant urban development Podgorica, Montenegro's capital, has a smaller economy with a strong focus on services, tourism, and energy, influenced by the country's transition to a market economy and EU integration efforts [89,90]. These economic differences may impact the adoption of circular and green building strategies, as investment capacity, policy incentives, and market demand vary between the two cities. However, they were not taken into account and represent a topic for further research.

H1: *The application of circular economy principles in green building projects differs across climate zones due to variations in locally available materials and technologies, influencing the extent to which climate-responsive and circular city objectives are achieved.*

The results of the study reveal no statistically significant differences in the application of circular economy principles between architects working in Belgrade and Podgorica, as evidenced by the independent sample *t*-test for the factor "Degree of application of circular economy principles in projects" ($p = 0.09$). This suggests that, despite the differing climates, architects in both cities approach the application of circular economy principles similarly. However, these findings contrast with the underlying expectation that locally available materials and technologies, which vary by climate zone, would have a distinct influence on how circular economy principles are implemented [88–90].

While the results do not fully support H1, it is important to consider that circular economy principles may be applied in other ways, beyond just the use of local materials and technologies. Additionally, factors such as economic conditions, policy frameworks, and awareness of sustainability issues might play a more significant role than initially anticipated. Further investigation into how these factors influence the implementation of circular economy practices could provide a more nuanced understanding of this relationship.

H2: *The urban form and architectural appearance of cities implementing circular economy principles in different climate zones vary significantly, as climate-responsive strategies shape building materials, design typologies, and spatial organization, leading to distinct esthetic and functional characteristics in the built environment.*

The findings regarding the factor "Degree of integration of greening strategies" indicate marginally significant differences ($p = 0.05$) between the two cities, although the Mann–Whitney test showed no significant results ($p = 0.08$). These results suggest that the integration of greening strategies, which is an important aspect of circular city objectives, may not differ significantly between Podgorica and Belgrade, even though the cities' climates are distinct.

However, the more pronounced differences in the factor "Degree of impact of the availability of local materials and technologies on adopting green building practices" ($p = 0.01$) indicate that the availability of locally sourced materials and technologies does influence the adoption of climate-responsive building practices. In Serbia, materials and technologies are more likely to facilitate the implementation of green building practices compared to Podgorica, which might suggest that the functional and esthetic outcomes of architectural projects in Serbia are shaped more by these local resources. This discrepancy could also be linked to the region's historical and cultural context, which might influence the way architects select materials and design buildings to respond to climate considerations [91].

In relation to H2, the study suggests that the architectural form and design elements, including greening strategies, do show some variation between the cities. This variation is likely shaped by both climate-responsive strategies and the available local resources, but the differences are not as pronounced as expected. Consequently, while some distinctions in the urban form may exist due to climate factors and local materials, the findings do not fully confirm that these cities exhibit significantly different urban and architectural forms purely due to climate-responsive strategies. Despite some sustainability initiatives, Serbia and Montenegro lack comprehensive policies that actively enforce circular economy principles in construction, with regulations primarily focusing on energy efficiency rather than material reuse or lifecycle sustainability [92]. Weak incentives for recycling construction waste and limited implementation of circular design guidelines further hinder the transition toward fully circular building practices in both countries.

Previous studies suggest that climate-responsive design significantly influences urban morphology, particularly in extreme climates. These findings align with newer research indicating that socio-economic and regulatory factors may play a larger role in shaping architectural outcomes [91,92].

In conclusion, while there are indications of differences in the way local materials and technologies influence the adoption of green building practices, the hypothesis regarding the distinct architectural characteristics in cities with differing climates remains inconclusive. Further research might explore other variables, such as policy interventions and socio-cultural factors, to gain a clearer understanding of how these factors interact and affect urban form and architectural design.

5. Conclusions

This study aimed to explore the integration of climate-responsive strategies and circular economy principles in green building projects across two cities with differing climates, Podgorica (Montenegro) and Belgrade (Serbia). Through a series of hypothesis-driven research questions, the study examined the impact of locally available materials, technologies, and climate conditions on the adoption of sustainable architectural practices.

The findings indicate that, despite the differing climates of the two cities, the application of circular economy principles in green building projects does not show significant differences ($p = 0.09$), suggesting that factors beyond local materials and technologies, such as economic conditions and policy frameworks, might influence the adoption of these principles. Thus, the first hypothesis (H1) is not fully supported, and further investigation into these additional factors is recommended.

Regarding the integration of greening strategies, the data showed marginal differences between the two cities, with no significant difference in the degree of greening strategy integration ($p = 0.08$), which questions the expected variance in architectural forms influenced by climate-responsive design. The second hypothesis (H2), proposing significant differences in the urban form and architectural appearance of cities in different climate zones, was also not fully supported. However, there were significant differences in the impact of locally available materials and technologies on green building practices ($p = 0.01$), especially in Serbia, where these materials had a greater influence.

Overall, this study demonstrates that, while some differences exist in how locally available materials shape green building practices, the broader integration of circular economy principles and greening strategies appears less influenced by climate zone differences than initially hypothesized. The findings suggest that a combination of local resources, regulatory frameworks, and other socio-economic factors play a significant role in shaping sustainable architectural practices. This study has significant theoretical value as it improves strategies for climate-smart green building in circular cities. This study provides valuable information for practitioners and policymakers focused on developing environmentally sustainable solutions in the building sector. The successful implementation of green building practices requires a strategic approach that allows for the creation of structured management practices that require the active participation of stakeholders, from government to local communities, organizations and architects themselves.

Future research should expand on these aspects to provide a more comprehensive understanding of the factors influencing circular city objectives and climate-responsive architecture.

Author Contributions: Conceptualization, M.M.; methodology, M.M.; software, M.M.R.; validation, D.K., M.M. and M.M.R.; formal analysis, D.K., M.M. and M.M.R.; investigation, M.M., M.M.R. and D.K.; resources, D.K., M.M. and M.M.R.; data curation, M.M., M.M.R. and D.K.; writing—original draft preparation, M.M., D.K. and M.M.R.; writing—review and editing, M.M., D.K. and M.M.R.; visualization, M.M., D.K. and M.M.R.; supervision, M.M. All authors have read and agreed to the published version of the manuscript.

Funding: This research is funding by the Ministry of Science, Technological Development and Innovation of the Republic of Serbia for funding the scientific research work, contract no. 451-03-18/2025-03/200155, with the support of the Faculty of Technical Sciences in Kosovska Mitrovica, University of Pristina and no. 451-03-136/2025-03, as well as the Institute of Economics, Belgrade, and University of Montenegro.

Institutional Review Board Statement: In the Republic of Serbia, the processing of personal data, including data collected through anonymous online surveys, is regulated by the Law on Personal Data Protection. This law applies to any processing of personal data carried out by a controller or processor with a registered office, residence, or domicile in the Republic of Serbia, regardless of where the processing activity physically takes place. Additionally, if the data are anonymous, meaning that no individual can be identified based on it, the law does not apply. Reference: https://www.paragraf.rs/propisi/zakon_o_zastiti_podataka_o_licnosti.html (accessed on 15 February 2025). Therefore, ethical review and approval were waived for this study due to legal regulations.

Informed Consent Statement: Informed consent was obtained from all subjects involved in the study.

Data Availability Statement: Data are contained within the article.

Conflicts of Interest: The authors declare no conflicts of interest.

References

1. Brglez, K.; Perc, M.; Lukman, R.K. The complexity and interconnectedness of circular cities and the circular economy for sustainability. *Sustain. Dev.* **2024**, *32*, 2049–2065. [CrossRef]
2. Brandsma, S.; Lenzholzer, S.; Carsjens, G.J.; Brown, R.D.; Tavares, S. Implementation of urban climate-responsive design strategies: An international overview. *J. Urban Des.* **2024**, *29*, 598–623. [CrossRef]
3. Lucchi, E.; Turati, F.; Colombo, B.; Schito, E. Climate-responsive design practices: A transdisciplinary methodology for achieving sustainable development goals in cultural and natural heritage. *J. Clean. Prod.* **2024**, *457*, 142431. [CrossRef]
4. Chen, M.; Chen, L.; Cheng, J.; Yu, J. Identifying interlinkages between urbanization and Sustainable Development Goals. *Geogr. Sustain.* **2022**, *3*, 339–346. [CrossRef]
5. NÆss, P. Urban planning and sustainable development. *Eur. Plan. Stud.* **2001**, *9*, 503–524. [CrossRef]
6. Williams, J. Circular cities. *Urban Stud.* **2019**, *56*, 2746–2762. [CrossRef]
7. Paiho, S.; Mäki, E.; Wessberg, N.; Paavola, M.; Tuominen, P.; Antikainen, M.; Heikkilä, J.; Rozado, C.A.; Jung, N. Towards circular cities—Conceptualizing core aspects. *Sustain. Cities Soc.* **2020**, *59*, 102143. [CrossRef]
8. Teng, J.; Mu, X.; Wang, W.; Xu, C.; Liu, W. Strategies for sustainable development of green buildings. *Sustain. Cities Soc.* **2019**, *44*, 215–226. [CrossRef]
9. Olubunmi, O.A.; Xia, P.B.; Skitmore, M. Green building incentives: A review. *Renew. Sustain. Energy Rev.* **2016**, *59*, 1611–1621. [CrossRef]
10. Darko, A.; Chan, A.P.C. Strategies to promote green building technologies adoption in developing countries: The case of Ghana. *Build. Environ.* **2018**, *130*, 74–84. [CrossRef]
11. Chan, A.P.C.; Darko, A.; Ameyaw, E.E. Strategies for Promoting Green Building Technologies Adoption in the Construction Industry—An International Study. *Sustainability* **2017**, *9*, 969. [CrossRef]
12. Ladu, L.; Blind, K. Overview of policies, standards and certifications supporting the European bio-based economy. *Curr. Opin. Green Sustain. Chem.* **2017**, *8*, 30–35. [CrossRef]
13. Morone, P.; Caferra, R.; D'Adamo, I.; Falcone, P.M.; Imbert, E.; Morone, A. Consumer willingness to pay for bio-based products: Do certifications matter? *Int. J. Prod. Econ.* **2021**, *240*, 108248. [CrossRef]
14. Kushch, E.; Grudina, O.; Sitokhova, T. Digital Solution in Circular Economy Frameworks. *Reliab. Theory Appl.* **2024**, *6*, 1202–1207.
15. Komatina, D.; Miletić, M.; Mosurović Ružičić, M. Embracing Artificial Intelligence (AI) in Architectural Education: A Step towards Sustainable Practice? *Buildings* **2024**, *14*, 2578. [CrossRef]
16. Butturi, M.A.; Neri, A.; Mercalli, F.; Gamberini, R. Sustainability-Oriented Innovation in the Textile Manufacturing Industry: Pre-Consumer Waste Recovery and Circular Patterns. *Environments* **2025**, *12*, 82. [CrossRef]
17. Zaniboni, L.; Albatici, R. Natural and Mechanical Ventilation Concepts for Indoor Comfort and Well-Being with a Sustainable Design Perspective: A Systematic Review. *Buildings* **2022**, *12*, 1983. [CrossRef]
18. Ching, F.D.K.; Shapiro, I.M. Green Building Illustrated. 2021, p. 326. Available online: https://books.google.com/books/about/Green_Building_Illustrated.html?hl=sr&id=I_UFEAAAQBAJ (accessed on 21 March 2025).
19. Yang, Y.; Haurie, L.; Wang, D.Y. Bio-based materials for fire-retardant application in construction products: A review. *J. Therm. Anal. Calorim.* **2021**, *147*, 6563–6582. [CrossRef]
20. Aghili, N.; Bin Mohammed, A.H.; Sheau-Ting, L. Key Practice for Green Building Management In Malaysia. *MATEC Web Conf.* **2016**, *66*, 00040. [CrossRef]
21. Saleh, R.M.; Al-Swidi, A. The adoption of green building practices in construction projects in Qatar: A preliminary study. *Manag. Environ. Qual. Int. J.* **2019**, *30*, 1238–1255. [CrossRef]
22. Milovanović, V.; Janošević, S.; Paunović, M. Quality management and business performance of Serbian companies. *Ekon. Preduz.* **2021**, *69*, 345–356. [CrossRef]
23. Paul, W.L.; Taylor, P.A. A comparison of occupant comfort and satisfaction between a green building and a conventional building. *Build. Environ.* **2008**, *43*, 1858–1870. [CrossRef]
24. Miletić, M.; Komatina, D.; Babić, L.; Lukić, J. Evaluating Energy Retrofit and Indoor Environmental Quality in a Serbian Sports Facility: A Comprehensive Case Study. *Appl. Sci.* **2024**, *14*, 9401. [CrossRef]
25. Kosanović, S.; Miletić, M.; Marković, L. Energy refurbishment of family houses in serbia in line with the principles of circular economy. *Sustainability* **2021**, *13*, 5463. [CrossRef]
26. Mukhtar, A.; Yusoff, M.Z.; Ng, K.C. The potential influence of building optimization and passive design strategies on natural ventilation systems in underground buildings: The state of the art. *Tunn. Undergr. Sp. Technol.* **2019**, *92*, 103065. [CrossRef]

27. Vijayaraghavan, K. Green roofs: A critical review on the role of components, benefits, limitations and trends. *Renew. Sustain. Energy Rev.* **2016**, *57*, 740–752. [CrossRef]
28. Bellini, E.; Macchi, A. Architectural Adaptability: Constructing Resilient Cities. 2022. Available online: https://flore.unifi.it/handle/2158/1282214 (accessed on 3 February 2025).
29. Houghton, A.; Castillo-Salgado, C. Analysis of correlations between neighborhood-level vulnerability to climate change and protective green building design strategies: A spatial and ecological analysis. *Build. Environ.* **2020**, *168*, 106523. [CrossRef]
30. Hoffman, M.; Shaw, R.; Santos, F.D.; Ferreira, P.L.; Strandsbjerg, J.; Pedersen, T. The Climate Change Challenge: A Review of the Barriers and Solutions to Deliver a Paris Solution. *Climate* **2022**, *10*, 75. [CrossRef]
31. Campbell-Lendrum, D.; Neville, T.; Schweizer, C.; Neira, M. Climate change and health: Three grand challenges. *Nat. Med.* **2023**, *29*, 1631–1638. [CrossRef]
32. Forster, P.M.; Smith, C.J.; Walsh, T.; Lamb, W.F.; Lamboll, R.; Hauser, M.; Ribes, A.; Rosen, D.; Gillett, N.; Palmer, M.D.; et al. Indicators of Global Climate Change 2022: Annual update of large-scale indicators of the state of the climate system and human influence. *Earth Syst. Sci. Data* **2023**, *15*, 2295–2327. [CrossRef]
33. Khaliq, W.; Mansoor, U. Bin Performance evaluation for energy efficiency attainment in buildings based on orientation, temperature, and humidity parameters. *Intell. Build. Int.* **2022**, *14*, 606–622. [CrossRef]
34. Mendes, A.M.; Monteiro, C.M.; Santos, C. Green Roofs Hydrological Performance and Contribution to Urban Stormwater Management. *Water Resour. Manag.* **2024**, *39*, 1015–1031. [CrossRef]
35. Cheval, S.; Amihăesei, V.A.; Chitu, Z.; Dumitrescu, A.; Falcescu, V.; Irașoc, A.; Micu, D.M.; Mihulet, E.; Ontel, I.; Paraschiv, M.G.; et al. A systematic review of urban heat island and heat waves research (1991–2022). *Clim. Risk Manag.* **2024**, *44*, 100603. [CrossRef]
36. Zakari, A.; Khan, I.; Tan, D.; Alvarado, R.; Dagar, V. Energy efficiency and sustainable development goals (SDGs). *Energy* **2022**, *239*, 122365. [CrossRef]
37. Akhimien, N.G.; Latif, E.; Hou, S.S. Application of circular economy principles in buildings: A systematic review. *J. Build. Eng.* **2021**, *38*, 102041. [CrossRef]
38. Ragheb, A.; El-Shimy, H.; Ragheb, G. Green Architecture: A Concept of Sustainability. *Procedia—Soc. Behav. Sci.* **2016**, *216*, 778–787. [CrossRef]
39. Elaouzy, Y.; El Fadar, A. Energy, economic and environmental benefits of integrating passive design strategies into buildings: A review. *Renew. Sustain. Energy Rev.* **2022**, *167*, 112828. [CrossRef]
40. Li, B.; Guo, W.; Liu, X.; Zhang, Y.; Russell, P.J.; Schnabel, M.A. Sustainable Passive Design for Building Performance of Healthy Built Environment in the Lingnan Area. *Sustainability* **2021**, *13*, 9115. [CrossRef]
41. Yang, L.; Fu, R.; He, W.; He, Q.; Liu, Y. Adaptive thermal comfort and climate responsive building design strategies in dry–hot and dry–cold areas: Case study in Turpan, China. *Energy Build.* **2020**, *209*, 109678. [CrossRef]
42. Zarzycki, A.; Decker, M. Climate-adaptive buildings: Systems and materials. *Int. J. Archit. Comput.* **2019**, *17*, 166–184. [CrossRef]
43. Owusu, P.A.; Asumadu-Sarkodie, S. A review of renewable energy sources, sustainability issues and climate change mitigation. *Cogent Eng.* **2016**, *3*, 1167990. [CrossRef]
44. Bhamare, D.K.; Rathod, M.K.; Banerjee, J. Passive cooling techniques for building and their applicability in different climatic zones—The state of art. *Energy Build.* **2019**, *198*, 467–490. [CrossRef]
45. Urban and Architectural Heritage Conservation within Sustainability. 2019. Available online: https://books.google.com/books/about/Urban_and_Architectural_Heritage_Conserv.html?hl=sr&id=Npn8DwAAQBAJ (accessed on 16 February 2025).
46. Scagnelli, S.D.; Ranjbari, M.; Fagone, C.; Santamicone, M.; Villa, V. Architecture Engineering and Construction Industrial Framework for Circular Economy: Development of a Circular Construction Site Methodology. *Sustainability* **2023**, *15*, 1813. [CrossRef]
47. Genovese, P.V.; Zoure, A.N. Architecture trends and challenges in sub-Saharan Africa's construction industry: A theoretical guideline of a bioclimatic architecture evolution based on the multi-scale approach and circular economy. *Renew. Sustain. Energy Rev.* **2023**, *184*, 113593. [CrossRef]
48. Gudekli, A.; Dogan, M.E.; Dogan, T.G.; Gudekli, D. Gender, Sustainability, and Urbanism: A Systematic Review of Literature and Cross-Cluster Analysis. *Sustainability* **2023**, *15*, 14994. [CrossRef]
49. Kanters, J. Circular Building Design: An Analysis of Barriers and Drivers for a Circular Building Sector. *Buildings* **2020**, *10*, 77. [CrossRef]
50. When not every response to climate change is a good one: Identifying principles for sustainable adaptation. *Clim. Dev.* **2011**, *3*, 7–20. [CrossRef]
51. Dodds, R.; Kelman, I. How Climate Change is Considered in Sustainable Tourism Policies: A Case of The Mediterranean Islands of Malta and Mallorca. *Tour. Rev. Int.* **2008**, *12*, 57–70. [CrossRef]

52. Childers, D.L.; Cadenasso, M.L.; Morgan Grove, J.; Marshall, V.; McGrath, B.; Pickett, S.T.A. An Ecology for Cities: A Transformational Nexus of Design and Ecology to Advance Climate Change Resilience and Urban Sustainability. *Sustainability* **2015**, *7*, 3774–3791. [CrossRef]
53. Sayigh, A. Sustainability, Energy and Architecture: Case Studies in Realizing Green Buildings. 2014, p. 551. Available online: https://books.google.com/books/about/Sustainability_Energy_and_Architecture.html?hl=sr&id=nkIobvJdjjwC (accessed on 17 February 2025).
54. Thermal, E.; Nunzi, J.-M.; El Ganaoui, M.; El Jouad, M.; Lourenço Niza, I.; Mendes da Luz, I.; Maria Bueno, A.; Eduardo Broday, E. Thermal Comfort and Energy Efficiency: Challenges, Barriers, and Step towards Sustainability. *Smart Cities* **2022**, *5*, 1721–1741 [CrossRef]
55. Saraiva, T.S.; de Almeida, M.; Bragança, L.; Barbosa, M.T. Environmental Comfort Indicators for School Buildings in Sustainability Assessment Tools. *Sustainability* **2018**, *10*, 1849. [CrossRef]
56. Reith, A.; Orova, M. Do green neighbourhood ratings cover sustainability? *Ecol. Indic.* **2015**, *48*, 660–672. [CrossRef]
57. Alyami, S.H.; Rezgui, Y. Sustainable building assessment tool development approach. *Sustain. Cities Soc.* **2012**, *5*, 52–62. [CrossRef]
58. Santos, P.; Cervantes, G.C.; Zaragoza-Benzal, A.; Byrne, A.; Karaca, F.; Ferrández, D.; Salles, A.; Bragança, L. Circular Material Usage Strategies and Principles in Buildings: A Review. *Buildings* **2024**, *14*, 281. [CrossRef]
59. Hsu, W.L.; Meen, T.H.; Yang, H.C.; Yu, W. Der Special Issue on Innovative Circular Building Design and Construction. *Buildings* **2023**, *13*, 1322. [CrossRef]
60. Li, M.; Ou, W.; Chai, X.; Khodabakhshi, H.; Fu, Z.; Yuan, J.; Horimbere, P.F.; Urban, E.D.L.J.; Makvandi, M.; Li, W.; et al. Urban Heat Mitigation towards Climate Change Adaptation: An Eco-Sustainable Design Strategy to Improve Environmental Performance under Rapid Urbanization. *Atmosphere* **2023**, *14*, 638. [CrossRef]
61. Ruíz, M.A.; Mack-Vergara, Y.L. Resilient and Sustainable Housing Models against Climate Change: A Review. *Sustainability* **2023**, *15*, 13544. [CrossRef]
62. Makvandi, M.; Li, W.; Li, Y.; Wu, H.; Khodabakhshi, Z.; Xu, X.; Yuan, P.F. Advancing Urban Resilience Amid Rapid Urbanization: An Integrated Interdisciplinary Approach for Tomorrow's Climate-Adaptive Smart Cities—A Case Study of Wuhan, China. *Smart Cities* **2024**, *7*, 2110–2130. [CrossRef]
63. Niazy, D.; Metwally, E.A.; Rifat, M.; Awad, M.I.; Elsabbagh, A. A conceptual design of circular adaptive façade module for reuse. *Sci. Rep.* **2023**, *13*, 20552. [CrossRef]
64. Hu, J.; Zhang, F.; Qiu, B.; Zhang, X.; Yu, Z.; Mao, Y.; Wang, C.; Zhang, J. Green-gray imbalance: Rapid urbanization reduces the probability of green space exposure in early 21st century China. *Sci. Total Environ.* **2024**, *933*, 173168. [CrossRef]
65. de la Joie Horimbere, E.; Chen, H.; Makvandi, M. *An Exploration of the Effects of Urban Block Design on the Outdoor Thermal Environment in Tropical Savannah Climate: Case Study of Nyamirambo Neighborhood of Kigali*; Springer: Berlin/Heidelberg, Germany, 2021; pp. 17–28. [CrossRef]
66. Dervishaj, A.; Gudmundsson, K. Parametric design workflow for solar, context-adaptive and reusable facades in changing urban environments. *J. Build. Perform. Simul.* **2025**, *18*, 161–190. [CrossRef]
67. Zhou, L.; Zhou, Y. Energy-resilient climate adaptation using a tailored life-cycle integrative design approach for national carbon abatement. *Cell Rep. Phys. Sci.* **2024**, *5*, 102306. [CrossRef]
68. Jelle, B.P. Accelerated climate ageing of building materials, components and structures in the laboratory. *J. Mater. Sci.* **2012**, *47*, 6475–6496. [CrossRef]
69. Nordgren, J.; Stults, M.; Meerow, S. Supporting local climate change adaptation: Where we are and where we need to go. *Environ. Sci. Policy* **2016**, *66*, 344–352. [CrossRef]
70. Bekier, J.; Parisi, C. Co-construction of performance indicators for a circular city and its relation to a local action net. *Account. Audit. Account. J.* **2023**, *ahead-of-print*. [CrossRef]
71. Balletto, G.; Ladu, M. The Role of Spatial Circular Planning in Urban Governance. A Set of Indicators to Evaluate Performance in Urban Regeneration. In *Lecture Notes in Computer Science (Lecture Notes in Computer Science)*; Springer: Berlin/Heidelberg, Germany, 2023; Volume 14111, pp. 104–118. [CrossRef]
72. Bîrgovan, A.L.; Lakatos, E.S.; Szilagyi, A.; Cioca, L.I.; Pacurariu, R.L.; Ciobanu, G.; Rada, E.C. How Should We Measure? A Review of Circular Cities Indicators. *Int. J. Environ. Res. Public Health* **2022**, *19*, 5177. [CrossRef]
73. Pegorin, M.C.; Caldeira-Pires, A.; Faria, E. Interactions between a circular city and other sustainable urban typologies: A review. *Discov. Sustain.* **2024**, *5*, 14. [CrossRef]
74. Falah, N.; Marrero, M.; Solis-Guzman, J. Identifying Circular City Indicators Based on Advanced Text Analytics: A Multi-Algorithmic Approach. *Environments* **2025**, *12*, 1. [CrossRef]
75. Paoli, F.; Pirlone, F.; Spadaro, I. Indicators for the circular city: A review and a proposal. *Sustainability* **2022**, *14*, 11848. [CrossRef]
76. Abdulateef, M.F.; Al-Alwan, H.A.S. The effectiveness of urban green infrastructure in reducing surface urban heat island. *Ain Shams Eng. J.* **2022**, *13*, 101526. [CrossRef]

77. Lundaev, V.; Solomon, A.A.; Le, T.; Lohrmann, A.; Breyer, C. Review of critical materials for the energy transition, an analysis of global resources and production databases and the state of material circularity. *Miner. Eng.* **2023**, *203*, 108282. [CrossRef]
78. Addas, A. The importance of urban green spaces in the development of smart cities. *Front. Environ. Sci.* **2023**, *11*, 1206372. [CrossRef]
79. Transformation of the New Belgrade Urban Tissue: Filling the Space Instead of Interpolation. Available online: https://raumplan.iaus.ac.rs/handle/123456789/187 (accessed on 20 March 2025).
80. Kibert, C.J. Green Building Design and Delivery. *Sustain. Constr.* **2016**, *1*, 1–36. Available online: https://books.google.rs/books?hl=sr&lr=&id=2xgWCgAAQBAJ&oi=fnd&pg=PR15&dq=80.%09Kibert,+C.J.+Green+Building+Design+and+Delivery.+Sustain.+Constr.+2016&ots=GaXp9Mc_pA&sig=tgRddmfMmtGex0ZSINgk6WDpw7U&redir_esc=y#v=onepage&q&f=false (accessed on 21 March 2025).
81. Project MUSE-Design with Climate. Available online: https://muse.jhu.edu/pub/267/edited_volume/book/64523 (accessed on 11 April 2025).
82. Bolger, K.; Doyon, A. Circular cities: Exploring local government strategies to facilitate a circular economy. *Eur. Plan. Stud.* **2019**, *27*, 2184–2205. [CrossRef]
83. Discovering Statistics Using IBM SPSS Statistics—Andy Field—Google Књиге. Available online: https://books.google.rs/books?hl=sr&lr=&id=83L2EAAAQBAJ&oi=fnd&pg=PT8&dq=Field,+A.+(2009),+Discovering+Statistics+Using+SPSS,+3rd+ed.,+SAGE+Publications,+London.&ots=UbLZFsFMHP&sig=yG_djoMqCzmDrP1YsBcC6ifaRgk&redir_esc=y#v=onepage&q=Field,%20A.%20(2009),%20Discovering%20Statistics%20Using%20SPSS,%203rd%20ed.,%20SAGE%20Publications,%20London.&f=false (accessed on 17 February 2025).
84. Curtis, F. Eco-localism and sustainability. *Ecol. Econ.* **2003**, *46*, 83–102. [CrossRef]
85. Calkins, M. A Complete Guide to the Evaluation, Selection, and Use of Sustainable Construction Materials. 2008, p. 464. Available online: https://books.google.com/books/about/Materials_for_Sustainable_Sites.html?hl=sr&id=FVqXqhvUhe0C (accessed on 17 February 2025).
86. Costa, C.; Cerqueira, Â.; Rocha, F.; Velosa, A. The sustainability of adobe construction: Past to future. *Int. J. Archit. Herit.* **2019**, *13*, 639–647. [CrossRef]
87. Godwin, P.J. Building Conservation and Sustainability in the United Kingdom. *Procedia Eng.* **2011**, *20*, 12–21. [CrossRef]
88. Philokyprou, M.; Michael, A. Environmental Sustainability in the Conservation of Vernacular Architecture. The Case of Rural and Urban Traditional Settlements in Cyprus. *Int. J. Archit. Herit.* **2021**, *15*, 1741–1763. [CrossRef]
89. Roadmap Towards the Circular Economy in Montenegro | United Nations Development Programme. Available online: https://www.undp.org/montenegro/publications/roadmap-towards-circular-economy-montenegro (accessed on 27 March 2025).
90. Roadmap for Circular Economy in Serbia | United Nations Development Programme. Available online: https://www.undp.org/serbia/publications/roadmap-circular-economy-serbia (accessed on 27 March 2025).
91. Steemers, K. Towards a research agenda for adapting to climate change. *Build. Res. Inf.* **2003**, *31*, 291–301. [CrossRef]
92. Ratti, C.; Baker, N.; Steemers, K. Energy consumption and urban texture. *Energy Build.* **2005**, *37*, 762–776. [CrossRef]

MDPI AG
Grosspeteranlage 5
4052 Basel
Switzerland
Tel.: +41 61 683 77 34

Sustainability Editorial Office
E-mail: sustainability@mdpi.com
www.mdpi.com/journal/sustainability

www.ingramcontent.com/pod-product-compliance
Lightning Source LLC
Chambersburg PA
CBHW041416210926
51850CB00013B/3287